D0164267

WATER PLANET

WATER PLANET

The Culture, Politics, Economics, and Sustainability of Water on Earth

Camille Gaskin-Reyes, Editor

An Imprint of ABC-CLIO, LLC
Santa Barbara, California • Denver, Colorado

Copyright © 2016 by ABC-CLIO, LLC

All rights reserved. No part of this publication may be reproduced, stored in a retrieval system, or transmitted, in any form or by any means, electronic, mechanical, photocopying, recording, or otherwise, except for the inclusion of brief quotations in a review, without prior permission in writing from the publisher.

Library of Congress Cataloging-in-Publication Data

Names: Gaskin-Reyes, Camille E., editor.
Title: Water planet : the culture, politics, economics, and sustainability of water on Earth / Camille Gaskin-Reyes, editor.
Description: Santa Barbara, California : ABC-CLIO [2016] | Includes bibliographical references and index.
Identifiers: LCCN 2016016137 | ISBN 9781440838163 (alk. paper) | ISBN 9781440838170 (ebook)
Subjects: LCSH: Water supply—Political aspects | Water-supply—Economic aspects. | Water resources development—Political aspects. | Water resources development—Economic aspects. | Water rights. | Marine ecosystem health. | Global environmental change. | Sustainable development.
Classification: LCC HD1691 .W3254 2016 | DDC 333.91—dc23
LC record available at https://lccn.loc.gov/2016016137

ISBN: 978-1-4408-3816-3
EISBN: 978-1-4408-3817-0

20 19 18 17 16 1 2 3 4 5

This book is also available as an eBook.

ABC-CLIO
An Imprint of ABC-CLIO, LLC

ABC-CLIO, LLC
130 Cremona Drive, P.O. Box 1911
Santa Barbara, California 93116-1911
www.abc-clio.com

This book is printed on acid-free paper ♾

Manufactured in the United States of America

Every reasonable effort has been made to trace the owners of copyrighted materials in this book, but in some instances this has proven impossible. The editor and publisher will be glad to receive information leading to more complete acknowledgments in subsequent printings of the book and in the meantime extend their apologies for any omissions.

CONTENTS

PREFACE

This book presents an integrated picture of the role of water in everyday existence, whether in our economic pursuits, social activities, or cultural life. Water has been a fundamental and indispensable part of human life for millennia and continues to be our most precious life-supporting liquid, but we have not been good stewards of this priceless resource. Flowing through this book is the theme that we are now faced with the daunting challenge of securing safe freshwater supplies for the masses of people on the water planet, and protecting oceans and marine ecosystems for current and future generations.

Chapters 1 and 2 examine water and climate change issues, including impacts of human-led greenhouse gas emissions since the industrial revolution, i.e., higher temperatures, ice melt, rising sea levels, storm intensity, and the vulnerability of low-lying areas. They delve into topics such as the increase of arid/water-stressed zones, water conflicts, climate equity for small island societies, and the urgency of climate change adaptation measures.

Both chapters focus on the linkages of climate change with population growth, economic development, energy, agricultural and industrial development, and disruption of the water cycle on a global scale. They make the point that since our shift to sedentary life and the growth of cities and industrial centers, we have become locked into an economic model based on fossil fuels, consumer-oriented growth, global trade, and competition for water.

The paradox discussed in these and other chapters is that on land we are increasingly facing problems of surface and groundwater quality, and on the oceans, pollution and overexploitation, yet we continue to draw down, pollute, compete, or increasingly fight for dwindling freshwater and marine resources. We have not been the best stewards of water since we stopped our hunting and gathering ways and moved into villages, cities, and industrial centers.

As freshwater becomes more and more threatened, humans can either turn to cooperation or conflict. Chapters 3, 4, and 5 address these themes as they relate to sharing freshwater and global use of the high seas. Chapter 3 discusses integrated water management and the concept of water as a public good or a commodity, as well as stellar and poor examples of shared lakes, seas, aquifers, waterways, and river basins. Chapter 4 explores threats and conflicts in ocean, coastal, and island environments as a result of pollution, acidification, coral reef destruction, tourism, overfishing, seabed mining, oil exploration, and lax enforcement of international laws on the high seas. Chapter 5 picks up this thread by discussing ocean and sea governance rules, people and drug trafficking, piracy on the high seas, and threats to peace and stability caused by competition for water.

Chapters 6 and 7 address the foundation of water for social and economic life and its significance for expansion of human settlements, i.e., through access to water sources, waterways, ports, lakes, and gateways to oceans. Chapter 6 zooms into the human overexploitation of water in areas such as agriculture, irrigation, manufacturing, energy, services, and the construction industry; the global problem of water haves and have nots; the conflicts caused by water privatization; and the public health risks of waterborne diseases for millions of people. Chapter 6 highlights the steady rise of the lucrative bottled water industry, different water consumption levels around the world, and the challenge of providing clean water to masses of people on the planet. Chapter 7 discusses technologies to conserve water in the construction industry, the historical development of cities, water problems in African cities, and drought and water shortages in California.

Chapter 8 turns to the vital role of water for exploration, warfare, land grabs, colonial expansion, the slave trade and the growth of globalized trade linkages among nations, including the development of transoceanic pathways to facilitate the passage of goods and people such as the Panama Canal. Chapter 9 reflects on the role of water in the generation of energy over millennia, controversies and conflicts of megadams, including impacts on communities and indigenous groups, environmental damage, reduction of water flow for downstream countries, and increased potential for regional conflicts.

Chapter 10 focuses on gender issues in water collection, distribution, allocation, and ownership rights, and changing gender roles parallel to the historical development of society. It emphasizes the situation of vulnerable groups and their access to water as well as the heavy water-carrying burden of women in many developing countries, but also the pace-setting roles of women explorers across oceans. Chapter 11 outlines the significance of water in our cultural, religious, and spiritual lives, its prevalence in rites and practices, indigenous legends and myths, and its significance in many creation stories and humans' relationships with deities throughout the ages.

Chapter 12 maps a possible path forward to cope with water management problems, including specific challenges and opportunities to preserve and better manage the world's water resources. It is a call to change. It emphasizes that even if we have not been the best stewards of water in the past or have undervalued its significance, we are now called upon to restore balance and face up to climate change, groundwater depletion and pollution, inefficiencies and inequity of water allocation, the gap between the global water haves and have nots—and the threats of a parched and water-polluted future.

INTRODUCTION

Water: the building block of life. The planet is awash in it. We marvel at a perfect snowflake, a roaring waterfall, a babbling brook, and ice-cold water on a sweltering day. But we also tremble at the wrath of a storm, the fury of a flood, and the terror of a tsunami. This book underscores the pervasiveness of water throughout our social, economic, and cultural lives. We cannot live without water, but we can perish if we have too much of it in too short a time. Globally, hundreds of thousands of people die or are displaced yearly in the wake of hurricanes, storms, landslides, torrential rains, floods, river overspills, and coastal storm surges.

WATER PLANET

A perspective of Earth from space makes it easy to see why we are called the water planet: 71 percent of Earth's surface area contains water. But 97.5 percent of this water is salt water, and only 2.5 percent freshwater. Two-thirds of this freshwater is frozen in snow, ice, and permafrost. The rest is surface water such as rivers, lakes and ponds, and groundwater stored in aquifers (Chellaney 2013).

This leaves relatively small amounts of freshwater for human consumption and other activities demanding water, creating the obligation for humans to be good stewards of this precious resource. This book is about how we have managed water in the past, how we are using or misusing it currently, and the challenges and opportunities we face in the future.

Water is in a state of continuous transfer called the hydrologic cycle. This cycle regulates the evaporation of water from oceans and the land to the air—and its return to the land or sea through precipitation in the form of rain, snow, ice, fog, or sleet. Its makeup is simple: two hydrogen atoms (H_2) bond with one oxygen atom (O) to form H_2O, which is water in liquid, gaseous, or solid form.

But while water's chemical composition is simple, water issues are complex. Around 7 billion people share the planet's limited freshwater resources, and they are not uniformly distributed around the world. Freshwater choices of the masses of people are mainly limited to aquifers below the ground, and springs, lakes, rivers, reservoirs, melted ice, or desalinated seawater on the surface. Many countries have an inadequate supply of freshwater, no matter the source; others are using up their water resources at fast rates.

Excessive sinking of wells by communities, industries, farmers, and urban water utilities has started to deplete underground aquifers, while the pollution of underground or surface water is also reducing freshwater supply. In some countries, water scarcity is provoking competition and sparking open conflicts; in others, insufficient water reserves are stimulating collaboration or sharing agreements among nations or regions to meet the increasing demand for water. Often, the problem is not a question of supply, but one of equitable water allocation and distribution means. These topics are addressed in this book.

Not only do we live on the water planet and depend on this precious resource; we mirror our connection to it through the water content of our bodies. Humans contain 50 to 70 percent of water, depending on body size. Water enabled the genesis and development of the first life forms. Humans are so dependent on water they cannot survive for more than a few days without it. Animals need water for drinking and body temperature regulation. Plants need it too for evapotranspiration, the flow of water and nutrients from roots to shoots and the release of water vapor into the air.

WATER AND EARLY CIVILIZATIONS

Water is vital for the development of early civilizations, which arose or thrived close to water sources and waterways. In the past, oceans and seas were key pathways for ocean-faring countries and intrepid explorers to discover, conquer, enslave, or trade with other lands and peoples. Curiosity and the human quest for knowledge also drew researchers to navigate across little-known waters for scientific expeditions. Alexander von Humboldt's five-year expedition to the Americas in 1799 accumulated valuable botanical data for posterity; and Darwin's ocean voyage to the Galapagos Islands, Ecuador, in 1831 advanced the theory of evolution and natural selection of the species.

Ancient water routes enabled raiding attacks, warfare, and plunder. In 43 CE, Romans sailed across the English Channel and conquered Great Britain. From the eighth century onwards, Vikings from Scandinavia traded with, looted, and colonized other regions, using their longboats and formidable navigational skills. Ocean powers such as Portugal, Spain, Holland, and England also developed distant trade routes and established (and were forced to defend) far-flung empires across the seas.

In the 16th century, Spain used its presence on two oceans, the Pacific and the Atlantic, to transport gold and silver from its Latin American colonies to the motherland. One major route was via the Isthmus of Panama. After loading the loot on ships in Peru on the Pacific, Spaniards sailed up the coast to Panama, transferred the cargo overland to Panamanian ports on the Atlantic, and then shipped it onward to Spain. This transoceanic route was so lucrative that it created possibilities for opportunistic seafarers, i.e., pirates. They ambushed Spanish galleons on the high seas and stashed the booty in coves and secret hideaways.

In the 18th and 19th centuries oceans provided a vital route for an infamous, global network: the slave trade. Slave brokers and trading companies developed a lucrative trade triangle between Africa, the Americas, and Europe, involving three legs: 1) enslaving Africans and transporting them in slave ships across the Atlantic to plantations in the Americas; 2) reloading the ships bound for Europe with plantation products such as sugar; and 3) completing the triangle by shipping European-made goods to Africa to trade for more slaves—then heading back across the Atlantic with human cargo to the Americas.

MODERN CHALLENGES

In modern times challenges on the high seas still abound, but the scale and nature of the problems vary. The rise in globalization and communications—including submarine cables that transmit global data across oceans—has intensified international trade and ocean traffic. Parallel to the explosion of global commerce, oceans have become pathways for new, sophisticated forms of high-seas piracy and illegal trafficking of people, arms, drugs, endangered species, stolen cultural artifacts, and other illicit products.

In addition, oceans and marine ecosystems are threatened by pollution, acidification, loss of marine biodiversity, overexploitation of fish stocks, and intensified seabed mining. In the Arctic Ocean the race for mineral resources is escalating with climate change and ice melt, which offers the prospect of ice-free shipping routes. Many nations and companies are vying for oil and mineral rights in this arena. This book reviews problems arising out of such competition, but also describes ocean-sharing and water cooperation practices.

GLOBAL OCEANS

Beyond a country's national 12- or 200-nautical-mile external limit and jurisdiction, the use of international ocean waters is up for grabs. Common ocean territory, called the high seas—more than 40 percent of the world's surface—belongs to everyone and to no one in particular. This situation offers both challenges and opportunities for collaboration and improved ocean governance.

Since 1958, international treaties and a consolidated 1982 United Nations Convention on the Law of the Sea (UNCLOS) have defined the rights and responsibilities of nations to use the oceans. They recognize the high seas as common international territory with free right of passage for all nations under the concept of *mare liberum* (Latin for "freedom of the seas"). Simply put, all nations and peoples have the right to use shared ocean resources, but they also have the duty to protect them for future generations. It is important to reconcile these two viewpoints to resolve the emerging crises of the oceans and achieve sustainable management of marine resources.

Parallel to the discourse on nations' and people's rights to the oceans as a global public good, there is an ongoing debate about freshwater supply and people's rights to freshwater on land. According to the United Nations (UN), the right to safe drinking water and sanitation is a basic human right for all. However, robust private water markets around the world have been developing due to rising demand and decreasing supply of freshwater.

As a result of the increasing monetary value of potable water and huge gaps in clean water provision in most developing countries, private companies have invested heavily in the global business of water. In some countries, the price of clean drinking water or bottled water is prohibitive for low-income communities. The costs and inaccessibility of clean water make it very difficult for the poorest to obtain clean water, which challenges the UN concept of water as a human right and a shared public good.

WATER INDUSTRY

The bottled water industry is among the fastest-growing and increasingly lucrative industries in the world. In 2000 just a few large water companies dominated more than 70 percent of the bottled water market in over 150 countries. These companies achieved global sales of about 22 billion gallons of bottled water for an approximate value of 23 billion dollars (Barlow and Clarke 2002, 142).

Free market proponents argue that water is a commodity, as any other good, that can be traded or privatized. The privatization of public water utilities in many countries has opened the door for water corporations to get into the business of potable water generation and distribution services, traditionally the domain of public utilities.

Since 1990 international water companies have been managing water supply and distribution networks in an increasing number of developing countries, which are privatizing or outsourcing these services. However, this trend has sparked conflicts and concerns of community leaders and nongovernmental organizations (NGOs) about social justice and water equity for the poor. Sections of the book examine this point in detail.

The already-limited supply of clean freshwater in many countries is progressively compromised through overpumping of aquifers, water pollution, and salt intrusion. Water shortages are building up due to population increase, poor water management, urban sprawl, and rising demand from agriculture, industry, and other sectors.

CLIMATE CHANGE

Climate change is compounding the problem. Higher global temperatures due to increased fossil fuel emissions since the industrial revolution are having an impact on freshwater supplies and ocean ecosystems. Increased evaporation, glacial melt, and the growing intensity of storms, floods, and droughts all aggravate global water issues. Accelerated evaporation of lakes and reservoirs is affecting both water availability and energy supply in countries dependent on hydroelectric power.

Vulnerable island states are experiencing rising ocean levels and associated threats. Some countries in the African Sahel and elsewhere are experiencing prolonged drought. Asian countries are feeling the effects of more intense typhoons and floods, and low-lying countries such as Bangladesh and the coastlines of most developing and developed countries are particularly endangered.

This book points out that climate change disruptions, erratic weather events and water shortages, and intensified use of water for irrigation, urban growth, industry, and energy are shifting economic, social, and cultural patterns. People are often forced to migrate from flood-prone or water-deprived zones to other regions, creating environmental refugees. When people from one community encroach upon the land or water resources of others, conflicts or skirmishes can erupt, particularly if migration stokes ethnic or other grievances, as in Darfur, Sudan.

Water stress situations are becoming more widespread around the world. This is a cause for global concern. At least eight global water danger zones (Sachs 2015) and 40 most-depleted freshwater sources (Richter 2014) have been identified. While this predicament is sparking global awareness about the need to conserve, recycle, and better manage water resources, worldwide action is still too slow to implement far-reaching changes.

History teaches us that human action has repeatedly threatened or even depleted plant and animal species, ecosystems, and water resources. Humans have been slow to address or reverse the negative impacts of their own actions or to learn from these experiences. This book discusses some lessons learned from the past as well as current efforts of local, regional, international, and nongovernmental (NGO) actors in water conservation and management.

Water is the one resource we cannot live without. Human collective action to safeguard water for the present and the future is an urgent requirement. Currently, one in eight people in the world—especially in low-income communities and water-deficit countries—lacks access to potable water, and about 3.3 million die every year from water-related health and sanitation problems (National Geographic 2010, 112). The task for humanity is daunting.

WATER CONSUMPTION

Access to and consumption of freshwater varies widely within the same country and among countries. A person needs at least 13 gallons a day for drinking, bathing, and toilet flushing (a toilet uses around 4.8 gallons per flush). On average, a U.S. resident consumes 79 gallons a day; a German, 32 gallons; a Somali, 2 gallons; and a Haitian, 0.79 gallons (Black 2004, 42; Institute Water for Africa).

Due to economic development and the growth of leisure and luxury activities, people use water for a wide spectrum of activities beyond drinking, basic sanitation, and routine household tasks. Agriculture and irrigation are important activities that consume water. With population growth and rising incomes, growing consumer demand for agricultural products and meat (livestock production is a big water user) in China and emerging economies has increased globally. City development, industrial growth, and high demand for services and energy have led to additional water stress.

As the book discusses, water is a vital aspect in economic, social, and cultural life. There is a wide range of activities that require water for everyday functioning of society. These include: construction, fishing, boating, cruising, gardening, irrigation of golf courses, firefighting, cleaning, and taking care of the sick. Many religions integrate water into spiritual rites. Water is present in art, design, architecture, and literature. There are also gender aspects of water ownership, decision-making, distribution, and allocation patterns.

WATER HAVES AND HAVE-NOTS

A main challenge facing the world is the growing global divide between water haves and have-nots. Location is a key factor for water endowment. It matters where countries or communities are located, whether in rainwater surplus or deficit areas, on rivers, in valleys, or in deserts. Other variables that influence water supply are weather patterns and climatic disruptions, irrigation, water contamination, damming of rivers, and saltwater intrusion into freshwater supplies. Population, migration pressures, and water pricing also affect water consumption and people's access.

In low-income countries, poor urban communities often lack access to clean drinking water and are forced to buy water from vendors at exorbitant prices. Roughly 90 percent of wastewater in developing countries goes untreated into rivers, jeopardizing freshwater sources (Black 2004, 46). In industrialized countries, the picture is different. Availability of water is a normal part of life—and often taken for granted. Most people turn on a tap and expect clean water or take a shower without much thought as to where water comes from. As a general rule, no one in a developed country—in contrast to some African countries—needs to walk for hours to fetch water. In North America, people can, at a click, even order bottled water online for prompt home delivery.

However, the water situation is changing rapidly for developed countries as well. For example, the 2014 and 2015 California droughts highlighted the importance of water conservation and placed water scarcity squarely at the forefront of U.S. public debate and policy action. Water shortages in water-stressed areas of the world are triggering competition among water users and pitting urban against rural residents, tourism industries against power companies, upstream against downstream communities, states against states, and nations against nations.

The spread of urbanization is also straining water supplies. More than half of the world's population lives in urban settlements; expanding cities are placing even higher demands on water resources. World population is expected to reach 8 billion by 2025 and 9 billion by 2040, with growth mostly occurring in cities (Sachs 2015, 53). Parts of this book describe water supply issues of city growth, and green design and building standards for efficient water use and wastewater recycling in urban design.

Water recycling, desalination technologies, and improvements in water management and efficiency are critical tools to address freshwater supply problems. While access to clean drinking water has improved on every continent over the last 30 years or so, severe problems remain. In 2014, 1 billion people on the water planet lacked safe drinking water and over 2 billion had no access to sanitation facilities (Chellaney 2013, 17). It is estimated that by 2100 over half of the world's population could face water and food shortages.

This book outlines how development aid agencies, international organizations, foundations, and NGOs are expanding efforts to address water supply and sanitation problems and improve local water transport methods in the poorest communities of the world, especially in Asia and Africa. Since women are usually the water carriers in rural areas of developing countries, such improvements would help relieve the physical and time burden on women and girls. The book's chapter on water and gender roles analyzes this topic in further detail.

Glacial melt and water loss are disrupting the supply of water for drinking and agriculture in farming and urban areas. This process is likely to exacerbate water and food shortages, extreme poverty levels, and conflicts in the future, given rising population and growing demand for food. Some development aid programs include climate change adaptation projects to help communities address problems caused by melting of ice caps and glaciers in the Andes, Himalayas, and other mountain ranges around the world. Demand outstrips availability of resources.

Global communities are drawing down on their water bank account more frequently and at higher rates, while at the same time paying inadequate attention to freshwater supplies for future generations. This is a complex, global challenge. Climate change, freshwater depletion, threats to biodiversity, and depletion of oceans and marine resources are daunting tasks that require coordinated approaches on many fronts to restore balance. The book describes many challenges and opportunities facing humankind and the need for effective strategies to address socioeconomic, political and cultural aspects of water use.

The final chapter pulls the key themes discussed throughout the book. It emphasizes global responsibility for water resources on the planet and the relationship between sound water management and more balanced economic and human development. It underscores the importance of consensus building, political commitment, and strong stakeholder engagement within and across nations and borders. It presents some windows of opportunity for future action such as fossil fuel reduction, technology development and transfer, improved governance, policy, regulatory and financing instruments, and coordination of international, regional, and local activities in water-related sectors.

REFERENCES AND FURTHER READING

Barlow, Maude, and Tony Clarke. *Blue Gold: The Fight to Stop the Corporate Theft of the World's Water*. New York: The New Press, 2002.

Black, Maggie. *The No-Nonsense Guide to Water*. Oxford, UK: New Internationalist Publications, 2004.

Chellaney, Brahma. *Water, Peace and War: Confronting the Global Water Crisis*. Lanham, MD: Rowman & Littlefield, 2013.

Institute Water for Africa. http://www.water-for-africa.org/en/water-consumption.html.

National Geographic. *A Special Issue: Water, Our Thirsty World*. Washington, DC: National Geographic Society, 2010.

Novaresio, Paolo. *The Explorers: From the Ancient World to the Present*. New York: U.S. Media Holdings, 1996.

Ponting, Clive. *A New Green History of the World: The Environment and the Collapse of Civilizations*. New York: Penguin Group, 2007.

Richter, Brian. *Chasing Water: A Guide for Moving from Scarcity to Sustainability.* Washington, DC: Island Press, 2014.

Sachs, Jeffrey D. *Common Wealth: Economics for a Crowded Planet*. New York: Penguin Press, 2008.

Sachs, Jeffrey D. *The Age of Sustainable Development*. New York: Columbia University Press, 2015.

1 WATER AND CLIMATE CHANGE

OVERVIEW

This chapter explores climate change, its relationship with human fossil fuel use, and its impacts on the planet, water, and people's livelihoods. In particular, the burning of fossil fuels such as coal and gas—once stored millions of years ago in rock formations—releases large quantities of greenhouse gases (GHG) such as carbon dioxide (CO_2) and methane (CH_4) into the atmosphere. These gases act like blankets thrown over the earth, trapping energy in the lower atmosphere and increasing the planet's temperature over time (the so-called greenhouse effect). Rising temperatures, thermal warming of the oceans, and melting ice, are leading in turn to the rise in sea level and other impacts discussed in this chapter.

Climate change represents serious challenges for societies and economies around the world. The signs of the planet's imbalances are already evident to the naked eye as well as to scientists: warming earth; melting ice; higher sea levels; extreme storms, floods and droughts; and growing water pollution and water cycle disruptions. These occurrences pose severe threats to human livelihoods and the preservation of marine, coastal, and freshwater ecosystems.

The case studies and examples in this chapter delve into a wide range of climate-related issues around the world. They show the conundrum of small, vulnerable islands in the Pacific in the wake of rising seas; climate change adaptation projects in some developed countries; and changes in the ecosystems of European lakes. The perspectives section discusses the emerging issue of climate justice. It zooms into ongoing demands for monetary compensation by vulnerable, low GHG-emitting nations, which are currently experiencing the impacts of high fossil fuel emissions discharged by other countries.

Melting iceberg in Iceland. (Maria Reyes Gaskin)

Climate Change: A Reality

Climate change—defined by the Intergovernmental Panel on Climate Change (IPCC) as significant modifications in the average state of the climate or its variability over a prolonged period—is already affecting freshwater regimes, ocean habitats, communities, coastal ecosystems, and islands around the world (IPCC 2014). Scientific research, documented in the reports of renowned, interdisciplinary scientists of the IPCC, has assembled landmark evidence that human-led climate change has accelerated in the last 50 years or so. The IPCC in its last five reports has sounded the alarm bell about the impacts of human activities on climate change and challenges for future development.

The IPCC is the world body charged with assessing the science of climate change. It was established in 1988 by the World Meteorological Organization (WMO) and the United Nations Environment Programme (UNEP). More than 800 scientists from 80 countries assessed and reviewed over 30,000 scientific papers in the preparation of the IPCC's latest Report, the Fifth Assessment Report, called AR5 (IPCC 2014).

AR5 provided further fact-based evidence of anthropogenic (human-led) influence on the climate system, and concluded that human release of increased amounts of GHG into the atmosphere since the Industrial Revolution is the main reason for earth's rising temperatures and climate disruptions. It warned of severe risks and potentially irreversible impacts for humanity if GHG emissions were not checked. It also underscored the urgent need for and the obligation of nations to agree on emissions reduction and adaptation measures for a sustainable future. However, it has not been easy for nations and stakeholders to negotiate joint solutions to this global problem.

Consistent with IPCC findings over the years, a World Bank (WB) publication (World Bank 2013) has also documented severe threats to the planet, and called for

Intergovernmental Panel on Climate Change (IPCC) chief Hoesung Lee, of South Korea, leaves the Elysee Palace after his talks with French President Francois Hollande in Paris, France, October 15, 2015. The new chief of the Nobel Prize–winning group of climate scientists was meeting Hollande in preparation for hosting major U.N. talks aimed at slowing global warming. (Jacques Brinon/AP Photo)

global action. This report indicated that the earth is about 0.8°C warmer than pre-industrial levels of the 18th century, and may already be locked into a warming path of about 1.5°C above these levels by the middle of the 21st century.

In line with IPCC scenarios, the WB report discussed the possibility that failure to reduce GHG emissions in the next decades could potentially lead to a worst-case increase of 4°C above pre-industrial levels by the year 2100 (World Bank 2013, 1). It outlined various warming scenarios and the range of potentially devastating impacts to global societies and economies. Similar to the IPPC findings, this report argued that humans risked pushing climate systems to an irreversible point—with severe consequences for the planet's, and human, survival. Such impacts include pervasive water and food shortages; health-threatening heat waves and droughts; hazardous storms and sea level rises; and the destruction of low-lying cities, coastal environments, islands, and ecosystems.

The IPCC and World Bank studies both pointed to the chain of events unleashed by the accelerated use of fossil fuels (oil, coal, and natural gas) for manufacturing and other human-led activities since the Industrial Revolution in the 19th century. Other studies have documented this process. For example, in 1958, scientists began to measure carbon dioxide in the atmosphere of the over 13,000-foot-high Mauna Loa summit in Hawaii. Findings indicate that since 1958, CO_2 concentrations have risen every year in line with economic growth and industrialization (Maslin 2014, 6).

The Mauna Loa study, combined with other information from ice core drillings in Greenland and other areas, confirmed that global temperature variations are linked to GHG content of the atmosphere in the past century or so (Maslin 2014, 7). In the last hundred years humans have put more carbon dioxide levels into the atmosphere than in the thousands of years before the Industrial Revolution (Maslin 2014, 6).

Greenhouse Gas Emissions

The IPCC 2014 report further indicated that about four-fifths of all global GHG emissions stemmed from energy production, industrialization, and transportation, while the remaining one-fifth came from land-use changes due to deforestation, conversion of forests to farming, and expansion of settlements (IPCC, 2014 Topic 1, Figure 1.7). The earth's land surface absorbs about half of all emissions coming from human-generated activities, the oceans about 25 percent, and the earth's lower atmosphere, called the biosphere, the remaining 25 percent, thus creating the greenhouse (blanket) effect mentioned earlier (Maslin 2014, 9).

However, there are differences in the contribution of specific countries or regions to global CO_2 emissions, reflecting varying degrees of industrial development. While developed countries historically have produced the most GHG due to their early start in industrialization, this situation has been changing in the last two decades. Larger, rapidly developing countries such as China, India, Brazil, Nigeria, and South Africa are also adding to the global emissions load. The main factors for emissions growth in these countries are rising population growth; industrial, agricultural, and energy production; unregulated deforestation; urbanization; and globalized trade and export networks.

North America, Europe, and some parts of Asia (mainly China) currently release over 90 percent of global, industrially emitted GHG (Maslin 2014, 7). While, as a whole, developed countries are still the largest emitters, China produced the highest levels of CO_2 in 2007, surpassing the United States for top position in terms of volume (Maslin 2014, 9). This situation reflects China's economic liberalization and its emergence as an economic and trade powerhouse in the international economy. China's high fossil fuel consumption—particularly its reliance on coal-fired plants—has fueled its rise as a top producer and exporter of consumer goods to the world, but also polluted its environment.

In recent years, China has become increasingly aware of climate change issues due to high urban pollution levels and escalating freshwater shortages and water contamination problems. Efforts are currently underway in China to address fossil fuel emissions and increase use of renewable energy sources, such as solar energy and hydropower. In 2014, China and the United States signed a landmark agreement in which both countries agreed to reduce carbon emissions. In this context China pledged that green energy, including solar and wind power, would provide 20 percent of its energy production by 2030 (Landler 2014).

GHG releases are also occurring in South America, Africa, and many parts of Asia, not only due to industrialization and accelerated city growth, but also as a result of wood and charcoal-based cooking practices, burning of rainforests, and forest conversion to farms and pastures. The latter is a big problem in tropical countries, since rainforests store huge amounts of carbon dioxide in trees and release them into the air when burnt. The ongoing destruction of tropical rainforests not only disrupts the water cycle locally, but can also disturb rain and seasonal weather patterns elsewhere.

The Heat Is On

Since 1990, all five IPCC reports have documented the impact of the progressive warming of the earth's surface on ice melting and glacial loss. The latest report underscored that the last 15 years were the warmest period of the previous 800 years in the Northern Hemisphere, with discernible impacts on ice and snow cover, and natural, human, and plant ecosystems (IPCC 2014, 1.1.3). It documented the rapid loss of sea ice, ice cover, and ice sheets in the Arctic, Greenland, and Antarctic (particularly West Antarctica); receding glaciers around the world; increased melting of the permafrost (frozen soil) in the Northern Hemisphere; and the associated release of higher amounts of methane, a greenhouse gas (IPCC 2014, 1.1.3).

Ice and glacial melt, together with ocean thermal expansion through warming, led to a global rise in sea levels of about eight inches between 1901 and 2010 (IPCC 2014, 1.1.4). Different warming scenarios estimate that global sea levels could rise up to four feet by 2100, if humanity fails to take effective action to reduce carbon emissions (Maslin 2014, 102).

The collapse of large parts of the West Antarctic Ice Sheet is already occurring due to ice melt and the impact of warm water welling up from ocean depths. The National Oceanic and Atmospheric Administration (NOAA) also documented in 2014 the speedy loss of Arctic snow cover and sea ice, the warming of Arctic Sea waters by approximately 0.17°C per decade since 1967, and the melting of 40 percent of Greenland's ice sheet during the summer of 2014 (Chang 2014).

It is estimated that Greenland has been losing 200 billion tons of ice per year, five times more than in the early 1990s (Maslin 2014, 102). Since Greenland started losing ice in the 1970s, the reduced weight of its melting, two-mile high ice sheet has caused the island to slowly rise. In 2004, the National Snow and Ice Data Center (NSIDC), affiliated with the University of Colorado Boulder, confirmed—based on analyses of satellite data—that an enormous (650-foot thick, 1,255–square mile section) ice mass had broken off from the Greenland landmass in 2002 (NSIDC 2004).

This mass of ice (called the Larsen B ice shelf) weighed billions of tons. It had formed part of the world's ice masses for 12,000 years; scientists and the world were astonished how quickly this ice shelf calved into numerous slabs and icebergs. In 2005, further melting was observed on all edges and ground levels of Greenland. In 2012, satellite images confirmed extreme melting at the surface of about 98 percent of the

View of the breakup of the northern section of the Larsen B ice shelf. The Larsen B ice shelf collapsed and broke away from the Antarctic Peninsula during February and March, 2002. (NASA)

Larsen ice sheet. In 2014, summer in Greenland lasted 70 days longer than it had 30 years earlier.

The pace and pattern of ice melting in Greenland is part of a feedback loop that is worrisome to scientists: as surfaces warm faster on the edge of ice sheets and the ice melts, more water penetrates into the interior of the ice mass through channels or vertical shafts, called crevasses, which slowly undermine the solidity of the frozen ice. Huge waterfalls of warmer water eventually form in the crevasses and make their way throughout the ice blocks, breaking up the ice from the inside out. As a result, the formation of lakes from melting ice continues to expand throughout Greenland.

More lakes in glaciers and ice-clad areas further accelerate melting and accentuate the feedback loop, because water surfaces are darker than white ice surfaces, retain more heat, and cause further widening of ice shafts. In Greenland, the breakup of ice sheets is expected to continue and change its culture and the life of indigenous groups. It will reshape Greenland's geography, expand growing seasons, and enable agricultural use of once frozen land.

But the thawing of frozen land (permafrost) in Greenland is releasing the greenhouse gas methane, and is expected to further reinforce climate change. According to worst-case estimates, potential melting of the entire Greenland sheet could raise global sea levels more than 20 feet in the coming century (Knauer 2007, 16). It should be pointed out that since Arctic Sea ice is suspended in ocean water (and already accounted for in global measurements), melting in the Arctic does not impact sea level rise.

Ice is disappearing fast in the Arctic Sea. In the summer of 2000, the Arctic was largely ice free; in 2005, ice cover was 20 percent less than in 1978 (Pearce 2007, 36). An additional contributing factor to Arctic melt is thought to be warm water pulses or currents called the Arctic Oscillation, which travel north into the Arctic. Photographs of toppled Inuit houses confirm what scientific research is indicating: the melting of the permafrost. Many indigenous communities are experiencing erosion of the foundations of their homes due to permafrost melting, and are relocating to more stable land.

Warming has raised the stakes for oil drilling and mineral prospecting in the Arctic, and also increased the potential use of year-round shipping lanes for trade and transport in the area around Greenland, Siberia, Alaska, Russia, and Canada. However, intensified natural resource extraction in the Arctic could be a double-edged sword. While the thaw improves access to formerly frozen reserves of oil, zinc, gold, diamonds, and uranium, there is a huge risk that mining and oil production could provoke conflicts, expedite permafrost melt, and release methane and carbon dioxide. Increased oil production also raises the likelihood of oil spills and threatens wildlife dependent on Arctic ecosystems.

Similar melting processes can be observed in tropical countries. Glaciers are disappearing fast on mountain peaks in highland zones. In 2014, the World Bank released a regional climate change report on Latin America and the Caribbean, documenting melting glaciers; reduced freshwater availability in the Andes; erosion, flooding, and agricultural losses; reduced water for drinking and hydroelectric production; and negative impacts on vulnerable, indigenous highland populations in Peru, Chile, Bolivia, and Ecuador, and on cities dependent on water from glacial runoff (World Bank 2014, 16). In Africa, most of the ice on Mount Kilimanjaro, Kenya has been lost in the last 90 years, and in Uganda, New Guinea, and the Himalayas, many highland glaciers are receding or have disappeared outright.

Health Impacts

Climate change impacts water, human health, and welfare in many ways: increased evaporation and drought; rapid snow and ice melt; water runoff and erosion; and floods, tidal surges, and saltwater intrusion. These events affect water supply and distribution regimes around the world. Inadequate or erratic rainfall and water scarcity threaten food production, drinking water supply, and ultimately human survival. Higher temperatures and extreme heat waves also pose health risks for people working outdoors, and the very young and elderly.

Specific health problems linked to climate change are expected to intensify in the future. Scientists estimate that warmer temperatures, higher humidity, and more frequent flooding episodes could increase mosquito habitats and the incidence of vector-borne diseases such as malaria. Widespread flooding, coupled with poor sanitation and contaminated drinking water in low-lying areas, could potentially lead to the spread of cholera and other diseases in developing countries.

Models based on global warming scenarios suggest that by 2080, millions of people currently living outside of malaria-endemic areas or in malaria-free higher altitudes could be severely threatened by direct exposure to this disease. Since the 1970s, due to warmer temperatures, we have seen the dramatic increase of malaria in highland cities such as Nairobi, Kenya. Nairobi is located 5,500 feet above sea level, and is normally outside of malaria's range. Other areas deemed most likely to suffer from expansion of malaria are Southeast Asia, the cooler Andean zones of South America, and other highland areas in Africa.

It is also estimated that in the future, worst-case warming scenarios of the earth could potentially create appropriate breeding conditions for malaria-carrying mosquitos in northern parts of North America, continental Europe, and the British Isles. In response, these areas would need to consider mitigation measures currently under use in tropical countries. They include draining of swamps and pools of water, pesticide application, and the use of mosquito nets to combat malaria. Higher pesticide use could potentially cause negative environmental health impacts and bring new challenges to the forefront.

Higher levels of carbon dioxide are also likely to promote the growth of ragweed and other plant pollen producers, provoking illnesses such as asthmatic or allergic episodes. Increased dust storms—due to droughts, growing desertification, and more frequent brushfires, as experienced in California or Australia—have the potential to exacerbate respiratory ailments worldwide. Outbreaks of diseases could disproportionately affect poor communities in developed countries or in vulnerable states already hit by climate change. In times of crises, public health facilities are heavily strained—and the poor are least able to afford quality or private medical care.

Impacts on Wildlife

The overall warming trend and variations in the timing of seasons are already affecting wildlife and causing local impacts and adaptations. A *New York Times* article (Gorman 2014) reported on a decades-long Arctic research project in the Hudson Bay, which assessed the impact of declining ice coverage on polar bears. The project findings indicated that polar bears—with less access to seals due to ice melt—have adapted by eating more snow geese and goose eggs. A new cycle has emerged: the bears come on shore earlier, because sea ice melts earlier, but the bears' arrival coincides with the snow goose breeding season on the land. This change in polar bears' eating habits compensates for less availability of seals, but may have potentially devastating consequences for snow goose survival. In spite of polar bears' adaptation responses, experts still warn that their nutritional status may be in jeopardy.

Alaska provides yet another example of climate change's impact on wildlife habitats. There, warming has increased the numbers of spruce bark beetles, traditionally a threat to Alaska's white spruce trees. In the past, these beetles were kept in check by colder winter temperatures, and mostly confined to the tree bark for most of the year. With higher temperatures and longer growing seasons, bark beetles are able to destroy larger swaths of white spruce forests. Due to forest loss and greater sun penetration on the ground, more grass has grown, which benefits moose, elk, and some birds. However, the loss of spruce trees is negatively affecting hawks, owls, voles, and red squirrels, which rely on the spruce as their habitat (Elert and Lemonick 2012).

Due to the melting of the ice, polar bears have to traverse larger areas of melting ice to hunt. (Shutterstock)

Down south in Antarctica, Magellan penguins are experiencing change as well. The increasing mortality of penguin hatchlings is linked to changing climate conditions. More intense storms and more days with heavy rains cause the plumage of penguin chicks to be soaked for longer periods, and reduce their survival rate. Penguins and their chicks are also finding it increasingly difficult to obtain food from the sea due to the growing masses of shifting ice and mega-icebergs breaking off from the melting land. In the Canadian Arctic, studies have also recorded the impact of more intense rainstorms on the mortality of peregrine falcon hatchlings (Fountain 2014).

Vulnerable Nations and Climate Equity

With the exception of China, the contribution of developing countries to global GHG still remains less than developed nations. This situation has sparked a debate on climate equity. Many developing countries are increasingly emitting GHG through deforestation, burning of biomass for cooking and heating, agricultural and industrial production, and methane discharges of livestock and vehicle emissions, but others are very low emitters. Most affected countries argue that they are disproportionately bearing the costs of climate change, caused in the first instance by industrialized countries.

The conundrum is that many developing countries—especially those with very low GHG emissions—are vulnerable to impacts resulting from the past (and increasingly the present), but lack resources to cope with their own development needs, let alone address additional climate change challenges. Low-lying countries such as Bangladesh and vulnerable islands and tropical highland countries with melting glaciers are already reeling from a range of problems. These problems include rising sea levels, flooding and displacement of coastal communities, unseasonal melt and excessive water runoff, and agricultural losses and freshwater shortages.

In some countries, the situation is complicated by additional factors such as high population density, low economic growth, lack of financial and technical resources, poor institutional capacity, and few alternative areas to relocate communities. The most vulnerable nations lack what is termed climate resilience, i.e., the ability to successfully adapt to climate change. The general rule of thumb is that the greater the financial, organizational, and technological resources of a country, the more adaptable and climate-change resilient it is likely to be.

Sea level rise is an alarming threat to many small islands and atolls in the Pacific Ocean. They are already experiencing severe coastal erosion, flooding, saline intrusion of the water table, crop failures, and the loss of groundwater reserves. To adapt, some islands have started to switch to less water-thirsty crops, while others have been forced to migrate to new settlements or build barriers to keep out the ocean.

For a few Pacific islands, it may already be too late for such prevention or mitigation measures; adaptation and exit strategies may be their best bet. However, they face two main challenges: 1) finding arable or uninhabited land on their own or neighboring islands to relocate displaced residents; and 2) addressing the reluctance of some developed nations to provide financial support or accept the displaced population as climate change refugees.

This problem has ignited discussions between affected states and industrialized countries in many United Nations climate change meetings and development aid conferences. To represent their interests, these nations have banded together under the Alliance of Small Island States (AOSIS), composed of 44 states and observers from the world, including the most vulnerable territories in the Indian and Pacific Oceans, and the Caribbean, Mediterranean, and South China Seas.

Most AOSIS states are also UN member states, representing their claim for financial support under the banner of a subgroup, called the Small Island Developing States (SIDS). Global nongovernmental organizations such as the Pacific Islands Association of Non-Governmental Organizations (PIANGO) and the Caribbean Policy Development Centre (CPDC) have been active in supporting the position of these vulnerable states (Commonwealth Foundation).

The most vulnerable islands are located in the Pacific and Indian Oceans. They include the Marshall Islands, Kiribati, Tuvalu, Tonga, the Federated States of Micronesia, and the Cook Islands. Islands in the Caribbean Sea and the Atlantic Ocean have also raised issues of climate justice, noting that they are located far from areas of melting ice and produce fewer emissions than most countries, yet are bearing the costs of rising seas and damaged ecosystems. Most insist that industrialized countries with the highest emissions since the Industrial Revolution owe them an ecological debt.

In response to these claims, developed countries—particularly the United Kingdom, Canada, Japan, Germany, and Scandinavian nations—and international aid organizations and lenders, are pledging some financial assistance to vulnerable states to help with climate change adaptation. However, many affected states consider these amounts to be inadequate to compensate for the potential loss of livelihoods, culture, and habitats. The perspectives section discusses this controversy in further detail.

Mitigation and Adaptation

Most experts agree that time is running out to prevent climate change, and there are mainly two principal courses of action to respond: a) mitigation, i.e., global agreements and national action at all levels to reduce greenhouse gases and limit the increase in average global temperature to no more than 2°C above preindustrial mean temperature; and b) adaptation, i.e., the implementation of specific measures to adapt to and live with the consequences of climate change.

Mitigation measures could be costly. They include, among others: the construction of engineering works to shore up coastal areas and cities against storm surges; research and development of technology to protect crops from higher temperatures and droughts; innovative agricultural and irrigation techniques; water storage and distribution practices to cope with water shortages; policies and programs to shift from fossil fuels to cleaner, renewable energy sources; the capture and storage of carbon; and global action to safeguard endangered freshwater resources and terrestrial, ocean, and marine ecosystems.

The main challenge is changing the current economic model. Oil, coal, and gas are the main engines of industry, agriculture, and other activities in developed countries and many developing countries. Moving away from fossil fuels to new patterns means that most countries, consumers, and businesses have to take politically difficult and costly decisions to retrofit their economic and social base.

Energy and transportation sectors are important opportunities for change. Potential areas include greater use of electric or hybrid cars, buses, and trains; redesigned cities; pedestrian, biking, and carpooling solutions; renewable energy such as wind, solar, geothermal, or wave power; more efficient irrigation, water management, and conservation techniques; and carbon capture and storage technologies.

In 1992, the world's nations held an Earth Summit in Rio de Janeiro, Brazil, and promised to take steps to curb carbon dioxide emissions. However, there were no firm, legally binding commitments to reduce GHG, then or in the numerous conferences held throughout the world over the last two decades. In fact, since the Rio summit, GHG emissions per year have continued to rise—partially due to China's economic growth and coal use, but also to the lack of coordinated action and follow up by nations.

Nonetheless, scientific research of the IPCC and the findings of many researchers and acclaimed institutions over the years have continued to advance the state of global knowledge on climate change and emphasize the importance of concerted action. A main obstacle to achieving more progress lies in the fact that climate change is a complex, multifaceted, multidimensional and multinational problem.

Potential solutions are complicated. They involve high stakes diplomacy, political consensus of the world's nations, and commitment of investors, business interests, agents of development, and individuals to change practices and attitudes. Stakeholders and policymakers across the world often have diverse (even short-term and conflicting) interests, dissimilar cultural and socioeconomic systems, and vastly different opinions and approaches, when negotiating strategies and measures to address a global problem.

Consensus building is also difficult, because shifting the economy of most countries is a politically, technologically and financially complicated undertaking. It requires time, money and unwavering commitment to action. Some countries such as Germany have made significant advances on retrofitting their energy profile, paving the way for other nations to follow their lead on future action. However, not all nations have the resources and the political will to follow the lead.

Spurred by the urgency of the situation, the most recent UN climate change conference, called the Conference of Parties (COP21), was held in December 2015 in Paris. It achieved some progress in obtaining voluntary, national GHG reduction pledges, and other commitments such as funding for vulnerable nations. Although there are mixed views on the effectiveness of these pledges to reduce emissions fast enough, COP21 has given rise to optimism and global expectations that countries might for the first time break their decades-long gridlock and follow through with collaborative agreements.

COP21 provides a huge window of opportunity for countries and communities to put the global good and collective long-term action ahead of short-term gains, or national or individual interests. Maintaining or increasing the momentum from Paris requires more than talk: it takes commitment, courage, creativity, and cooperation. The final chapter of this book touches upon the outlook for concerted global action on climate change and improved water management and protection strategies, including attention to equity considerations in the most affected and vulnerable countries in the world.

Camille Gaskin-Reyes

REFERENCES AND FURTHER READING

Ackerman, Diane. *The Human Age: The World Shaped by Us.* New York, Norton & Company, 2014.

Chang, Kenneth. "Snow is Down and Heat is Up in the Arctic, Report Says." *New York Times*, December 17, 2014. http://www.nytimes.com/2014/12/18/science/snow-is-down-and-heat-is-up-in-the-arctic-report-says.html?_r=0.

Commonwealth Foundation. Interview with H. E. Marie-Pierre Lloyd, Seychelles High Commissioner to the UK, about the Challenges and Opportunities that Face Small Island Developing States (SIDS). http://www.commonwealthfoundation.com/updates/looking-beyond-horizon-resilience-small-island-states.

Elert, Emily, and Michael D. Lemonick. *Global Weirdness: Severe Storms, Deadly Heat Waves, Relentless Drought, Rising Seas, and the Weather of the Future.* New York: Penguin Random House, 2012.

Fountain, Henry. "For Already Vulnerable Penguins, Study Finds Climate Change is Another Danger." *New York Times*, January 29, 2014. http://www.nytimes.com/2014/01/30/science/earth/climate-change-taking-toll-on-penguins-study-finds.html.

Gillis, Justin, and Kenneth Chang. "Scientists Warn of Rising Oceans from Polar Melt." *New York Times*, May 12, 2014. http://www.nytimes.com/2014/05/13/science/earth/collapse-of-parts-of-west-antarctica-ice-sheet-has-begun-scientists-say.html.

Gorman, James. "For Polar Bears, A Climate Change Twist." *New York Times*, September 22, 2014. http://www.nytimes.com/2014/09/23/science/for-polar-bears-a-climate-change-twist.html?mtrref=query.nytimes.com&gwh%20=82EA7B87E4E908E94215490EE3887184&gwt=pay.

Intergovernmental Panel on Climate Change (IPCC). "Climate Change 2014: Synthesis Report." http://www.ipcc.ch/report/ar5/syr/.

Jopson, Barney, and Pilita Clark. "Compensation Measures Key to UN Climate Change Summit." *Financial Times*, November 25, 2015.

Klein, Naomi. *This Changes Everything: Capitalism vs. the Climate*. New York: Simon & Schuster, 2014.

Knauer, Kelly. *Global Warming: The Causes, the Perils, the Solutions, the Actions, What You Can Do*. New York: Time, 2007.

Kolbert, Elizabeth. *The Sixth Extinction: An Unnatural History*. New York: Henry Holt and Company, 2014.

Landler, Mark. "U.S. and China Reach Climate Accord After Months of Talks." *New York Times,* November 11, 2014. http://www.nytimes.com/2014/11/12/world/asia/china-us-xi-obama-apec.html.

Maslin, Mark. *Climate Change: A Very Short Introduction*. Oxford: Oxford University Press, 2014.

McDonnell, G., J. D. Ford, B. Lehner, L. Berrang-Ford, and A. Sherpa. *Climate-related Hydrological Change and Human Vulnerability in Remote Mountain Regions: A Case Study from Khumbu, Nepal*. Oxford: Springer-Verlag, 2012.

National Snow & Ice Data Center (NSIDC). 2014. "Larsen Ice Shelf Breakup Events." https://nsidc.org/news/newsroom/larsen_B/index.html.

Pearce, Fred. *With Speed and Violence: Why Scientists Fear Tipping Points in Climate Change*. Boston: Beacon Press, 2007.

Sachs, Jeffrey D. *Common Wealth: Economics for a Crowded Planet*. New York: Penguin Press, 2008.

World Bank. *4 Degrees: Turn Down the Heat: Climate Extremes, Regional Impacts and the Case for Resilience*. Washington, D.C.: World Bank Publications, 2013.

World Bank. *Climate Change Impacts in Latin America: Confronting the New Climate Normal.* Washington, D.C.: World Bank Publications, 2014.

CASE STUDIES

The following case studies focus on the impacts of climate change on spaces, places, and communities in different geographical areas and economic levels of development. They include the dilemma of vulnerable, low-lying islands and their adaptation strategies; planned or executed protection works of developed countries against rising sea levels and storm surges; and the impact of climate change on lakes in Europe.

Case Study 1: Small, Vulnerable Islands: Fallout of Climate Change

This case study discusses the vulnerability of small islands in the Pacific and Indian Oceans facing rising sea levels and extreme storm events. It also looks at mitigation and adaptation strategies and the ongoing efforts of such island states to rally against the impacts of climate change on their societies and economies.

The Carteret Islands belong to this group. They are low-lying coral islands that are part of Papua New Guinea in the South Pacific. Since the 1960s, they have been progressively experiencing more severe storms, and are currently subject to erosion and invasion of the sea. Carteret Island residents are rapidly losing their way of life, their ancestral homes, and their livelihoods.

The situation worsened after 2001, when strong winds and heavy seas cut off parts of their atolls and hindered fishing activities. The islands became more and more narrow through encroaching seawater, and crops started to die as a result of saltwater intrusion. In 2005, the government of Papua New Guinea made arrangements for the Carteret islanders to move to central Papua New Guinea. For these inhabitants, outmigration had become the last and only option to respond to climate change (Sherbinin et al. 2011).

Threatened Communities on the Move

The 10,000 citizens of the nine inhabited islands of Tuvalu are either on the move or preparing to move. Over the last decade, these residents have been suffering from the encroachment of high tides and salt water on their fields and coconut groves. Calling climate change the new weapons of mass destruction, in 2007 the authorities of Tuvalu created a National Adaptation Programme of Action (NAPA) to formulate a strategy to adapt to climate change. One of Tuvalu's adaptation responses was to sign an agreement with New Zealand, 1,800 miles away, to accommodate the anticipated migration of Tuvalu's inhabitants to New Zealand in the future.

Kiribati, another island nation, is one of the most severely threatened Pacific Islands. With a total land area of about 313 square miles, Kiribati is composed of 33 coral atolls, of which 21 are inhabited. It has limited resources and relies on severely threatened activities such as the production of copra (dried coconut fiber) and fishing

(Kennedy 2015). Kiribati lost two entire islands to the ocean in 1999. Since then, inhabitants of the smaller outer atolls have been moving to larger islands within the island group in preparation for a potential, final move to other countries. In May 2014, Kiribati purchased land in Fiji to ensure food security for its inhabitants and possible resettlement of communities in the future. Kiribati's limited tourism activities are expected to decline and peter out as the sea advances (Davenport 2014).

While Fiji, another Pacific Island, is open to helping its neighbor, Kiribati, Fiji itself is under climate stress. It too has suffered from the effects of the invading ocean. The government of Fiji has started to relocate inhabitants from the outer islands of the archipelago to the safety of other areas. In many coastal villages of Fiji, salt water has seriously damaged soils and rendered them useless for agriculture. Fiji's adaptation response to threats to its freshwater has been to construct desalination plants to convert seawater to freshwater and also build huge water containers to store freshwater reserves.

The Maldives, an island chain in the Indian Ocean, off the tip of Sri Lanka, is also threatened. It risks losing a huge chunk of its land to sea intrusion and flooding. Home to 350,000 people, it is a well-known tourist destination, with tourism accounting for about 28 percent of its economic output. However, the foundation of life and human welfare in the Maldives is likely to be undermined. Malé, its capital, houses 100,000 people on less than a square mile, making it one of the most densely populated capitals in the world. Freshwater reserves are becoming more and more depleted in low rainfall periods. In addition, coral bleaching, declining biodiversity, and accelerated beach erosion are posing serious threats to tourism activities and people's livelihoods (AOSIS 2015).

One adaptation response of the Maldives to the growing crisis has been to construct seawalls to reduce its vulnerability to ocean flooding and storm surges. The Great Malé Wall was constructed at a cost of 60 million dollars. Maldivians are also building additional stone structures to protect their homes and reclaim new land to house people forced to migrate from other areas of the island.

Global Organization and International Protests

Government representatives of the Maldives continue to be tireless speakers on behalf of themselves and climate change-affected areas around the world. In a 2010 interview the then-president of the Maldives stressed that sea levels were rising at a rate of about 0.03 inches per year, and warned that a potential three-foot rise in sea level under worst case scenarios would be devastating for the country (Bowermaster 2010, 65). The government is also establishing a sovereign wealth fund to set aside reserves for future generations of Maldivians to cope with the impacts of the crisis.

The government of the Maldives, which once convened an underwater cabinet meeting to raise global consciousness of the country's inundation problems, has lamented that many developed nations around the world had not yet taken sufficient action to curb their fossil fuel emissions. The authorities of the Maldives (and other vulnerable island states) have emphasized accountability of all nations and outlined their own plan to become largely carbon neutral within the next decade. This plan foresees greater use of renewable energy sources such as wind, solar, and biomass use.

Like Kiribati, Tuvalu, and the Maldives, the Marshall Islands (MI) are experiencing difficulties associated with tidal floods and coastal erosion. Located halfway between Australia and Hawaii, the MI are composed of two archipelagic island chains of 29 atolls and five single islands. The land area is about 45 square miles with a coastline of 230 miles. In the past, economic assistance from the United States and lease payments for the use of Kwajalein Atoll as a U.S. military base have played a major role in the economy. Currently, the MI are less reliant on U.S. payments and depend more on production of coconuts and breadfruit, two products threatened by climate change. The government has approached the Australian government for help to address the impending crisis, but so far no official agreement has been finalized.

Vanuatu is an example of another Pacific Ocean Island subject to destruction. In 2015, a devastating cyclone hit the island. This cyclone brought to the forefront the danger of extreme storms and natural disasters in low-lying islands. Cyclone Pam destroyed about 90 percent of Vanuatu's housing and caused the deaths of at least 24 inhabitants. It damaged food sources throughout the island, destroyed social infrastructure such as schools and clinics, and wiped out a range of economic activities across the island (AOSIS 2015).

All the islands discussed above are part of the Small Island Developing States (SIDS) group, which—represented by AOSIS within the United Nations system—has become increasingly outspoken in world climate change discussions and disaster risk reduction conferences. Using what is termed climate change diplomacy, they have consistently argued that for many of them, climate change prevention mechanisms are no longer possible. Most of their efforts are related to seeking funding from development institutions and industrial countries to finance mitigation and adaptation projects.

The members of the SIDS group and their sympathizers contend that assistance is not a matter of charity but survival, and high GHG emitters have a moral obligation to help low emitters stuck with the problems. In support of these claims, India and other countries have floated a proposal for climate justice reparations that would require developed countries to earmark 0.5 percent to 1 percent of economic output to assist affected nations. This idea has not had traction among industrialized nations (Jopson and Clark 2015).

Camille Gaskin-Reyes

REFERENCES AND FURTHER READING

Aginam, Obijofor. *Climate Change Diplomacy and Small Island Developing States*. Tokyo: United Nations University, 2011.

Alliance of Small Island States. http://aosis.org/.

AOSIS. "Maldives Environment Minister Addresses Sendai Conference in Wake of Cyclone Pam." March 17, 2015. aosis.org/maldives-environment-minister-addresses-sendai-in-wake-of-cyclone-pam/.

Bowermaster, Jon, ed. *Oceans: The Threats to Our Seas and What You Can Do to Turn the Tide*. Philadelphia: Public Affairs Books, 2010.

Davenport, Coral. "Rising Seas." *New York Times,* March 28, 2014. http://www.nytimes.com/interactive/2014/03/27/world/climate-rising-seas.html.

Jopson, Barney, and Pilita Clark. "Compensation Measures Key to UN Climate Change Summit." *Financial Times,* November 25, 2015.

Kennedy, Warne. "Against the Tide: Rising Seas Threaten to Swamp Kiribati, but the Spirit of the Islanders Is Resolute." *National Geographic* 228, no. 5 (November 2015).

Sherbinin, Alexander et al. "Preparing for Resettlement Associated with Climate Change." *Science* 334 (2011): 456–457.

Tuvalu National Adaptation Programme of Action (NAPA). www.adaptation-undp.org/projects/Tuvalu-napa.

Case Study 2: Climate Change Adaptation Measures in Developed Countries

The previous case studies on problems facing low-lying islands in the Pacific and Indian Oceans outlined their vulnerability to the ravages of climate change and their efforts to protect themselves against threats to their existence. Most developing nations lack funds and technology to protect freshwater reserves, ensure food security, and implement adaptation projects such as seawalls, stone barriers, and large-scale community relocation—while coping with the developmental challenges of poverty, inequity, and inadequate social and health services.

While United Nations conferences and other global fora continue to debate funding for projects to assist vulnerable states, some developed countries with low-lying coastlines and exposure to growing storm surges and other impacts have already set aside funds to cover their own projects. This case study takes a look at examples of planned or already operational high-cost barrier projects in England, Holland, and Italy to address rising sea levels and flood risks in their environments.

The huge scale and financial, technical, or technological requirements of such projects make it difficult for lower-income countries to replicate these efforts. However, they provide valuable lessons for the future. One key lesson is that planning for climate change takes foresight, engineering, technical know-how, organization, governance capacity, stakeholder engagement, and considerable human and financial resources. The reality is that most poor, vulnerable countries cannot afford such costs or may not be able to quickly acquire the technical capacity and expertise to plan and execute major works.

Protecting London from Storm Surges

The idea to build a River Thames barrier to protect London arose after a massive storm in the 1950s led to the flooding of London and the death of 58 people. However, it was not until 1983 that the floodgates on the River Thames became operational. These protective gates, located about 11 miles from central London, are designed to close and open, either fully or partially, to protect the city from flooding due to tidal waves and surges coming from the Thames estuary.

The planners of the Thames project estimated that the barriers might need to close once or twice a year. However, due to extreme storms along the North Sea and the English Channel in the past decade, the floodgates have closed an average of 10 times per year. Modeling scenarios from 2004 estimate that a combination of rising temperatures, sea-level increases, and more intense storm events in the future could cause tidal surges to reach even further inland than planned, exposing more people, agricultural land, and property to flooding (Royal Geographical Society).

Under the highest GHG emission scenarios, it has been estimated that the risk of floods to England would be 30 times greater in 2080 than at present levels. The barrier gates were designed to provide for a high level of protection against a one-in-2000-year storm, but more recent projections have indicated a scenario by which sea level rise by 2030 could produce a one-in-1000-year (a higher incidence) intensity of storms. In fact, there are ongoing discussions about the possibility of adding an additional 12 inches of protection on top of the existing gates (Knauer 2007, 68).

Protecting the Netherlands from Flooding

Another example of flood protection is the Delta Works barrier in the Netherlands along the North Sea coast. Traditionally, the Dutch have always had to address storm surges, floods, and water intrusion through a complicated system of dykes, sluices, dams, and polders to mitigate the risks of their low-lying geographical location. Dutch legal provisions for coastline protection traditionally foresee building sea defense installations to withstand the odds of one-in-100-year flooding events.

Due to climate change, accelerated snow melt in the Alps, and increased water flow of the Rhine River (which empties into the North Sea in Holland), Dutch planners are now rethinking flood management scenarios, coming from both the sea and the land. The new strategy is to adapt to projected increases of flooding through providing predesignated areas to store additional floodwaters from all sources, until they recede over time.

While there are increasing risks from the potential overflow of the Rhine River, which the Dutch are addressing in the planning of new protection works, the greatest flood dangers to the Netherlands in the future are still expected to come from the North Sea. Maintaining the highest levels of protection and adaptation to increased flooding through fortified dams, barriers, and massive engineering works is projected to cost about 1.3 billion dollars annually (Knauer 2007, 67).

The Dutch are willing to bear these costs, because adaptation to climate change is a major priority for the government and the entire society. State-of-the-art information on climate change scenarios and the growing threats to the Netherlands are well known to the Dutch population and discussed throughout all sectors and communities. The Dutch are not only committed to protecting coastal cities, low-lying agricultural areas, and their very existence, but also to reducing GHG through using sources of clean energy and a policy of widespread public transport and bike use.

Safeguarding the Legacy of Venice

Another example of an adaptation project currently underway in a developed country is the proposed Venetian Lagoon project (also called the Moses Project after the biblical reference). This is a megaproject intended to protect Venice from further floods. The lagoon project is still in the construction and testing stage and is slated for completion in 2016. Since record flooding in Venice in 1966, the city has increasingly experienced frequent floods and water overflows into inhabited and culturally important areas.

Numerous photographs of tourists wading through floodwaters in St. Mark's Square have appeared time and time again on the front pages of newspapers in Italy and the world (*Guardian* 2015). Venice has been sinking over the last thousand years, but due to rising sea levels in the last century or so, the average water level in the Venetian Lagoon is nine inches higher than in the last century. In 2000, the city's most popular and impressive square, St. Mark's Square, flooded about 50 times. Since then, flooding has become a regular occurrence (Nadeau 2001).

This development is literally eroding the cultural and physical base of Venice's 16th-century palaces, and submitting priceless pieces of art and architectural gems to humidity, mold, and decay. It is also threatening the city's tourism lifeline. There is an additional complication. Since the 1966 landmark flood, the local population has been steadily declining. Since 2006, Venice's population has practically halved from 121,000 to 62,000 people. Since about 20 million tourists visit Venice each year (roughly 55,000 on average per day), it is likely that on some days tourists may outnumber the local population (Kington 2009).

The Venetian Lagoon project started construction in 2003; after severe delays and financial setbacks, the first four modules of the project were installed at the Lido inlet in 2013, one of the three waterways that connect the Venetian Lagoon to the Adriatic Sea. The protective barriers will be comprised of 78 independently operated gates of varying sizes. Unlike the Thames floodgates, which are located above the river level to stop sea or water surges from entering, the Venice gates are designed to lie on the seafloor, and to rise as needed through pumping with compressed air (Guarino 2014).

Historically, the early development of Venice was based on its city-state commercial function and location on major trade routes. Today, it is still a major port that handles a sizeable amount of cruise ship and cargo traffic. The proposed barriers are designed to include a deeper canal to allow large tanker and cruise ships to enter the lagoon, and thus preserve Venice's mercantile function. Nonetheless, some environmental groups have opposed the lagoon project, claiming that the engineering works could cause negative impacts on the lagoon and the inlets, and potentially increase erosion through the creation of an artificial island at the inlet.

Critics have suggested alternatives to the megaproject such as the construction of smaller-sized projects within lower parts of the city to protect buildings and shore up walls at the banks of the historic Venice canals. Some have maintained that the problem of flooding in Venice results less from rising sea levels and more from the lack of maintenance of canals and the inadequate cleaning and dredging of garbage and sediments suspended in Venice's numerous waterways. Other critics have raised more general objections about the increasing size of cruise ships and tankers currently entering the Venice lagoon and their potential impact on marine pollution, lagoon erosion, and the city's historical and cultural integrity.

Until the lagoon project is completed and in full operation, controversy on the effectiveness or lack thereof of the Venice barriers is expected to continue. At this point, Venice is at an important crossroads: failure to arrest the problem of a sinking city subject to more frequent floods would condemn it to becoming a beautiful but empty museum-like theme park that would increasingly be submerged under water.

Completion of the barrier project would afford some protection, but it appears the design would be more acceptable to environmentalists and other groups if it integrated concerns about the equilibrium of the lagoon ecosystem, the preservation of Venice's culture, and the implementation of more robust erosion control measures.

This situation represents a paradox of sorts: Venice may need tourist flows and cruise ships to survive economically and arrest the decline of its population and economy, but an accelerated increase in mass tourism has the potential to undermine the city's future, destroy its beauty, and spur further departure of Venetians.

Camille Gaskin-Reyes

REFERENCES AND FURTHER READING

Guardian. "A History of Flooding in the Sinking City of Venice—in Pictures." June 16, 2015. http://www.theguardian.com/cities/gallery/2015/jun/16/history-flooding-sinking-city-venice-in-pictures.

Guarino, Alessandro. "Venice Flood Barrier: MOSE Project Keeps the Sea at Bay," *Engineering and Technology Magazine* 9, no. 8 (August 11, 2014). http://eandt.theiet.org/magazine/2014/08/venice-master-of-water.cfm.

Kington, Tom. "Who Now Can Stop the Slow Death of Venice?" *Guardian*, February 28, 2009. http://www.theguardian.com/world/2009/mar/01/venice-population-exodus-tourism.

Knauer, Kelly. *Global Warming: The Causes, the Perils, the Solutions, the Actions, What You Can Do*. New York: Time, 2007.

Nadeau, Barbie. "The Plan to Refloat Venice." *Newsweek*, March 11, 2001. http://www.newsweek.com/plan-refloat-venice-148687.

Royal Geographical Society. 21st Century Challenges. "The Thames Barrier." http://21stcenturychallenges.org/the-thames-barrier/.

Case Study 3: Impacts of Climate Change on European Lakes

Lakes have been termed sentinels of climate change because they collect water that drains from the landscape and integrate water flow in watersheds. Warming primarily alters the physical properties of lakes and reservoirs, which in turn influences chemical and biological processes. Long term monitoring of lakes in many parts of the world has already revealed clear evidence of the impacts of climate change, and provided a glimpse of what can be expected in an even warmer future. This case study describes such effects in European lakes and reservoirs, with specific examples from Lake Stechlin and Lake Arend in Germany.

Effects of Warming on Lakes

One of the effects of climate change is an increase in surface water temperature. In general, average surface water temperature closely follows average air temperature. For example, air and surface water temperatures have increased by about 1.3–2°C over the last 35 years in Lake Stechlin and Lake Arend, causing an increase in the length of the swimming season. Whereas the number of days with a water temperature greater than 22°C in Lake Arend has increased from four to nine since 1980, projections suggest it may increase to 55 days by the end of the 21st century. Higher temperatures also increase the burden on water quality in lakes.

Temperature affects almost all chemical and biological processes. Warming of lakes and reservoirs increases recycling of nutrients and primary productivity, thus mimicking eutrophication (Jeppesen et al. 2010). Eutrophication is the increase in growth of plants and algae, usually as a result of pollution from fertilizers and wastewater. Germany has taken steps to manage eutrophication, including improved wastewater treatment and bans on phosphate (a powerful fertilizer) from household detergents. While these measures have largely been successful in recent decades, climate change threatens to roll back this progress.

Stratification and Mixing of Lake Waters

Climate change also alters stratification, the layering and mixing of water in lakes. Lakes stratify when layers of warm and cold water form and mingle. In temperate zones, shallower lakes may stratify intermittently in summer, whereas deeper lakes (more than 30 feet) tend to stratify continuously from spring through summer, when warmer, less dense surface water floats above colder deep water. Deeper lakes mix in the fall and spring, a process known as overturn, and can also stratify in winter, if air temperatures fall below freezing, or if a colder layer of ice or near-freezing water floats above warmer water.

Since climate warming raises surface water temperatures higher than deep-water temperatures, it increases the strength and duration of stratification. Recent records show that the duration of stratification has increased by three to eight days per decade in a number of German lakes and reservoirs (Shatwell et al. 2013). This means that the length of the vegetation period or growing season is also increasing due to climate warming.

Stratification is very important for lake ecology and affects many physical, chemical, and biological processes. Cyanobacteria or blue-green algae profit from stable stratification with warm surface waters (Paerl & Huisman 2008). Many species of cyanobacteria contain gas vesicles, which allow them to rise to the water surface during calm weather and form green scum. These species are often toxic, a nuisance to bathers as well as a threat to humans and animals if they enter potable water supply systems.

Droughts, heat waves, and storms are expected to increase due to climate warming, but more specific effects on lakes and freshwater ecosystems are still poorly understood. Changes in rainfall regimes, for instance, may lower water levels in many small, shallow lakes. The loss of water volume and the increase of evaporation would concentrate nutrients and promote algae growth. Reservoirs are particularly vulnerable, because maintaining high water levels, and thus better water quality, must be balanced with water supply needs.

Oxygen Depletion Problems

The European heat waves in 2003 and 2006 caused severe oxygen depletion and large algae blooms in lakes (Jankowski et al. 2006; Jöhnk et al. 2008). In a curious cascade of events in Lake Stechlin in the summer of 2011, a large storm stirred surface waters and brought deep-dwelling algae to the surface, which initiated a bloom and a chemical reaction that turned an ultra-clear lake into a milky green. This incident showed that algae are the general winners in a warmer climate.

Another consequence of altered stratification and mixing is decreased oxygen. Oxygen is vital for the survival of many aquatic organisms and for maintaining a chemical balance in lakes. During stratification, deep water is disconnected from surface water, and consequently from the supply of fresh oxygen in the air. An increase in the duration of stratification increases oxygen depletion and leads to the oxygen-free "dead zones" being reported in lakes around the world as a result of climate change (North et al. 2014).

Winters have become milder under climate change, decreasing the length of ice cover and winter stratification in temperate lakes (Magnuson et al. 2000). Since ice cover also restricts oxygen supply, shorter ice duration improves oxygen and reduces the occurrence of winter fish kills in shallow lakes due to asphyxiation. However, the effect in deep lakes is less clear, because warming may also decrease the duration and intensity of mixing during overturn, which is necessary to replenish depleted oxygen (Livingstone 2008).

Lake Stechlin in Germany provides an interesting experimental insight: cooling water pumped into the lake from a nuclear power plant has effectively eliminated ice cover. However, the net effect was not an increase in oxygen but a strong decrease, probably because mixing during winter was too weak to carry oxygen down to deep waters.

Altogether, it is obvious that climate warming appears to have strong negative effects on oxygen in lakes. The absence of deep-water oxygen can accelerate the loss of water quality by altering the sediment chemistry to cause the release of more nutrients (phosphorus) into the water. Higher levels of phosphorus often lead to a downward spiral into oxygen loss (eutrophication) because increased nutrients fuel the production of organic matter, which depletes oxygen even further (Carpenter 2003). This loss of water quality has already been observed in Lake Tahoe in the United States and Lake Zurich in Switzerland. It appears to be a major threat to Lake Stechlin, where nutrient levels and oxygen consumption have increased drastically in recent years (Sahoo et al. 2013; North et al. 2014).

Climate change therefore poses a threat to health and biodiversity of lakes and reservoirs and a number of challenges. Major challenges include managing diffuse sources of nutrients (e.g., from the air and groundwater), solving usage conflicts in reservoirs (e.g., managing water quality and quantity), and controlling blooms.

Long-term monitoring has also shown how warming counteracts management efforts to address eutrophication (Schindler 2006). Eutrophication in the 21st century will be climate induced and mainly dependent on the condition of the watershed. Managing climate change consequences thus requires complex political decisions supported by the development and use of ecological models and long-term monitoring.

Tom Shatwell

REFERENCES AND FURTHER READING

Carpenter, S. *Regime Shifts in Lake Ecosystems: Pattern and Variation*. Oldendorf/Luhe: International Ecology Institute, 2003.

Jankowski, T, D. M. Livingstone, H. Bührer, R. Forster, P. Niederhauser. "Consequences of the 2003 European Heat Wave for Lake Temperature Profiles, Thermal Stability, and Hypolimnetic Oxygen Depletion: Implications for a Warmer World." *Limnology and Oceanography* 51 (2006): 815–819.

Jeppesen, E, B. Moss, H. Bennion, et al. "Interaction of Climate Change and Eutrophication." In *Climate Change Impacts on Freshwater Ecosystems*, edited by M. Kernan, R. Battarbee, and B. Moss, pp. 119–151. Oxford, UK: Wiley-Blackwell, 2010.

Jöhnk, K. D., J. Huisman, J. Sharples, B. Sommeijer, P. M. Visser, J. M. Stroom. "Summer Heat Waves Promote Blooms of Harmful Cyanobacteria." *Global Change Biology* 14 (2008): 495–512.

Livingstone, D. M. "A Change of Climate Provokes a Change of Paradigm: Taking Leave of Two Tacit Assumptions About Physical Lake Forcing." *International Review of Hydrobiology* 93 (2008): 404–414.

Magnuson, J. J., D. M. Robertson, B. J. Benson, et al. "Historical Trends in Lake and River Ice Cover in the Northern Hemisphere." *Science* 289 (2000): 1743–1746.

North, R. P., R. L. North, D. M. Livingstone, O. Koster, R. Kipfer. "Long-term Changes in Hypoxia and Soluble Reactive Phosphorus in the Hypolimnion of a Large Temperate Lake: Consequences of a Climate Regime Shift." *Global Change Biology* 20 (2014): 811–823.

Paerl, H. W. and J. Huisman. "Climate—Blooms Like It Hot." *Science* 320 (2008): 57–58.

Sahoo, G. B., S. G. Schladow, J. E. Reuter, et al. "The Response of Lake Tahoe to Climate Change." *Climatic Change* 116 (2013): 71–95.

Schindler, D. W. "Recent Advances in the Understanding and Management of Eutrophication." *Limnology and Oceanography* 51 (2006): 356–363.

Shatwell, T, G. Ackermann, M. T. Dokulil, et al. "Langzeitbeobachtungen zum Einfluss von Klimawandel und Eutrophierung auf Seen und Talsperren in Deutschland." *Korrespondenz Wasserwirtschaft* 12 (2013): 729–736.

ANNOTATED DOCUMENT

Intergovernmental Panel on Climate Change (IPCC) Fifth Report 2014: Summary for Policymakers (PM)

Background

In 1988 the World Meteorological Organization (WMO) and the United Nations Environment Programme (UNEP) established the Intergovernmental Panel on Climate Change (IPCC) to assess climate change and to enable policymakers around the world to make decisions on reducing GHG and curbing the impacts of climate change.

Following a series of four reports over the years, the latest global IPCC report of 2014 (called AR5) is a compendium of state-of-the-art scientific information on climate change, its impacts, future risks, and options for adaptation and mitigation. Over 800 scientists from 80 countries prepared and/or reviewed over 30,000 scientific papers to assemble the report.

The bulk of the IPCC work was based on three technical Working Groups: I: Physical Science Basis; II: Impacts, Adaptations, and Vulnerability; and III: Mitigation of Climate Change. A synthesis report integrates main research findings into one document. The final IPCC report consisted of a main document with 16 chapters, various Annexes, a Technical Summary, and a Summary for Policymakers (SPM).

The SPM contains three main sections: Section A: observed climate change impacts, vulnerability, exposure, and adaptive responses; Section B: future risks and potential benefits; and Section C: principles for adaptation to climate change and relationships among adaptation, mitigation, and sustainable development measures. The following section summarizes the content of the SPM and its implications for policymakers. Excerpts from the document are also included.

Summary

The gist of the SPM is that nations of the world have a sound, factual basis to understand the science of climate change and its impacts. Such impacts fall under three main categories: physical systems (glaciers, snow, ice, permafrost, aquifers, rivers, lakes, and sea levels); biological systems (weather events, terrestrial and marine ecosystems); and human-managed systems (food, livelihoods, health, and economic development).

Based on the findings presented by researchers in various disciplines across the world, the IPCC compressed the wealth of scientific information in the SPM, in a format that is easier for top decision-makers and policy makers to grasp and act upon.

Implications of the SPM: Addressing Vulnerability to Climate Change in a Complex and Changing World

Scientific evidence from the IPPC and the SPM overwhelmingly illustrates that climate change is linked to the effects of human (anthropogenic) interferences with the global climate system, and related impacts on natural ecosystems and human environments throughout the world. The main risks associated with these changes are the vulnerabilities of specific locations, societies, and communities, and differences in the capacity of countries and regions to address threats and to cope with the fallout of climate change.

According to the SPM, variability in precipitation regimes and increased snow and ice melt are modifying the planet's hydrological systems, with far-reaching impacts on water volume, water quality, and sea levels. These changes have been exacerbated by the rise in population growth, the rapid pace of urbanization, and increasing human demands on freshwater and marine resources.

Terrestrial, freshwater and marine species are increasingly under stress, and some organisms are forced to adapt to changes in ecosystems through adjusting their seasonal or breeding activities, migration patterns, and interactions with other species.

Some observed or potential occurrences as a result of climate change include: extreme storms, sea level rise, and coastal flooding; disrupted livelihoods of coastal inhabitants; and increased mortality due to heat, food insecurity, spread of disease vectors, and freshwater shortages. Additional risks are related to the outright loss of some marine, coastal, terrestrial, and inland water species.

The SPM indicates that more frequent heat waves, cyclones, droughts, and wildfires are also leading to water and food supply disruptions in many regions of the world; low-lying nations, low income countries, or those undergoing political or armed conflicts are more vulnerable to these developments.

While it is still hard to quantify the exact impact of climate change on human health, IPCC reports point to the fact that temperature and rainfall changes are altering the distribution of some waterborne diseases. Heat-related mortality is on the rise, while cold-related mortality cases appear to be declining. These changes are most likely to jeopardize the poorest rural and urban communities as well as highland and lowland inhabitants in developing countries, and potentially increase the incidence of poverty.

Managing Risks and Building Resilience

The likelihood and severity of climate change and associated risks to humanity varies under IPCC scenarios, which warn of considerable impacts under a 1° to 2°C increase of global mean temperatures, and irreversible impacts under a 4°C or more increase above preindustrial levels.

To prepare for and cope with such impacts, the IPCC urges nations and communities to build resilience, i.e., the ability to adapt. Some countries and communities are more aware of this skill and consequently better prepared than others to address threats by embedding adaptation measures in their planning process and implementing engineering works and water and disaster risk management measures. The case studies in this chapter show how Holland, for example, has already started to integrate climate change planning into coastal zone and water management and environmental and land use practices, while other countries are still struggling to cope with rising sea levels.

Due to lack of engagement with the severity and timing of anticipated climate change scenarios and impacts, inadequate political consensus internally, or lack of funds, some nations and policymakers have not yet fully grasped the implications of climate change or made specific decisions to curb greenhouse gas emissions and implement other measures. The SPM report aims to raise consciousness and assist policy makers to take action. On a national scale, decisions to commit to concrete action require engagement of the entire society, financial resources, and the development of innovative, technological strategies.

On a global scale, responses to climate change threats are even more complicated, since they involve political consensus and cooperation among hundreds of nations to act together for the common good. This includes the ability to break through institutional gridlock and get to robust commitments to curb fossil fuel emissions within countries and key sectors and across countries. In addition to political will, countries also need to actually make use of the available scientific information (in IPCC and other reports) and use decision-making criteria and the analysis of issues and options to understand the risks of inaction as well as the benefits and trade-offs of taking action.

Managing risks is crucial to reduce vulnerability and build climate change resilience, i.e., the ability to cope with climate change on all fronts—from communities to individual citizens to policy makers to politicians in power. How well nations mitigate climate change and manage risks is also affected by political and economic interests and different levels of economic development. Social and cultural contexts also matter in a diverse world, as do differences in global environmental consciousness, public attitudes, and perceptions of risk.

The development of climate-resilient pathways includes improvements in legal, financial, policy, and governance structures; food and water security; resolution of water conflicts and management of natural disasters; protection of water and terrestrial ecosystems; and greater use of green technology for the energy and transportation sectors. The SPM report underscored the need for nations to take mitigation and adaptation measures. It emphasized good governance as crucial to better manage and coordinate responses of governments, decision makers, communities, stakeholders, and organizations to cope with all aspects of climate change.

Camille Gaskin-Reyes

Excerpts from the IPCC Summary for Policy Makers (SPM) Document

SPM 1. Observed Changes and Their Causes
Human influence on the climate system is clear, and recent anthropogenic emissions of greenhouse gases are the highest in history. Recent climate changes have had widespread impacts on human and natural systems.

SPM 1.1. Observed Changes in the Climate System
Warming of the climate system is unequivocal, and since the 1950s many of the observed changes are unprecedented over decades to millennia. The atmosphere and oceans have warmed, amounts of snow and ice have diminished, and sea level has risen.

SPM 1.2. Causes of Climate Change
Anthropogenic greenhouse gas emissions have increased since the pre-industrial era, driven largely by economic and population growth, and are now higher than ever. This has led to atmospheric concentrations of carbon dioxide, methane and nitrous oxide that are unprecedented in the last 800,000 years. Their effects, together with those of other anthropogenic drivers, have been detected throughout the climate system and are extremely likely to have been the dominant cause of the observed warming since the mid-20th century.

SPM 1.3. Impacts of Climate Change
In recent decades, changes in climate have caused impacts on natural and human systems on all continents and across the oceans. Impacts are due to observed climate change, irrespective of its cause, indicating the sensitivity of natural and human systems to changing climate.

SPM 1.4. Extreme Events
Changes in many extreme weather and climate events have been observed since 1950. Some of these changes have been linked to human influences, including a decrease in cold temperature extremes, an increase in warm temperature extremes, an increase in extreme high sea levels, and an increase in a number of heavy precipitation events in a number of regions.

SPM 2. Future Climate Changes, Risks and Impacts
Continued emission of greenhouse gases will cause further warming and long lasting changes in all components of the climate system, increasing the likelihood of severe, pervasive, and irreversible impacts for people and ecosystems. Limiting climate change would require substantial and sustained reductions in greenhouse gas emissions, which, together with adaptation, can limit climate change risks.

SPM 2.1. Key Drivers of Future Climate
Cumulative emissions of CO_2 largely determine global mean surface warming by the late 21st century and beyond. Projections of greenhouse gas emissions vary over a wide range, depending on both socioeconomic development and climate policy.

SPM 2.2. Projected Changes in the Climate System
Surface temperature is projected to rise over the 21st century under all assessed emission scenarios. It is very likely that heat waves will occur more often and last longer, and that extreme precipitation events will become more intense and frequent in many regions. The ocean will continue to warm and acidify, and global mean sea level to rise.

SPM 2.3. Future Risks and Impacts Caused by a Changing Climate
Climate change will amplify existing risks and create new risks for natural and human systems. Risks are unevenly distributed and are generally greater for disadvantaged people and communities in countries at all levels of development.

SPM 2.4. Climate Change Beyond 2100, Irreversibility and Abrupt Changes
Many aspects of climate change and associated impacts will continue for centuries, even if anthropogenic emissions of greenhouse gases are stopped. The risks of abrupt or irreversible changes increase as the magnitude of the warming increases.

SPM 3. Future Pathways for Adaptation, Mitigation and Sustainable Development
Adaptation and mitigation are complementary strategies for reducing and managing the risks of climate change. Substantial emissions reductions over the next few decades can reduce climate risks in the 21st century and beyond, increase prospects for effective adaptation, reduce the costs and challenges of mitigation in the longer term, and contribute to climate-resilient pathways for sustainable development.

SPM 3.1. Foundations of Decision-Making About Climate Change
Effective decision-making to limit climate change and its effects can be informed by a wide range of analytical approaches for evaluating expected risks

and benefits, recognizing the importance of governance, ethical dimensions, equity, value judgments, economic assessments, and diverse perceptions and responses to risk and uncertainty.

SPM 3.2. Climate Change Risks Reduced by Mitigation and Adaptation
Without additional mitigation efforts beyond those in place today, and even with adaptation, warming by the end of the 21st century will lead to high to very high risk of severe, widespread, and irreversible impacts globally (high confidence). Mitigation involves some level of co-benefits and of risks due to adverse side effects, but these risks do not involve the same possibility of severe, widespread, and irreversible impacts as risks from climate change, increasing the benefits from near-term mitigation efforts.

SPM 3.3. Characteristics of Adaptation Pathways
Adaptation can reduce the risks of climate change impacts, but there are limits to its effectiveness, especially with greater magnitudes and rates of climate change. Taking a longer-term perspective in the context of sustainable development increases the likelihood that more immediate adaptation actions will also enhance future options and preparedness.

SPM 3.4. Characteristics of Mitigation Pathways
There are multiple mitigation pathways that are likely to limit warming to below 2°C relative to pre-industrial levels. These pathways would require substantial emissions reductions over the next few decades and near zero emissions of CO_2 and other long-lived greenhouse gases by the end of the century. Implementing such reductions poses substantial technological, economic, social and institutional challenges, which increase with delays in additional mitigation and if key technologies are not available. Limiting warming to lower or higher levels involves similar challenges but on different timescales.

SPM 4. Adaptation and Mitigation
Many adaptation and mitigation options can help address climate change, but no single option is sufficient by itself. Effective implementation depends on policies and cooperation at all scales and can be enhanced through integrated responses that link adaptation and mitigation with other societal objectives.

SPM 4.1. Common Enabling Factors and Constraints for Adaptation and Mitigation Responses
Adaptation and mitigation responses are underpinned by common enabling factors. These include effective institutions and governance, innovation and investments in environmentally sound technologies and infrastructure, sustainable livelihoods, and behavioral and lifestyle choices.

SPM 4.2. Response Options for Adaptation
Adaptation options exist in all sectors, but their context for implementation and potential to reduce climate-related risks differs across sectors and regions.

Some adaptation responses involve significant co-benefits, synergies, and trade-offs. Increasing climate change will increase challenges for many adaptation options.

SPM 4.3. Response Options for Mitigation
Mitigation options are available in every major sector. Mitigation can be more cost-effective if using an integrated approach that combines measures to reduce energy use and the greenhouse gas intensity of end-use sectors, decarbonize energy supply, reduce net emissions, and enhance carbon sinks in land-based sectors.

SPM 4.4. Policy Approaches for Adaptation and Mitigation, Technology, and Finance
Effective adaptation and mitigation responses will depend on policies and measures across multiple scales: international, regional, national, and subnational. Policies across all scales supporting technology development, diffusion and transfer, as well as finance for responses to climate change, can complement and enhance the effectiveness of policies that directly promote adaptation and mitigation.

SPM 4.5. Trade-offs, Synergies and Interactions with Sustainable Development
Climate change is a threat to sustainable development. Nonetheless, there are many opportunities to link mitigation, adaptation, and the pursuit of other societal objectives through integrated responses (high confidence). Successful implementation relies on relevant tools, suitable governance structures, and enhanced capacity to respond (medium confidence).

Source: IPCC, 2014: Summary for policymakers. In *Climate Change 2014: Impacts, Adaptation, and Vulnerability*. Part A: Global and Sectoral Aspects. Contribution of Working Group II to the Fifth Assessment Report of the Intergovernmental Panel on Climate Change [Field, C. B., V. R. Barros, D. J. Dokken, K. J. Mach, M. D. Mastrandrea, T. E. Bilir, M. Chatterjee, K. L. Ebi, Y. O. Estrada, R. C. Genova, B. Girma, E. S. Kissel, A. N. Levy, S. MacCracken, P. R. Mastrandrea, and L. L. White (eds.)]. Cambridge University Press, Cambridge, United Kingdom and New York, NY, USA, pp. 1–32. Available online at https://ipcc-wg2.gov/AR5/images/uploads/WG2AR5_SPM_FINAL.pdf. Used by permission.

PERSPECTIVES

Should Developed Countries Pay Monetary Reparation for the Ecological Debt Caused by Their Fossil Fuel Emissions?

Overview

Discussions on international collective action towards the mitigation of and adaptation to the impact of climate change caused by human activity have proceeded at many levels and in a number of fora from as early as the late 1970s. The multiplicity of conferences and meetings is not surprising since climate change is a complex problem. It requires contributions from many disciplines and fields of research as well as consensus, if the goal of reducing the emission of greenhouse gases (GHGs) to levels that would limit midcentury global temperatures to less than 2°C is to be achieved.

To give cohesion and focus to their collective efforts, countries joined the United Nations Framework Convention on Climate Change (UNFCCC) in 1992 to consider what they could do to limit global temperature increases (mitigation) and to cope with the increase of global average temperatures (adaptation). There are now 195 Parties to the Convention and 192 Parties to the Kyoto Protocol, which legally binds developed countries to GHG emission reduction targets. The commitment period of the protocol started in 2008 and ended in 2012. The second commitment period began on January 1, 2013 and will end in 2020.

Costs Associated with Reducing GHG Emissions

There are costs associated with reducing GHG emissions, at least in the short run. These costs include the expense of developing and deploying low-emission and high-efficiency technologies as well as the costs to consumers of switching spending from emissions-intensive to low-emission goods and services (mitigation). There is a relationship between economic growth and GHG emissions, and it stands to reason that so far developed countries have been the major contributors to climate change.

In addition, the countries with the technological capacity to make a significant contribution to the reduction in GHG emissions are for the most part developed countries. A 2007 estimate of mitigation costs suggests that they could amount on average to 1 percent of their gross domestic product (GDP) or total economic output, but would vary according to how and when emissions are cut (Stern 2006, 211).

There are also costs associated with coping with the consequences of climate change that cannot be avoided because of GHG emission levels already in the atmosphere. To respond to these GHG levels, adaptation measures are needed. There is, not surprisingly, a wide variation in coping capacities among countries. Countries that are least capable of adapting to the impact of climate change are described as environmentally vulnerable.

The projected global consequences of climate change for economic development and human living conditions continue to outstrip available funding for both mitigation and adaptation. These impacts cover a broad range of areas such as the reduction of tourism activities due to coastal and beach erosion, coral reef destruction, water shortages, intrusion of salt water into water tables and groundwater aquifers, and reduction of water flows for potable water, agriculture, and energy production.

Charles Skeete

Perspective 1: Yes, Compensate the Most Vulnerable Countries

Reduced to its essential element, there is a strong case for monetary compensation of countries that are adversely impacted by climate change events. It is simply the following logical implication: those who emit the most should compensate the most, and those who suffer the most, relative to their ability to mitigate (prevent) and cope with (adapt to) adverse impacts, should be compensated the most.

This approach represents a form of distributive justice applied at the global level. Statistics of GHG emissions on a per capita basis (excluding land use and forestry) by country show that China, the European Union, India, Japan, Russia, and the United States accounted for 55 percent of GHG emissions in 2010 (compiled from World Resources Institute data, 2014). The significant difference in responsibility for GHG emissions between a relatively small group of countries and the rest of the world is supported by other sources, as will be stated below.

The Intergovernmental Panel on Climate Change (IPCC) is the international body that was set up in 1988 to assess the science related to climate change. IPCC assessments provide a scientific basis for governments at all levels to develop climate-related policies, and they underlie negotiations at the UN Climate Conference—the United Nations Framework Convention on Climate Change (UNFCCC). The IPCC does not tell policy makers what actions to take; it presents projections of future climate change outcomes, based on different scenarios and risks, and discusses the implications of response options.

In its Fifth Assessment Report (AR5), the IPCC projected that: (a) surface temperature would rise over the 21st century *under all emission scenarios* (emphasis added); (b) heat waves were likely to occur more often and last longer; and (c) extreme precipitation events would become more intense and frequent in many regions. It also estimated that the ocean will continue to warm and acidify, and global mean sea level will rise (IPCC 2014, 10). AR5 pointed out that affected countries have little option but to take adaptation measures to reduce adverse climate change impacts.

Not for the first time, the IPCC has reported that many countries most vulnerable to climate change have contributed little to GHG emissions. The IPCC report added that insufficient adaptation responses of these countries to emerging impacts are already eroding the basis for their sustainable development (IPCC 2014, 17). These findings corroborate those of other studies to the effect that climate change poses a real threat to the developing world, including difficulties to combat poverty (Stern 2006, 92).

Difficulty of Obtaining Collective Action

Given the disparities in contributions to GHGs, there are some principles of distributive justice that could be applied to support the case for monetary compensation of the most vulnerable countries by those countries that contribute most to climate change impacts. Three of these principles are proportionality, fairness, and equality. Discussion of their relative merits is beyond the scope of this essay. What is relevant is how large a difference there is in responsibility for GHG emissions between major emitting countries and countries most vulnerable to climate change.

Although there is a wealth of fact-based information and scientific consensus on climate change causes and impacts, there are considerable difficulties that stand in the way of collective action to mitigate climate change and to share the burden of coping with its impact. The scientific evidence is unequivocal in its conclusion that developed countries are largely responsible for the greenhouse gases *already accumulated in the atmosphere.* Their higher level of development can be accounted for by past energy use and high GHG emissions.

However, on the other hand, not all *current* emitters are developed countries; a number of *current* major emitters are developing countries (for example, China and India). These major emitters argue, however, that, given the disparity in their levels of development, past major emitters and current emitters should shoulder differential burdens for mitigation going forward.

If left unchecked, climate change threatens the basic elements of life—access to water, food production, health, and use of land and the environment. The costs for mitigation and adaptation are beyond the capacity and resources of the most vulnerable countries.

Therefore, suitable and fair monetary arrangements have to be worked out to assist these countries to cope with the impending crises. Because the gap between available resources and need is not likely to be closed in the short—or even the medium—term, criteria for the allocation of funding between the competing goals of mitigation and adaptation, and among the affected countries, should be worked out with rigor and fairness.

A World Bank study estimates the cost of adapting to climate change at an average of $70 billion to $100 billion a year at 2005 prices between 2010 and 2050 (World Bank 2010 (a)). Compared to current development assistance levels, these are considerable sums of money. And yet, without adaptation, the survival, let alone the development, of many developing countries is at stake. The questions, then, are: which are the most vulnerable countries and what are the most efficient means of adaptation?

Making the Case for Monetary Compensation

A definition of vulnerability has been the subject of much study and controversy. The difference between the degree of vulnerability becomes evident when vulnerability is defined to take at least two factors into account: *exposure* to the consequences of climate change; and the *capacity to cope* with its consequences. This approach makes it possible to divide countries into those whose existence and survival are threatened by the adverse impacts of climate change (the most vulnerable countries) and those who have the capacity to withstand such events without suffering significant damage to their economic viability and survival (the least vulnerable countries). For these purposes, a ranking of vulnerability would be enough. This approach also makes it possible to rank vulnerability threats at the sectoral level (for example, food production threats vs. health threats).

The case of Bangladesh and the United States is often used to illustrate the importance of distinguishing between the frequency and intensity of climate change events on the one hand, and the degree of damage to ecosystems, property, and human life on the other. Climate change events are far more frequent in the United States than in Bangladesh, yet the loss of life in Bangladesh is many times greater than in the United States. Bangladesh is therefore considered to be more vulnerable. The case for compensation of the most vulnerable rests on who has responsibility for emission levels *already* in the atmosphere. It also rests on the evidence that, if unchecked, climate change will overwhelm, and in some cases could even reverse development.

In other words, financing sustainable development in the most vulnerable countries on as timely a schedule as possible is the most effective way of achieving adaptation goals. Without adaptation there will be little or no development in vulnerable countries. There will be an increase in poverty. And without development, there will be no adaptation.

Getting to Yes

But even though we can make the case for monetary compensation, there is still the matter of actually getting to consensus building on levels and amounts of payments by high–GHG-emitting countries. Some European countries such as the United Kingdom, Germany, and Scandinavian states have been funding adaptation projects in some lower-income and most vulnerable countries, but the amounts are woefully inadequate to address the magnitude of the problem. In the multitude of conferences, there has been lots of talk but not enough action. More effective international diplomacy is needed for GHG-emitting countries to step up to the plate to address global climate change injustice.

Charles Skeete

Perspective 2: No Direct Monetary Compensation for the Most Vulnerable Countries

While there is no doubt that most vulnerable nations are experiencing the impacts of climate change, it is not clear how direct monetary compensation of more developed countries would *automatically* lead to improvements in climate change mitigation and adaptation actions of affected states. Even if there were global consensus on the provision of payments to affected countries, there is no guarantee that these funds would be properly allocated to the right or most appropriate national programs. This is because problems of corruption, inadequate governance, and weak financial management structures as well as insufficient technical expertise of vulnerable governments could stand in the way of the most appropriate use of these funds for the desired outcomes.

Technical Assistance Rather Than Dollars

Developed countries should provide specific technical assistance—rather than direct, non-targeted compensation payments—to developing countries to enable them to choose their own customized pathways to respond to climate change. It is also crucial for vulnerable states to change their *own* policies and practices to better manage land use changes and deforestation, reduce their GHG build up, and switch to available sources of renewable energy such as hydropower, sun, wind, waves, and geothermal resources.

How effective would it be for vulnerable nations to receive huge sums of compensatory payments from industrialized countries without being able to direct those payments efficiently to affected sectors and areas? Should they at the same time be making internal efforts to curb their own GHG emissions? Should vulnerable countries not first develop the instruments to manage these sums of money before becoming overwhelmed by different streams of funds? There is another question that comes to mind: should developing countries such as China and India, which are increasingly emitting greenhouse gases, compensate other developing countries, which are low emitters?

These questions expose the complexity of the compensation problem. Some developing countries have argued that developed countries were able to industrialize and develop their economies based on fossil fuels and exploitation of the resources of past colonies (now developing countries). Therefore, they contend, they too should have the opportunity to develop industries and services and grow their economies, lest they be relegated to the role of exporting primary commodities. However, this argument does not hold water, because scientific evidence is already available to document the negative impacts of past fossil fuel use of industrialized economies: developing economies have a valuable opportunity to learn from them and not repeat the lessons of the past.

Time for Policy Shifts in Developing Countries

Many developing countries have had the option of using their abundant (nonfossil) renewable energy resources to propel economic growth for at least three decades, but have not opted to change their policies and make the shift to these resources, even while continuing to import costly fossil fuels. This reluctance is partially due to their insertion in the globalized economy, either as providers of commodities, tourism and other services, and/or due to convenience or the perceived benefits of following established coal, oil, and gas based economic models.

In addition, some island states have aggressively developed tourism-based economies without much regard to their environmental impacts on their ecosystems or the pursuit of sustainable ecotourism models. Yet other developing countries have had lax policies on curbing deforestation of tropical forests and have even actively promoted policies to expand livestock production—with huge consequences for GHG emissions.

Through unsustainable mass tourism practices, some developing countries, especially island economies, have unwittingly provoked some degree of environmental and climate change–related impacts in their own societies. These include the carbon footprint of massive tourism; degradation of beaches, mangroves, and coral reefs; excessive construction of resorts in coastal areas; lax legal, regulatory, and environmental protection standards; and ongoing contamination of coastal ecosystems.

It can be argued that when industrialization started and progressed on the basis of fossil fuels—first coal, then oil and natural gas—countries such as the UK were not fully aware of the impacts of growing GHG emissions. In other words, at the time, their actions did not have the express purpose of harming themselves or other countries of the world. During the Industrial Revolution and for many decades thereafter, there was not enough scientific information on the potential impacts of fossil fuel use in industrial activities.

The Question of Willful Damage

Nonetheless, there are recent concerns that in the last three decades, oil companies may have been carrying out scientific research on climate science without disclosing potential risks of fossil fuel emissions to investors. In 2015, the New York attorney general began an investigation into whether oil companies might have funded groups seeking to undermine state-of-the-art scientific research on climate change effects (Gillis and Krauss 2015).

However, in their quest for monetary compensation, it has been difficult for affected countries to make the case that climate change is a result of the *willful* misconduct of industrialized GHG emitters from the outset. Therefore, demands for monetary reparation based on liability of industrialized countries have not gained traction in intentional negotiations.

If we look at some examples across social and economic life, we can observe other areas of alleged willful damage. In the tobacco industry, for example, it has been demonstrated that tobacco companies—through their own research or the research of others—were aware early on of the health dangers and risks of cigarette smoking (Oreskes 2015). However, due to bottom-line considerations, corporate secrecy, or the lack of stringent public health regulations at the time, this information was not available to the public.

The slave trade is another example of willful damage: for economic reasons, colonial nations and individual agents engaged in the slave trade and enslaved other peoples, directly inflicting physical, cultural, and emotional harm on others for economic gain. For this reason, there have been calls from some African American groups in the United States for reparation due to cultural and other damages incurred as a result of the slave trade.

Returning therefore to the question of climate justice, it is highly unlikely that vulnerable countries could make any progress on the reparation argument, since it is difficult to single out any one country or group to which to assign blame. However, the increasing severity of their situation and their capacity to organize within the United Nations system are pushing them to plead the case for financial assistance more forcefully. One reason for the reluctance of industrialized countries to provide more funds is that they—now faced with climate change fallout in their countries—are more apt to use national budgets to finance their own mitigation and adaptation measures.

Vulnerable nations should request or press for targeted technical assistance to develop their human resource and technical capabilities to address climate change—instead of fighting long, drawn-out battles for monetary compensation that may continue to fall on deaf ears. Technical support and increased know-how would enable these countries to be better negotiators to lobby for emissions reductions in a global setting, to pursue more effective long-term strategies to adjust their use of fossil fuels, to take adaptation measures, and reengineer their economies towards a more sustainable path.

In an uncertain future, it can be argued that immediate, strategic shifts of vulnerable states to cleaner energies and more sustainable social, economic, and transportation development strategies would place them in better standing to compete in a future green world economy. In this manner, they could jump-start or frog-leap their way to green industries and exports of low GHG products, rather than continuing along a fossil fuel–laden path to destruction.

Pending Problems

There is still the problem of what vulnerable countries can do now to address some of the funding gaps for climate change impacts. To obtain flows of financial resources for immediate mitigation and adaptation measures—and at the same time fast-forward efforts to make the necessary economic shift—these countries could establish new internal tax policies together with stricter environmental regulations.

For example, they could levy green taxes on corporations, mining, and tourism industries, incentivize the use of renewable energy sources, and raise sales taxes on luxury (high emissions) consumer goods. They should also speedily institute policy and regulatory measures to curb forest destruction, water depletion, and GHG releases. These measures could provide some quick gains.

At the same time, vulnerable countries should work with national private enterprises and international investors to enforce regulations on water-polluting and GHG-producing enterprises. They should provide strong economic incentives for green investments and the retrofitting of enterprises. Finally, they should direct most of their international negotiation efforts to seeking overall financial debt relief and green grants from developed countries, while implementing pressing internal policy actions.

Camille Gaskin-Reyes

REFERENCES AND FURTHER READING

Gillis, Justin, and Clifford Krauss. "Exxon Mobil Investigated for Possible Climate Change Lies by New York Attorney General." *New York Times*, November 6, 2015. http://www.nytimes.com/2015/11/06/science/exxon-mobil-under-investigation-in-new-york-over-climate-statements.html.

Intergovernmental Panel on Climate Change (IPCC). "Fifth Assessment Report (AR)," 2014. https://www.ipcc.ch/report/ar5/.

Oreskes, Naomi. "Exxon's Climate Concealment." *New York Times*, October 19, 2015. http://www.nytimes.com/2015/10/10/opinion/exxons-climate-concealment.html.

Stern, Nicholas. *The Stern Review on the Economics of Climate Change*. Washington, DC: World Bank Publications, 2006.

United Nations Environment Programme. "Assessing Human Vulnerability to Environmental Change: Concepts, Issues, Methods and Case Studies." Nairobi: United Nations Press, 2003.

World Bank. "Economics of Adaptation to Climate Change." Washington, DC: World Bank Publications, 2011.

2 GLOBAL WATER CYCLE DISRUPTIONS AND INTENSE WEATHER EVENTS

OVERVIEW

This chapter explores the processes related to weather patterns around the world, disruptions to the water cycle, and the increasing manifestations of severe global weather events such as storms and floods, heat waves, and droughts. The case studies examine the issues of dust bowls and droughts in the Great Plains and the Aral Sea; the impacts of major hurricanes such as Mitch in Honduras, Katrina in New Orleans, and Sandy in New Jersey; and the dilemma of water supply bottlenecks in China and its neighboring region. The perspectives section looks at social equity and flood protection in New Orleans and factors behind the vulnerability of the city in withstanding megastorms.

Life and ecological processes depend on water. Every country, every region, every community has a water bank account, but the assets in this account—and people's livelihoods—depend on factors such as water supply and demand, location, altitude, latitude, temperature, wind, and rainfall regimes. Available surface and groundwater is a part of the dynamics of the hydrologic cycle and also linked to climate change. These elements determine national and regional differences and why some places are water-rich or water-poor.

The Hydrologic Cycle

The hydrologic cycle is the general term used to describe the processes by which water moves or recycles through various areas of the planet. How does it work? It begins with evaporation, the upward movement of water in the form of vapor from the land and the oceans through heating. As the surface warms, water vapor rises into the atmosphere and cools in higher levels of the earth's surface, where it condenses, forms clouds, and falls back to land or sea as precipitation, i.e., fog, rain, ice, sleet, or snow. Some vapor from the ocean also drifts over land, where it may fall in various forms of precipitation.

While there is a transfer of water vapor from ocean to land, there is also movement of surface water in the opposite direction, from land to ocean. Rivers usually start in higher altitudes and flow down to lower land, eventually into seas or oceans (except when extreme heat dries them up underway). Some precipitation stays on the land surface, but moisture also percolates downward into the soil or bedrock, concentrating in aquifers. Groundwater may also escape into the sea and seawater may intrude into aquifers, continuing the circulation process within the water cycle.

Latitude and Altitude Matter

The regions with the highest temperatures are located close to the equator. They receive rain all year due to intense evaporation and the continuous return of water vapor to the earth. This picture changes north and south of the tropics, as year-round rains of the equatorial zones change to seasonal rainfall, usually in the summer months.

Further south or north of the tropics (within 20 to 30° latitude) in the Northern and Southern Hemispheres there is a zone called the subtropics. This area is marked by a noticeable reduction in precipitation and the growing incidence of deserts such as the Sahara, Gobi, Kalahari, Australian, and the Atacama. Deserts usually receive less than four inches of annual rainfall because they absorb down currents of dry air that descend from the subtropics.

The poles are located north and south of the subtropics, depending on the hemisphere: these are the temperate regions of the world, where parts of the United States, Canada, Europe, Japan, and China are found. Precipitation patterns are more complex at the poles. Cold air masses from the North and South Poles interact with relatively warmer air from the tropics and subtropics and create pockets of weather turbulence.

Not only does latitude (location close to the equator, tropics, or poles) determine the degree to which countries are endowed with water; altitude matters as well. Mountain ranges affect both precipitation and temperature levels. One side of a mountain usually gets more rain than the other side (called the rain shadow) because mountains cause air currents to rise, condense, and drop their moisture before descending as drier air. On the big island of Hawaii, for example, annual rainfall varies from 200 inches on the side facing the trade winds to less than 9 inches on the downwind coast. The higher the mountain range, the cooler the temperature at the top.

Water and Human Settlements

Human settlements that are close to rivers and other surface water, or positioned on groundwater aquifers, have the obvious advantage of tapping available surface or ground water in their proximity. Some settlements on the coasts of oceans may receive more water than inland areas, depending on prevailing winds and rainfall patterns, while communities in inland, semiarid/arid zones may not have enough rain, rivers, or groundwater aquifers to support inhabitants all year round. Most countries depend on seasonal rainfall and some may store water for irrigation and other uses. Many arid regions are already experiencing water stress and Dust Bowl–like conditions, as we explore later in the chapter.

Some countries in coastal areas also experience desert-like conditions due to the upwelling of cold ocean currents. In Chile or Peru, for example, the Pacific coastal areas are so dry that water for settlements depends heavily on the flow of rivers from the highlands to the coast. Though it hardly rains on the Pacific coast, pockets of dense fog continuously form due to these cold ocean currents. To capture moisture, Peru and Chile are extracting fog vapor from the air using fog catchers or high nets, around which fog condenses as water and is collected in containers (Collyns 2009; Mulder 1976).

Changing Patterns

Precipitation and overall meteorological and water flow patterns of the planet have become more erratic and less predictable since the end of the preindustrial area. Temperatures have been rising steadily, as discussed in chapter 1 (0.85°C in global average temperature rise since 1880). More intense or extreme weather disruptions such as storms, floods, heat waves, wildfires, and droughts are now associated with the ongoing process of climate change.

While research on how climate change directly affects the entire meteorological and water cycle is still ongoing, it is already evident that higher temperatures are accelerating the precipitation and evaporation cycle in many areas of the world. More heat means that more water evaporates from oceans, lakes, rivers, reservoirs, and soils. As more water vapor rises to the atmosphere, the land becomes drier and affects farming negatively. With more rapid condensation, rain episodes tend to be more torrential and pose higher risks of flooding. Another more frequent phenomenon is the spread of wildfires: land gets parched due to higher temperatures combined with drought, forests become dry, wildfires break out more frequently.

Increasing Groundwater Extraction

Some countries and areas have been experiencing declining rainfall and water shortages since the beginning of the 21st century. To compensate, they are increasingly extracting water from aquifers below the surface or from rivers and lakes on the surface, even exceeding sustainable limits on such extraction.

In a general sense, humans are drawing heavily on their water bank accounts, leaving fewer assets for the future. A rate of groundwater withdrawal that is higher than its recharge through rainfall will eventually lead to depletion, potential collapse of aquifers, or saline intrusion (the seeping in of salt water). The latter can occur frequently, if groundwater sources are located close to oceans or floodplains.

The Ogallala Aquifer in the United States sharply illustrates the problems of aquifer drawdown. Water levels in the Ogallala Aquifer, a huge underground reservoir beneath the Great Plains, have declined since 1950 by over 200 feet in parts of Kansas and Texas because demand is outstripping supply (Elert and Lemonick 2012, 49). Users are pumping this water—stored underground in rock layers over 1,000 years ago—for crop irrigation at very fast rates, creating the potential for very serious impacts (Barlow and Clarke 2002, 161).

Rancher sits on a flowing well on his property, in Sand Hills, Nebraska, April 18, 2012. The rancher is fighting to move the Keystone XL pipeline away from the Ogallala Aquifer that lies below his land. (Lucas Oleniuk/Toronto Star via Getty Images)

Declining Surface Water Supply

In addition to the depletion of underground water, there are increasing surface water problems. The availability of fresh surface water is declining through population pressure, overuse, pollution, drought, excessive evaporation, and erratic rainfall. Water levels of some lakes and reservoirs in the United States, such as Lake Mead, have been dropping in recent years. Many rivers that used to flow all the way to the sea often dry out before they get there as a result of overuse or rapid evaporation of parts of their courses. This has happened with the Yellow River in China, the Colorado River in the United States, and the Ganges River in India, among others (Elert and Lemonick 2012, 49).

More Intense Storm Events

While some parts of the planet are experiencing severe drought, heat waves, and fires, others are undergoing frequent episodes of heavy precipitation, storm surges, and dangerous floods. Climate models suggest that shorter bouts of torrential rainfall and more intensive storms are likely to provoke more severe flooding in the future.

From 1980 to 2013, the United States experienced 33 coastal storms. Each caused more than $1 billion in damages, accounting for about 75 percent of globally insured losses from natural catastrophes. Damages from coastal storms on the East and Gulf Coasts have tripled since 1980 due to extreme weather events (Dean 2014).

In 1960 about 14 million inhabitants in the United States lived in coastal areas and in the potential path of hurricanes and storm surges; in 2009 this number reached over 32 million (Elert and Lemonick 2012, 140). Settlement trends indicate that this number is expected to increase. Population growth and expansion of settlements in low-lying coastal areas such as New York, New Jersey, the Carolinas, and the Atlantic and Gulf Coasts of Florida, Louisiana, and Texas, have placed more people in immediate harm's way.

Hurricanes receive their energy and intensity from heat in tropical ocean waters. They usually emerge at sea temperatures above 26°C. Rising surface temperatures in the Atlantic—the birthplace of most tropical storms and hurricanes in Central America and the Caribbean—have been a major factor in storm intensity along the Atlantic Seaboard, the Gulf Coast, and the Caribbean in recent years.

Vulnerable Coastlines and Islands

Island and coastline dwellers are becoming increasingly concerned about the risks of severe hurricanes, storms, and dangerous storm surges. Surges are walls of seawater pushed up to islands and low-lying coastal shorelines. They accelerate beach erosion and increase the exposure and vulnerability of coastal and island communities to the onslaught of the seas.

Global sea levels are about eight inches higher now than in 1900, and could potentially rise up to three feet in the next hundred years, further endangering island and coastal communities, including major metropolitan areas around the world such as Boston, New York, Miami, Rotterdam, London, and Manila (Elert and Lemonick 2012, 27).

In the United States, urban planners, policy makers, and natural disaster–management agencies are faced with the urgency of addressing accelerated coastal flooding on the Atlantic coast from north to south. In 2014, inhabitants of Fire Island, a barrier island on a sandbar off the metropolitan area of New York, initiated specific measures to address the rapid loss of the island's sand and coastal ecosystems due to more intense storms and wave erosion (Dean 2014).

Shoreline projects to stabilize inlets, replenish lost sand, and build protective sea walls and rock jetties are also underway on other barrier islands and coastal areas from Massachusetts to the Outer Banks of North Carolina. Barrier islands and mangroves are very important parts of coastal ecosystems. They are usually the first lines of defense—like coral reefs for tropical areas—against storms and storm surges. But they are losing their protective functions due to rising seas and more damaging storm surges. In many coastal areas, residents have moved their houses or are raising them up by 20 feet in anticipation of more intense storms and flooding.

In 2012 Hurricane Sandy brought home the risks. It was a devastating hurricane in New York and New Jersey on the East Coast, and was the most destructive storm of the

2012 Atlantic hurricane season. Sandy inflicted severe damage in a metropolitan area far north of the regular hurricane areas. In Manhattan and New Jersey, Sandy destroyed thousands of homes, killed 73 people, forced thousands of residents into shelters, and damaged vital infrastructure such as transmission lines, roads, railways, and water and sewage treatment facilities. It also revealed the shortcomings of meteorological and natural disaster warning and management systems to address super storms in large metropolitan areas.

Coastal erosion is also increasing in Florida, now considered a very vulnerable area. Miami is particularly exposed to rising sea levels, and in some parts is already experiencing the impacts of coastal road flooding and antiquated sewage systems. Since Miami lies above a porous limestone foundation, higher levels of seawater are able to seep into the bedrock and rise to the surface through storm drains. It is anticipated that in the future the limestone bedrock could become more saturated and crumble, as the ocean encroaches further upon the land.

Densely populated islands and coastal areas in the Caribbean and Central America are traditionally affected by hurricanes, since they lie directly in the storm path of hurricanes coming from the Atlantic Ocean. However, they are even more threatened by climate change and increasingly violent storms. The force of destructive Category 4 and 5 hurricanes (wind speeds over 130 miles per hour) is packing a bigger punch not only in the Atlantic but also in the Pacific, where they are called typhoons. Megastorms pose higher risks to life and cause extreme damage to property, livestock, farms, hospitals, and schools, requiring high reconstruction costs. The case study on hurricanes in this chapter explores this topic in greater detail.

To illustrate this point, in November 2013, Typhoon Haiyan slammed into the coastal settlement of Tacloban in the Philippines, leaving a death toll of 6,000 and widespread damage in its wake. Community relocation costs were estimated at millions of dollars (*Economist* 2013). Vulnerable areas such as Tacloban are in double jeopardy; they have to adapt to rising sea levels and storms, and at the same time address the threat of climate change. As Tacloban shows, severe storms require huge rebuilding efforts that syphon off valuable financial resources from economic development and climate change adaptation activities.

Droughts, Heat Waves, and Floods

In 2014, a National Climate Assessment study of the United States, released by the White House, reported the growing trend of erratic, regional weather variations: frequent water scarcity in dry regions, torrential rains in wet regions, extended periods of unusually high heat waves, and intense wildfires due to droughts and forest dryness. The study concluded that climate change was a contributing factor to such changes, and warned of future trends such as more water stressed areas; longer, hotter summers; torrential downpours in some areas; and drought in others. The study also projected rising seas, greater inland flooding, and more loss of coastal arable land.

The report reflected the situation already experienced by many states. In 2006 drought in North Dakota destroyed 10 percent of the state's wheat crop (*Economist* 2014). In 2014, drought severely affected agricultural production in California and availability of water for residential and other purposes. It caused the Rio Grande to become a trickle, affecting water supply in many states (Wines 2015). On the opposite side of the spectrum, torrential rains in 2015 crippled the city of Houston, Texas, and caused the deaths of four people (Fernandez and Pérez-Peña 2015).

Some parts of the United States have been receiving more rainfall over the last 50 years: 71 percent higher in the Northeast, 37 percent in the Midwest, and 27 percent in the South. Intensive torrential rains have caused higher frequencies of devastating floods. For example, in 2010, Nashville, Tennessee, flooded after 20 inches of rain fell in 48 hours; in 2014, the Florida Panhandle flooded after withstanding 24 inches of rain in 24 hours; and in 2015, parts of Texas flooded after 10 inches of rainfall fell in 24 hours (Gillis 2014).

In the United States 2015 was reported as being the hottest year on record since 1880, when data tracking first began. Of the 14 hottest years in the history of U.S. record keeping, 13 were registered in the 21st century; the fall of 2015 was more than three degrees warmer than in the 20th century, on average (Bromwich 2015). In 2014, drought declarations were issued in California, and the federal government announced millions of dollars in drought relief.

On the other side of the world in Asia, more intense storms and floods are occurring in the monsoon belt. Monsoon seasons are the most important rainfall events in Asia, and two-fifths of the world's population lives in the monsoon belt. Monsoons bring life-giving rains to areas such as India and Bangladesh, but they are causing more serious floods that endanger the lives of millions. They develop every year when there are huge differences in temperature between the ocean and the continent. With rising temperatures, the ocean heats up faster, and hotter air holds more water vapor. Monsoon winds coming from more heated oceans dump more rain onto the land, causing more flooding. In a monsoon belt, an entire year's supply of rainfall can fall during one, or at most two, seasons.

On the other side of the spectrum, increased heat accelerates evaporation and dries up water reservoirs, wells, irrigation canals, and lakes more quickly. Like floods, water shortages and agricultural losses pose serious risks for human health and food security. Intense, sustained heat waves are also life threatening, particularly for vulnerable groups such as the very young and elderly.

In Europe, summer average temperatures at the beginning of the 21st century were warmer than in the previous century (Pearce 2007, 202). In the summer of 2003, temperatures climbed steeply in Europe and led to the deaths of at least 35,000 people. Average day temperatures soared to 47°C, while night temperatures hovered around 30°C, bringing no relief to residents (Pearce 2007, 201). Many elderly people without access to air conditioning died in their homes.

2003 was recorded as Europe's hottest summer in half a millennium. Crops died, forest fires broke out, power shortages occurred, and river levels sank to very low levels. A second heat wave occurred in Europe in 2006, but produced fewer casualties due to the establishment of heat warning systems and other preventive measures instituted after the 2003 crisis. Heat waves also occurred in other places in 2013 and 2014. A dangerous one struck Australia in 2013 and continued into 2014, briefly shutting down the Australian Open tennis tournament after the temperature registered 43°C.

Studies have shown that when temperatures exceed 37°C the supply of available labor for farming, construction, and landscaping jobs falls by an hour per day, compared with relatively milder days of 24° to 27°C. It is also estimated that labor productivity of outdoor workers falls by 3 percent for every 27 to 50 extra days that record daytime temperatures of over 35°C. A week of outside temperatures over 32°C is estimated to cause a drop of factory production of cars by 8 percent (*Economist* 2014).

In many developed countries, prolonged water scarcity disrupts some agricultural production and other activities, but in developing countries the situation is even more serious. In the poorest areas of the world, droughts cause famine, out-migration, humanitarian crises, even armed conflicts. People fleeing drought-stricken areas may die of hunger, end up in refugee camps, or encroach upon the land of other communities. In vulnerable villages, even relatively short periods of drought can deplete drinking water supply, damage the local economy, destroy crops and livestock, and push inhabitants off the edge of viable existence.

Water-Stressed Areas and Danger Zones

The African Sahel, the Horn of Africa, and many Middle Eastern and Central Asian countries are examples of arid zones undergoing more prolonged drought and severe water stress. In these zones, the decline in rainfall has provoked severe losses of agricultural and livestock production, economic hardships, famine, loss of lives, and the eruption of political conflicts.

Sub-Saharan Africa is particularly affected. Water shortages are provoking deprivation in thousands of water-poor refugee communities. In the driest parts of the African continent, the Sahel, the farming of food crops without reliable water sources is almost impossible. Communities in the Sahel are mostly pastoral, i.e., they raise herds of goats, camel, sheep, and cattle on pasturelands. But herders are also suffering. Due to limited water and the need to find grassland for animals, they are forced to move across wide swaths of land following short, seasonal rains that replenish grass for the herds. Since rains are failing more frequently, animals are dying, and the existence of herders is becoming more precarious. As a result of this vicious cycle, herders are also fleeing to other areas.

There is a connection between lack of water and poverty. The African countries rated as having the lowest human development in the world are also water-deficient with a precarious economic base. Counties such as Mali, Chad, Ethiopia, Burkina Faso, Somalia, and Sudan have high poverty levels and lack sufficient rainfall, rivers, or reliable water sources year-round. In addition, many have low water storage capacity and irrigation potential. Climate change is making this situation worse, as pastoralists are competing with farmers and urban dwellers for the few water resources still available.

Sudan and Chad are in a group of countries and regions around the world characterized as water danger zones. These areas are: The Sahel, the Horn of Africa, Israel-Palestine, the Middle East, Pakistan, Central Asia, the Indo-Gangetic Plain, the North China Plain, the Southwest and California in the United States, and the Murray-Darling Basin in Australia (Sachs 2008, 126). This list could get longer in the future.

These water danger zones share similar characteristics: they are drawing down on their groundwater (and their future), they are experiencing declining precipitation, and/or suffering from loss of glacial melt or prolonged droughts. Some are located in more developed countries such as the United States (Southwest and California) or Australia; others are in poverty-stricken regions with high political and ecological instability.

There are many dry land belts across the world that have been further affected by climate change. They include the already mentioned Sahel Zone (Senegal, Burkina Faso, Mali, Niger, and Chad); the Horn of Africa (Ethiopia, Eritrea, Somalia, and Sudan); East Africa (Northern Uganda, Northern Kenya); the Middle East (Yemen); and Central Asia (Afghanistan, Pakistan, Uzbekistan, and Tajikistan). In addition, there are at least 40 rivers and aquifers around the world experiencing water depletion (Richter 2014, 15).

Example of Darfur

Darfur, Sudan, exemplifies drought, poverty, and conflict. Darfur is a landlocked, arid region in the west of Sudan, located on the border with Chad. It suffers from extreme water stress and dire levels of impoverishment. Rainfall has declined during the last 30 years. This situation is compounded by a sevenfold increase in population in the same period and extreme pressure on dwindling water resources. Darfur's population cannot meet its most basic needs (Sachs 2008, 248).

A Somali family relocating to the United Nations High Commissioner for Refugees (UNHCR)'s Ifo Extension camp in eastern Kenya, August 5, 2011. Drought and famine in the horn of Africa have pushed over 1,000 newcomers a day to the camp. (Jerome Delay/AP Photo)

Darfur lacks safe water and sanitation services as well as access to irrigation, food, and housing. The conflict has reflected the increasing desperation of inhabitants and intense competition for land and water. Nomadic, pastoral communities from northern Darfur and from Chad have started to move further south with their livestock in the search for water, encroaching upon areas of the more sedentary farmer communities of southern Darfur. This situation has intensified already existing ethnic and linguistic differences between the two groups and led to violent conflicts.

The deteriorating water situation and the process of climate change and desertification (encroachment of the desert upon farmlands and communities) in Darfur, Chad, and other Sahel areas require integrated approaches and strategies. These could include the building of water storage systems to improve communities' access to water, the cultivation of more water-resistant crops, community education programs, and the development of improved land, water management, and environmental practices.

Example of the Murray-Darling Basin

The Murray-Darling Basin, Australia's breadbasket, is one of the areas reported to be experiencing water stress. The basin covers one-seventh of Australia and produces about a half of its agricultural output. It contains 20 major river systems and numerous wetlands, which supply drinking water to more than two million people. It is a key area for sheep and livestock herding, and a pillar of Australia's export economy. Two hundred years of clearing native vegetation for agriculture, cattle-herding, and extensive irrigation projects have increased erosion and soil damage in the basin area.

Many of the basin's problems stem from drought and irrigation: irrigation canals have raised the water table, damaged root zones of crops through evaporation, built up salts, and polluted rivers and fields across the watershed. More than 1.5 million acres of cropland were seriously damaged by salinity and soil erosion, a process aggravated by severe drought starting in the 1990s. Many of the basin's rivers and streams stopped flowing due to extreme evaporation (Richter 2014, 117).

In response to the problem, Australian officials formed the Murray-Darling Basin Commission in 1985 to put in place water conservation and land management measures and reduce "boom-and-bust" agricultural production attitudes between wet and dry years. The commission introduced drainage and water-saving drip irrigation techniques, limited water extraction, reduced soil salinity, increased sustainable water storage mechanisms, and supported the participation of farmers and herders in programs to curb erosion and restore environmental and hydrological balance.

These efforts have started to bear fruit. The main success factors were strong community participation, sound legal and regulatory frameworks, and the consensus of affected stakeholders on how to address the problems. This positive experience stands out in contrast to the situation observed in water danger zones in the Sahel area of Africa. There, political or ethnic conflicts, lack of financial and technical resources, governance weaknesses, and the grinding poverty of the region make it difficult for communities to pursue effective, collective responses to water and land use crises.

Camille Gaskin-Reyes

REFERENCES AND FURTHER READING

Barlow, Maude, and Tony Clarke. *Blue Gold: The Fight to Stop the Corporate Theft of the World's Water*. New York: New Press, 2002.

Bromwich, Jonah. "A Fitting End for the Hottest Year on Record." *New York Times*, December 23, 2015. http://www.nytimes.com/2015/12/24/science/climate-change-record-warm-year.html.

Collyns, Dan. "How Peru is Netting Water Supplies." *BBC News*, October 20, 2009. http://news.bbc.co.uk/2/hi/8297276.stm.

Dean, Cornelia. "Growing, and Growing Vulnerable: Barrier Islands Feeling the Effects of Climate Change." *New York Times*, September 29, 2014. http://www.nytimes.com/2014/09/30/science/barrier-islands-feeling-the-effects-of-climate-change.html.

The Economist. "Climate Change and the Economy: The Cost of Doing Nothing." June 28, 2014. http://www.economist.com/news/united-states/21605936-scorched-farms-flooded-homes-and-lower-productivity-cost-doing-nothing.

The Economist. "The Philippines and Typhoon Haiyan: The Winds of Change." December 21, 2013. http://www.economist.com/news/asia/21591908-month-later-worst-hit-town-looks-future-winds-change.

Elert, Emily, and Michael D. Lemonick. *Global Weirdness: Severe Storms, Deadly Heat Waves, Relentless Drought, Rising Seas, and the Weather of the Future*. New York: Penguin Random House, 2012.

Fernandez, Manny, and Richard Pérez-Peña. "Rain Spreads Destruction in Houston, Killing Four." *New York Times*, May 26, 2015. http://www.nytimes.com/2015/05/27/us/texas-rains-bring-flooding-to-houston-area.html.

Gillis, Justin. "U.S. Climate Has Already Changed, Study Finds, Citing Heat and Floods." *New York Times*, May 6, 2014. http://www.nytimes.com/2014/05/07/science/earth/climate-change-report.html.

Maslin, Mark. *Climate Change: A Very Short Introduction*. Oxford: Oxford University Press, 2014.

Mulder, Michelle. *Every Last Drop: Bringing Clean Water Home*. Victoria: Orca Book Publishers, 1976.

Pearce, Fred. *With Speed and Violence: Why Scientists Fear Tipping Points in Climate Change*. Boston: Beacon Press, 2007.

Richter, Brian. *Chasing Water: A Guide for Moving from Scarcity to Sustainability*. Washington, DC: Island Press, 2014.

Sachs, Jeffrey D. *Common Wealth: Economics for a Crowded Planet*. New York: Penguin Press, 2008.

Wines, Michael. "Mighty Rio Grande Now a Trickle Under Siege." *New York Times*, April 12, 2015. http://www.nytimes.com/2015/04/13/us/mighty-rio-grande-now-a-trickle-under-siege.html?_r=0.

CASE STUDIES

The case studies in this section address the factors underlying the past Dust Bowl in the Plains states of the United States and those contributing to an emerging dust bowl in the Aral Sea; the impacts of more intensive hurricanes and the lessons learned; and the dilemma of water shortages in China and their impacts on the water supply of downstream countries in the region.

Case Study 1: Old and New Dust Bowls

This case study compares and contrasts the Dust Bowl of the 1930s with a new dust bowl in another part of the world, the Aral Sea. It exemplifies how, in both examples, unsustainable, self-destructive, and market-oriented land use and agricultural production practices, combined with weather variations, have caused environmental destruction and severe social and economic impacts on society.

The U.S. Dust Bowl

The 1930s Dust Bowl of the U.S. Great Plains ranks as one of the major ecological disasters in North American history. For many people who did not experience or even know about this event, grainy black-and-white film clips showing dust clouds invading homes and entire families fleeing the area tell some of the story. This period was also captured in novels such as John Steinbeck's *The Grapes of Wrath*, which depicted the plight of migrants leaving the devastation of the Dust Bowl to migrate to California.

The Dust Bowl was a six-year period of dust storms that occurred as a result of extreme drought, overcultivation of marginally productive land, and inappropriate farming methods. The problem originated after settlers cleared the Great Plains for agriculture. In doing so, they replaced natural prairie grasses—better at holding the soil intact—with food crops.

Periods of drought caused the soil to dry up, and the winds on the prairie contributed their part by whipping up soil particles throughout the Panhandles of Texas and Oklahoma and the adjacent states of Kansas, Colorado, and New Mexico. In 1934, dust storms swirled more than 300 million tons of topsoil across the continent, destroying more than 7 million hectares of agricultural land, and provoking the flight of thousands of people from the area. Over 5,000 people died of heat and drought at the time (Ponting 2008, 254).

Lessons Learned

The events leading up to the Dust Bowl provide important lessons for today. Human removal of prairie grasses in the Great Plains for crop agriculture destroyed the natural cover that bound water in the soil and watersheds. Extension of agriculture into marginal, unsuitable lands triggered erosion and disrupted the water cycle. Loss of water vapor due to drought and high evaporation reduced moisture in the area and expanded the normal dry season, prolonging aridity. Human action exacerbated seasonal dry cycles and turned them into long-term drought, slowly creating an agricultural disaster for the states, farmers, and nonfarming residents.

Before the Dust Bowl disaster, the official policy of the United States in the 1930s had been to foster increased settlement of the Great Plains; at that time, individuals' attitudes and official promotion strategies were focused on the benefits of people settling the land to dominate the natural environment. Most parties paid little attention to or were simply unaware of the natural limits of the Plains' ecosystem and water supply boundaries.

There was also high land speculation—a process fueled by the promise of quick, short-term economic gains. The lure of prosperity led to a rush of settlers and tenant farmers to work land owned by many nonresident landowners. The practice of the settlers was to strip more grass and expand agricultural production to respond to growing market demand for crops such as wheat. As the crisis unfolded, nonresident owners abandoned or sold the land first, leaving tenants and resident farmers to fend for themselves. By then, however, it was too late to put in place serious conservation efforts.

In the aftermath of the Dust Bowl, the U.S. Bureau of Reclamation, the Works Progress Administration (WPA), and the National Resources Committee completed in 1936 a report on the Dust Bowl catastrophe and the soil erosion problems. The report indicated that a major factor for the crisis was the prevailing attitude to nature at the time, i.e., emphasizing human domination of nature. This approach promoted the notion that land resources were virtually inexhaustible. Land speculation and land sale agents promised owners and tenant farmers the American dream of land ownership and prosperity, which continued to lure settlers to the Great Plains until disaster struck.

The 1936 assessment led to an official examination of the lessons learned and produced recommendations for conservation programs to restore water and land resources in the Great Plains and curb unchecked expansion of farming on marginal lands. In particular, the report's recommendations emphasized the importance of preventive measures such as terracing and the reduction of cropping on marginal land to address the vulnerability of semiarid ecosystems (Ponting 2008, 255).

Unfortunately, people's memories were short. In the period after World War II, agricultural production and exports—based on large-scale farming—continued to expand in the Great Plains in response to the growing demand for food and livestock fodder and (in the last decade) for ethanol from corn.

Even though there is greater awareness of land and water management conservation principles in Plains states since the Dust Bowl, there are ongoing concerns that renewed intensity of crop cultivation—coupled with climate change and seasonal rainfall variations—could lead to further problems. The Southwest and Great Plains have a high likelihood of experiencing further droughts in the future, if warming levels and further water drawdown from reservoirs and aquifers continue unchecked (Richter 2014, 28).

Despite the lessons of the U.S. Dust Bowl experience, similar processes of overstepping natural boundaries are occurring in other parts of the world. In tropical countries, the burning of rainforests has stripped the soil of natural forest cover, disrupted the limits of fragile self-contained rainforest ecosystems, and produced greenhouse gases. This process is already disrupting the water cycle, causing severe erosion, and reducing precipitation in tropical forest areas.

Tropical forests, stripped of protective tree cover, are also subject to increased wind and soil erosion, a process that is occurring in the Amazon basin. In the Amazon basin, short-term thinking (as during the U.S. Dust Bowl) is propelling farmers and squatters—some acting on behalf of absentee large landowners—to go after quick gains. Speculators have cut down tropical forests for lumber and cleared land for unsustainable and inappropriate land uses such as cattle ranching and the cultivation of commercially important crops such as soybean and sugar cane for ethanol production and exports.

New Dust Bowl: The Aral Sea

Across the world in Central Asia, and notwithstanding the lessons of history, a new dust bowl has developed around the Aral Sea. The Aral Sea is located in Central Asia, between the southern part of Kazakhstan and northern Uzbekistan. It was once the world's fourth largest inland sea on earth, but since 1960 it has shrunk to 74 percent of its original area and 90 percent of its volume (Chellaney 2013, 166). Its ongoing destruction mirrors the Great Plains experience of the 1930s. The Aral Sea's degradation is a result of the unsustainable expansion of agriculture, accelerated erosion, and rapid depletion of water reserves in a vulnerable, dry region.

How did this happen? In the 1950s, the then government of the Soviet Union built irrigation canals near the border with Afghanistan and diverted the flow of two rivers (the Amu Darya and Syr Darya) that flowed from the Himalayas into the Aral Sea. The rechanneled water was used to grow cotton in an adjacent desert, which absorbed a high portion of the water. Some of the transported water evaporated before even arriving at the cotton fields. Over time, the rerouting of water away from the Aral Sea created an ecological imbalance that caused the sea to dry up over four decades and increased the salinity of its soil (Black 2004, 20).

The destruction of the Aral Sea killed fish stocks, and by the 1980s the sea's demise had wiped out the fishing industry for good. In the same way that old photos of the Dust Bowl depicted haunting images of dust storms and people fleeing from the dust, contemporary photos of the Aral Sea immortalize stranded fishing boats in the middle of a dry, cracked wasteland, which was once a large, sparkling, inland sea.

Other effects of the Aral Sea crisis have been the lowering of the water table due to evaporation, the destruction of oases in the area, the disappearance of farms and wildlife, the leaching of salts over the Sea's surface, and the accumulation of pesticides and poisonous sediments over the exposed seabed. Similar to the Dust Bowl, dust storms around the Aral Sea are constantly blowing exposed soil throughout the area, provoking respiratory ailments and other health issues in the regional population.

Ecological Crisis

As a result of this ecological crisis, the local and regional climate around the Aral Sea has drastically changed. Historically, the sea had cooled the area in summer and warmed it in winter. With the change, the area became more extreme, exhibiting wider temperature ranges between the two main seasons. Less rain fell, and there was a shorter, more intense summer growing season, making it hard for the ever-shrinking cotton crop to mature. In recent years, farmers have made the decision to switch from cotton to rice, a more lucrative but thirstier crop—with further disastrous impacts. The Aral Sea ecological calamity is now difficult to reverse.

Five Central Asian nations are located around the shrunken, damaged Aral Sea. In 1992, they set up Interstate Councils of the Aral Sea basin to address the problems jointly. Despite some initial momentum on collaboration, joint water management has been hampered due to the lack of funds and technical resources, institutional difficulties, and weak commitment of the convening countries to implement agreements.

The Aral Sea remains in jeopardy, providing yet another lesson of the consequences of human action in a vulnerable area. It is ironic that this disaster has attracted visitors, who come to the Aral Sea to photograph stranded ships in an arid wasteland. Some scientists believe that in the next two decades, the sea might totally disappear, making it an eerie memory for future generations, similar to the Dust Bowl of the 1930s.

Camille Gaskin-Reyes

REFERENCES AND FURTHER READING

Barlow, Maude. *Blue Future: Protecting Water for People and the Planet Forever*. New York: New Press, 2013.

Black, Maggie. *The No-Nonsense Guide to Water*. Oxford: New Internationalist Publications, 2004.

Burns, Ken. *The Dust Bowl*. PBS, 2012.

Chellaney, Brahma. *Water, Peace, and War: Confronting the Global Water Crisis*. Lanham: Rowman & Littlefield, 2013.

Ponting, Clive. *A New Green History of the World: The Environment and the Collapse of Civilizations*. New York: Penguin Group, 2008.

Richter, Brian. *Chasing Water: A Guide for Moving from Scarcity to Sustainability*. Washington, DC: Island Press, 2014.

Steinbeck, John. *The Grapes of Wrath*. New York: Penguin Books, 2002.

Case Study 2: Hurricanes and Intense Storms

The ferocity of Hurricanes Mitch in Central America, Katrina in New Orleans, and Sandy in the New Jersey/New York area in the past years has highlighted the intensity of storms and their damaging or life-threatening impacts. This case study focuses on comparing and contrasting the experiences, impacts, and lessons learned from these devastating events.

Hurricane Mitch in Central America

Hurricane Mitch was a megastorm that affected most of Central America in October 1998, but Honduras bore most of its fury. It left 10,000 Hondurans dead in its path, 3 million people homeless, and billions of dollars of damage to the economy. In addition to wind speeds of up to 180 miles per hour and torrential rains and floods, Mitch caused devastating landslides on denuded slopes, and wiped out entire communities (Flannery 2005, 137).

The year 1998 was the warmest of the 20th century. Hurricane Mitch in its early stages absorbed extraordinarily huge amounts of overheated Atlantic waters and developed enormous strength. The storm's impact was heightened by widespread deforestation of hillsides and slopes in Honduras, unregulated building of settlements on steep and treeless slopes, and the general lack of natural disaster preparedness efforts.

Mitch destroyed 60 percent of Honduras's bridges, 25 percent of its schools, and half of its agricultural production, including most of its banana plantations. Observers commented that the hurricane had set the country back 20 years in its development, and Central America by about a decade (Maslin 2014, 78). Many of the severely affected communities were poor people eking out livelihoods in vulnerable areas such as urban squatter settlements on bare slopes or low-lying coastal zones.

These communities had little or no surplus food or cash reserves to weather the catastrophe. This is one of the most important lessons learned from Mitch: poor countries such as Honduras lack resources and planning systems to set up warning, evacuation, and response strategies; and poor communities have few defense and response mechanisms, and very little resilience to cope with the aftermath of natural disasters.

Hurricane Katrina in Louisiana

In contrast to Mitch, Hurricane Katrina occurred in a developed country, the United States. Katrina struck on August 28, 2005, a summer remembered for the destructive force of hurricanes that shattered many records. Surface waters of the northern Gulf of Mexico reached about 30°C. In the Atlantic basin, 27 named storms and hurricanes developed in that season, beating the 1933 record of 21 hurricanes. Katrina was the first and most powerful of a string of three powerful Category 5 hurricanes (Katrina, Rita, and Wilma) within the space of a two-month period.

During its four-day passage through Gulf waters, Katrina progressed from a mere thunderstorm first sighted in the Bahamas to a Category 1 storm of 75 miles an hour in Florida, then to a full-fledged Category 5 with winds of over 175 miles an hour, finally slamming into New Orleans on August 28, 2005. The floods covered 80 percent of the city with water; more than 1,500 people lost their lives. The economic impact of Hurricane Katrina on the U.S. economy exceeded 150 billion dollars (Maslin 2014, 78).

Hurricane Rita developed in the Gulf of Mexico 23 days after Katrina, becoming the fourth most powerful hurricane ever recorded in the Atlantic basin. It packed winds of 175 miles an hour and wind gusts topping 235 miles an hour. Three weeks after Rita, Hurricane Wilma came to life off the coast of Jamaica. It zigzagged its way across Mexico, Florida, and the Bahamas, and eventually reached sustained winds of over 175 miles per hour at high atmospheric levels. Of the three storms, however, Katrina was the most memorable hurricane.

Much in the same way as Hurricane Mitch affected the poverty-stricken in Honduras, Hurricane Katrina's impact was hardest on the poorest, most vulnerable communities in the low-lying wards of the city of New Orleans. Katrina provided many lessons on the factors responsible for the devastation of New Orleans and the disruption and loss of so many lives.

One main lesson of Katrina is that the stage for the devastation was set long before 2005. Prior to the storm, New Orleans had already been subsiding—as had most of the Louisiana coastline—due to land loss, coastal erosion, sinking of the shoreline and sea intrusion. Thousands of miles of canals, built by the oil industry, crisscrossed the Louisiana shoreline, adding to the subsidence problem. These factors, and the ongoing loss of sediments from the Mississippi River, were major contributors to erosion of the land and the vulnerability of the coastline. The perspectives section in this chapter further examines the issues regarding the vulnerability of New Orleans and the levee systems.

Before Katrina, many experts had warned that the New Orleans levees needed reinforcing in low-lying areas, but this problem was not adequately addressed. With Katrina, many factors came together to create the perfect storm: land loss, erosion, the destruction of Louisiana's barrier islands and mangroves (which traditionally had offered storm protection), the strength of the hurricane, the weak levees (which did not hold at the most critical moment), and poor disaster management.

New Orleans had always been a classic case of a divided city. In what has been termed the racial geography of the city, a white urban population with higher incomes generally occupied the higher, better-drained areas of the city, while low-income African Americans lived in the swampier, lower-lying Ninth Ward (Colten 2005, 81; Dyson 2006, 5).

Over the years, levee and drainage system construction enlarged the city's habitable territory, but in the end levee development proved to be insufficient to protect the vulnerable low-lying wards from the onslaught of Katrina. In addition, measures taken to prepare for and respond to the storm were woefully inadequate. New Orleans before the storm was divided; during and after the storm it became even more divided.

Two factors amplified Katrina's impact: flaws in the levee design and construction in the poorest areas of the city, and flaws in the city's and the federal government's responses once the storm struck. Inadequate preparedness, evacuation, and storm response actions aggravated the already existing social and economic gap in the city.

Thousands of low-income, vulnerable people, including the sick and elderly, were left behind to fend for themselves, suffer the ravages of the storm, and bear the indignities of its aftermath. Evacuation plans assumed that most people had access to cars or transportation to leave the city. But poor residents had few cars, and not enough public buses were mobilized to evacuate the poor, the sick, the elderly, and the disabled.

Television images of Katrina remain imprinted in the memories of people who watched the horror unfold on live television. Prior to the storm, the images showed traffic jams with cars full of people fleeing to safety. During the storm, the images showed the raw suffering of the sick, elderly, disabled, and poor trapped in a flooding city. Who could forget these terrifying scenes of helpless and dying people witnessed by viewers all over the world?

Hurricane Sandy on the East Coast

Hurricane Sandy, which struck the East Coast of the United States in October of 2012, also demonstrated the effects of warming temperatures and super storms. Sandy was the most destructive hurricane of the 2012 Atlantic season. Reaching as far south as New Jersey, it damaged thousands of homes and killed over 100 people. It forced thousands into shelters and damaged vital infrastructure, including power lines, transportation arteries, and water and sewage facilities. Sandy fully disrupted parts of Manhattan, flooded tunnels and subway stations, and wreaked havoc on businesses and city travel, plunging large parts of the city into total darkness.

There are multiple lessons to be gleaned from Sandy. It exposed the vulnerability of the New York and New Jersey coastlines as well as the need to improve the resilience of U.S. coastlines and for communities to adapt to a new breed of super storms. Sandy brought the issue of climate change home to a major metropolitan area by highlighting the urgency of shoring up sea defenses and coastal zones. The storm also exposed inadequate regulations for construction in coastline areas, the vulnerability of coastal drainage and sewerage systems, and the imperative of additional storm barriers and natural disaster–management strategies.

Similar to that of Mitch and Katrina, Sandy's impact was disproportionately felt in the poorer boroughs of the New York metropolitan area and public housing projects, home to the most vulnerable groups such as the elderly, the sick, and low-income residents. An important lesson learned was the importance of coordinating all responsible jurisdictions and agencies to enable disaster management planning and responses.

Camille Gaskin-Reyes

REFERENCES AND FURTHER READING

Brinkley, Douglas. *The Great Deluge.* New York: Harper Collins, 2005.

Cobb, Jelani. "Race and The Storm," *The New Yorker*, August 24, 2015. http://www.newyorker.com/magazine/2015/08/24/race-and-the-storm.

Colten, Craig E. *An Unnatural Metropolis: Wresting New Orleans from Nature.* Baton Rouge: Louisiana State University, 2004.

Cumming-Bruce, Nick. "U.N. Disaster Chief Warns of More Natural Catastrophes to Come." *New York Times,* December 23, 2014. http://www.nytimes.com/2014/12/24/world/un-disaster-chief-warns-of-more-natural-catastrophes-to-come.html.

Dyson, Michael Eric. *Come Hell or High Water: Hurricane Katrina and the Color of Disaster.* New York: Basic Civitas, 2006.

Flannery, Tim. *The Weather Makers: How Man Is Changing the Climate and What It Means for Life on Earth.* New York: Grove Press, 2005.

Heerden, Ivor van, and Mike Bryan. *The Storm.* New York: Viking, 2006.

Li, Wei, Christopher A. Airriess, Angela Chia-Chen Chen, Karen J. Leong, and Verna Keith. "Katrina and Migration: Evacuation and Return by African Americans and Vietnamese Americans in an Eastern New Orleans Suburb." *The Professional Geographer* 62, no. 1 (February 2010): 103–118.

Maslin, Mark. *Climate Change: A Very Short Introduction.* Oxford: Oxford University Press, 2014.

Winterfeldt, Detlef von. "Using Risk and Decision Analysis to Protect New Orleans Against Future Hurricanes." In *On Risk and Disaster: Lessons from Hurricane Katrina*, edited by Ronald J. Daniels, Donald F. Kettl, and Howard Kumreuther. Philadelphia: University of Pennsylvania Press, 2006.

Case Study 3: Water Issues in China and Neighboring Countries

China's rapid progress into an international economic powerhouse has been nothing less than astonishing. It is the world's most populated nation, with slightly over 1.3 billion people. Its economy has grown almost eightfold since the mid-1990s, and has lifted millions of Chinese out of poverty, based on a global export model. This study explores how this model—in spite of these impressive achievements—has impaired the quantity and quality of the country's natural resources and freshwater supply, and is affecting neighboring countries as well.

Water Woes

Historically, China has suffered from both droughts and floods. Over lengthy periods of China's history, the waters of the Yellow and Yangtze Rivers have burst their banks many times, causing deaths and famine. Currently, the country is undergoing severe water stress. China has 22 percent of the world's people, but only 6 percent of its freshwater (Black 2004, 23). More than a quarter of China's land is classified as desert; it is one of the planet's most arid regions in general. Most of its rivers are either overpolluted or too silted to provide adequate supplies of freshwater. Every spring, China's capital, Beijing, experiences raging dust storms from Mongolia, where hectares of grasslands are undergoing desertification due to prolonged drought and overuse of the land.

Water scarcity and contamination has become a major developmental and environmental issue for China. The overuse of water resources to support population, urban growth, and coal-based industries and exports is a problem. While widespread famine and the deprivation of the past centuries have been resolved, the country remains vulnerable to new threats such as droughts, floods, natural disasters, health effects of air and water pollution, and additional risks posed by climate change.

Increasing Vulnerability

In the Yangtze River basin of China, the conversion of floodplains and flood diversion zones into farmland to support population growth has increased the country's water vulnerability and exacerbated water scarcity. Land reclamation for farming has shrunk many lakes and reduced available water supplies. In 1998, the Yangtze floods were the highest ever recorded in China's flood records. This disaster required the mobilization of millions of soldiers and citizens to build up dykes and levees to save downstream cities in the path of floodwaters. These floods affected 16 million people, killed 4,000 persons, and degraded 9 million hectares of land (Black 2004, 94).

North-South Water Differential

The flipside of the intensification of floods in China is increasing water scarcity and the high incidence of droughts in some areas. One of the main problems of China's water supply and water management is related to the fact that there is a hydrological North-South divide. The arid north has just 19 percent of China's water but 50 percent of the country's population and 64 percent of its cultivated land. In contrast, the less-populated south is more humid, with 80 percent of rainfall and snowmelt as well as the immense water resources of the Yangtze and other river basins (Chellaney 2013, 237).

Due to the past policies, the more arid north of China was traditionally the country's food basket between 1950 and 1980—with disastrous environmental consequences. Large-scale irrigation projects continuously drew down water resources from the Yellow River, which eventually became one of the world's heavily dammed and sediment-laden rivers. Since then, the Yellow River has lost most of its flow and no longer reaches the ocean. In northern China, many reservoirs that supplied water to farms were also diverted to households and factories, aggravating water problems for agricultural production.

Megaprojects and Dam-Building

China is planning the construction of an enormous canal project to bring water from the wetter south to the drier north at a cost of $60 billion (Chellaney 2013, 239). To address the north-south water differential, the country is also considering an $80 billion project to reroute the Brahmaputra River and create three new artificial rivers to carry water from the Tibetan Plateau to the drier north (Schneider and Pope 2008). Neighboring India, which also relies on the waters of the Brahmaputra, has strongly opposed this proposal. Other critics contend that China is relying too much on big engineering solutions to resolve structural problems of water misuse, pollution, runoff, and water management policies (Hornby 2014).

Furthermore, in order to address seasonal rainfall variations, droughts, and floods, and to generate electricity for its economy, the country launched a number of massive dam-building programs in upstream areas of the watersheds of rivers flowing from the Tibetan Plateau. One example of such a megaproject is the controversial Three Gorges Dam with a cost of over $25 billion and the displacement of millions (Chellaney 2013, 212).

Potential Conflicts in the Neighborhood

In attempting to resolve its own problems, China could potentially be contributing to water shortages further downstream. Some analysts have made the point that China's numerous dam and river projects have the potential to create tensions with neighboring countries such as India, Burma, Cambodia, Laos, Thailand, and Vietnam. There are concerns that the Tibetan Plateau could become an epicenter of regional water conflicts in the future. This plateau is the world's third-largest reserve of freshwater after the Arctic and Antarctica, and the source of Asia's large rivers such as the Brahmaputra, Indus, Mekong, Arun, and others.

The Tibetan Plateau is home to glaciers, freshwater lakes, and waterfalls, which form the headwaters of large Asian rivers. Half of the world's population lives in their watershed areas. Therefore, China's actions regarding the use of the headwaters or upstream river courses—whether through dam building or rerouting rivers—have to be viewed in terms of potential hydrologic and political ramifications for the entire region.

A serious problem affecting the Tibetan Plateau is climate change. Glaciers are receding at a rate of three feet per year; rapid melting and evaporation of freshwater is provoking additional concerns about regional water supply and the need for collaborative sharing of water resources among China, Tibet, and other countries.

In addition to the Brahmaputra River, the Mekong River has the potential to become a source of conflict. Researchers report that Chinese dams upstream on the Mekong are already affecting water levels in Vietnam, Cambodia, Laos, and other downstream neighbors. It was reported that for the first time in several thousand years the Mekong Delta in Vietnam, home to 18 million people, was shrinking (Barlow 2013, 271).

In the Tibetan Plateau area, millions of people could be affected in the future if potential problems from glacial melt, shrinking lakes, rapid runoff, and sedimentation are not adequately addressed. Some experts have warned that further water scarcity and conflicts in Asia could result in millions of future environmental refugees (Piper 2014, 133).

These examples pose challenges but also opportunities for regional cooperation and sustainable management of common water resources in Asia. Cross-border collaboration is a crucial element to develop a win-win strategy to reduce or mediate political conflicts, stem the loss of agricultural production, and curb the expansion of regional poverty.

China's Industrialization and Water Scarcity

China's accelerated industrialization has caused fierce competition among national water users and led to pressures to pull water away from farmers to satisfy the demand of urban consumers and industrial development zones. The water table below China's capital, Beijing, has dropped over 121 feet in the past four decades (Black 2004, 45). The very industries that are using freshwater at a fast pace are at the same time discharging industrial pollutants into rivers and rapidly degrading their quality.

It is estimated that around 80 percent of China's major rivers are so polluted that they no longer support fish (Black 2004, 101). In the Yellow River, pollutants from tanneries, paper and pulp mills, textile factories, oil refineries, and chemical plants are reported to have released so many heavy metals that the water is unfit for irrigation purposes. Pollution is also affecting lakes. Lakes Dianchi and Taihu are covered by dense algae blooms, which have intensified loss of oxygen and the abandonment of fish farming (Barlow 2013, 164).

Because China's total water resources are estimated to have dropped 13 percent since the beginning of the century, the country is aware that its economic and industrial expansion model is reaching choke points due to inadequate freshwater and groundwater reserves, poor air quality, and energy bottlenecks.

Coal Production and Climate Change

There is an additional complication. China uses coal-fired plants for energy, and there is an important nexus between water and coal production, and between coal production and the release of GHG. Freshwater is needed for mining and processing coal, the largest area of industrial water use in China. To address its water problems, the country has adopted some water conservation measures, including efforts to improve efficiency of water use, retrofit sewage treatment systems to recycle wastewater, and mechanisms to capture rainfall from rooftop installations. However, water scarcity remains a major challenge.

Parallel to efforts to address water scarcity and water pollution, China needs to step up efforts to reduce greenhouse gas emissions, air, and soil pollution levels. China's coastal cities will be dramatically affected by climate change (the country produces more carbon dioxide than any other country), if fossil fuel emissions are not reduced, and temperatures continue to rise. The country will also need to spend huge sums of money on shoreline barriers and flood control systems to protect its coasts and cities. Its largest city, Shanghai, with a population of about 24 million located on the East China Sea, would face almost complete inundation by 2100 under the worst-case scenario of 4°C warming. Tianjin and Shantou are other cities that would be severely threatened by rising oceans and encroaching seas (Watkins 2015).

Camille Gaskin-Reyes

REFERENCES AND FURTHER READING

Barlow, Maude. *Blue Future: Protecting Water for People and the Planet Forever*. New York: New Press, 2013.

Black, Maggie. *The No-Nonsense Guide to Water*. Oxford: New Internationalist Publications, 2004.

Chellaney, Brahma. *Water, Peace, and War: Confronting the Global Water Crisis*. Lanham: Rowman & Littlefield, 2013.

Hornby, Lucy. "Concern Mounts in China Over Yangtze Diversion Project." *Financial Times*, October 15, 2014. http://www.ft.com/cms/s/0/a39d9c7c-4fa2-11e4-a0a4-00144feab7de.html#axzz48Zhev2xf.

Piper, Karen. *The Price of Thirst: Global Water Inequality and the Coming Chaos*. Minneapolis: University of Minnesota Press, 2014.

Schneider, Keith. "Choke Point China—Confronting Water Scarcity and Energy Demand in the World's Largest Country." *Circle of Blue*, February 15, 2011. http://www.circleofblue.org/2011/world/choke-point-chinaconfronting-water-scarcity-and-energy-demand-in-the-worlds-largest-country/.

Schneider, Keith, and C. T. Pope. "China, Tibet, and the Strategic Power of Water: Pollution and Global Warming Threaten Asia's Most Important Freshwater Source." Circle of Blue, May 9, 2008. http://www.circleofblue.org/2008/china/china-tibet-and-the-strategic-power-of-water-pollution-and-global-warming-threaten-asiass-most-important-freshwater-source/.

Watkins, Derek. "China's Coastal Cities, Underwater." *New York Times*, December 11, 2015. http://www.nytimes.com/interactive/2015/12/11/world/asia/Chinas-Coastal-Cities-Underwater.html.

ANNOTATED DOCUMENT

United Nations Convention to Combat Desertification in the Countries Experiencing Serious Drought and/or Desertification, Particularly in Africa (1994)

Background

The 1994 UN Convention addressed the problem of desertification around the world, particularly in African countries experiencing drought and degradation of dry lands. It established a framework and regional action plans to be supported through international cooperation and partnership arrangements among developed and developing countries.

The Convention excluded natural deserts, and focused on dry lands, defined as arid, semi-arid, and dry sub-humid areas. Desertification therefore refers to land degradation in non-desert areas, which is caused by factors such as climatic variations and human activities. It emphasized that this process has created desert-like conditions and intensified the encroachment of natural deserts upon arid lands.

The Convention recognized that achieving its objective to combat desertification would require long-term integrated strategies. It focused on addressing affected areas through activities such as improved land productivity, rehabilitation, conservation and sustainable management of water resources, and other concerted efforts to alleviate poverty.

The Convention gave priority to mitigating the effects of drought in the most vulnerable communities of the world. It underscored the promotion of community participation, especially of women and the youth; the involvement of pastoralist and farmer groups and non-governmental actors; and the fostering of an enabling environment, i.e., more robust legislation and policies to combat desertification.

Summary

The Convention contains a number of Parts and their respective Articles as follows:

Part I contains an introduction, and three articles on the use of terms, objectives, and principles. Part II refers to general provisions, general objectives, obligation of affected country parties, obligations of developed country parties, and relationship with other conventions. Part III includes action programs, scientific and technical cooperation and supporting measures, including thirteen articles on multiple topics such as: national, sub-regional, and regional programs; international cooperation; information collection, analyses and exchanges; research and development; technology transfer, acquisition, adaptation, and development; capacity building, education and public awareness; and financial resources and financial mechanisms;

Part IV addresses institutions and is comprised of four articles on institutions, agencies, and bodies, while Part V contains procedures, with seven articles on resolution of questions on implementation, the settlement of disputes, and amendments to the convention; and Part VI includes final provisions, including seven articles to regulate implementation of the convention.

The four Annexes of the Convention outline specific regional Action Plans as follows:

Annex I addresses the African Region and commitments and obligations of African and developed countries; strategic frameworks for sustainable development; a timetable for actions; technical assistance and transfer of technology; financial resources and mechanisms; and partnership agreements.

Annex II contains the Regional Implementation Plan for Asia, taking into account sub-regional, national, and joint action programs; regional projects; financial resources and mechanisms; and cooperation and coordination mechanisms.

Annex III contains the Regional Implementation Plan for Latin America and the Caribbean, focusing on technical cooperation and scientific arrangements; institutional frameworks; and financial resources and mechanisms.

Annex IV includes the Action Plan for the Northern Mediterranean to address specific drought and dry-land conditions of the Northern Mediterranean region; a strategic planning framework for sustainable development; and co-ordination of subregional, regional, and joint action programs.

All Regional Action Plans include the following: the establishment and/or strengthening of early warning systems at the subregional and regional levels; mechanisms for assisting displaced persons; drought preparedness, management and drought contingency plans based on seasonal to interannual climate predictions; and food security systems, including the construction of storage and marketing facilities in rural areas.

Postconvention Follow-Up: Decade for Deserts

We can see from the above that in the 1990s, the problems of drought, soil erosion, and water scarcity were already of concern to water-stressed nations around the world. Since the 1994 convention, desertification has intensified with climate change and more extended periods of droughts in the Sahel Zone in Africa, the North American continent, the Aral Sea, the Indian subcontinent, and other parts of Asia.

With more countries around the world facing further land and soil erosion, water shortages, and expansion of the desertification process, the United Nations declared a "Decade for Deserts" from 2010 to 2020 to accelerate efforts to combat desertification.

The new Decade for Deserts' declaration recognized that the original convention had not made enough progress in the fight against desertification. It called for more cooperation among dry land and desertification-affected countries and developed nations, and for greater financial and technical support to countries undergoing the challenges of climate change.

Some objectives of the declaration are related to the United Nations Millennium Development Goals, such as the eradication of poverty and environmental sustainability objectives. Using the original convention as a backdrop, the UN developed a 10-year Strategic Plan and framework to implement the Action Plans of the four regions included in the convention's Annexes.

The new 10-year Strategic Plan includes efforts to mobilize financial and technical support to the Convention Secretariat, monitor results of programs and prepare progress reports for the UN Secretary General's United Nations Report to the General Assembly. Experience of the first convention showed that due to adequate funding support from developed countries and technical assistance to states impacted by desertification, the objectives remained largely declarations of intent. It is important that nations around the world take the threat of further desertification seriously and commit to action so that affected countries and regions receive the required financial and technical assistance.

Camille Gaskin-Reyes

Excerpts from the Convention Document

Preamble

The Parties to this Convention,

Affirming that human beings in affected or threatened areas are at the center of concerns to combat desertification and mitigate the effects of drought,

Reflecting the urgent concern of the international community, including States and international organizations, about the adverse impacts of desertification and drought,

Aware that arid, semi-arid, and dry sub-humid areas together account for a significant proportion of the Earth's land area and are the habitat and source of livelihood for a large segment of its population,

Acknowledging that desertification and drought are problems of global dimension in that they affect all regions of the world and that joint action of the international community is needed to combat desertification and/or mitigate the effects of drought,

Noting the high concentration of developing countries, notably the least developed countries, among those experiencing serious drought and/or desertification, and the particularly tragic consequences of these phenomena in Africa,

Noting also that desertification is caused by complex interactions among physical, biological, political, social, cultural, and economic factors,

Considering the impact of trade and relevant aspects of international economic relations on the ability of affected countries to combat desertification adequately,

Conscious that sustainable economic growth, social development, and poverty eradication are priorities of affected developing countries, particularly in Africa, and are essential to meeting sustainability objectives,

Mindful that desertification and drought affect sustainable development through their interrelationships with important social problems such as poverty, poor health and nutrition, lack of food security, and those arising from migration, displacement of persons, and demographic dynamics,

Appreciating the significance of the past efforts and experience of States and international organizations in combating desertification and mitigating the effects of drought, particularly in implementing the Plan of Action to Combat Desertification which was adopted at the United Nations Conference on Desertification in 1977,

Realizing that, despite efforts in the past, progress in combating desertification and mitigating the effects of drought has not met expectations and that a new and more effective approach is needed at all levels within the framework of sustainable development,

Recognizing the validity and relevance of decisions adopted at the United Nations Conference on Environment and Development, particularly of Agenda 21 and its chapter 12, which provide a basis for combating desertification,

Reaffirming in this light the commitments of developed countries as contained in paragraph 13 of chapter 33 of Agenda 21,

Recalling General Assembly resolution 47/188, particularly the priority in it prescribed for Africa, and all other relevant United Nations resolutions, decisions and programs on desertification and drought, as well as relevant declarations by African countries and those from other regions,

Reaffirming the Rio Declaration on Environment and Development which states, in its Principle 2, that States have, in accordance with the Charter of the United Nations and the principles of international law, the sovereign right to exploit their own resources pursuant to their own environmental and developmental policies, and the responsibility to ensure that activities within their jurisdiction or control do not cause damage to the environment of other States or of areas beyond the limits of national jurisdiction,

Recognizing that national Governments play a critical role in combating desertification and mitigating the effects of drought and that progress in that respect depends on local implementation of action programs in affected areas,

Recognizing also the importance and necessity of international cooperation and partnership in combating desertification and mitigating the effects of drought,

Recognizing further the importance of the provision to affected developing countries, particularly in Africa, of effective means, inter alia substantial financial resources, including new and additional funding, and access to technology, without which it will be difficult for them to implement fully their commitments under this Convention,

Expressing concern over the impact of desertification and drought on affected countries in Central Asia and the Trans-Caucasus,

Stressing the important role played by women in regions affected by desertification and/or drought, particularly in rural areas of developing countries, and the importance of ensuring the full participation of both men and women at all levels in programs to combat desertification and mitigate the effects of drought,

Emphasizing the special role of non-governmental organizations and other major groups in programs to combat desertification and mitigate the effects of drought,

Bearing in mind the relationship between desertification and other environmental problems of global dimension facing the international and national communities,

Bearing also in mind the contribution that combating desertification can make to achieve the objectives of the United Nations Framework Convention on Climate Change, the Convention on Biological Diversity and other related environmental conventions,

Believing that strategies to combat desertification and mitigate the effects of drought will be most effective if they are based on sound systematic observation and rigorous scientific knowledge and if they are continuously re-evaluated,

Recognizing the urgent need to improve the effectiveness and coordination of international cooperation to facilitate the implementation of national plans and priorities,

Determined to take appropriate action in combating desertification and mitigating the effects of drought for the benefit of present and future generations,

Source: United Nations Convention to Combat Desertification in the Countries Experiencing Serious Drought and/or Desertification, Particularly in Africa (1994). A/AC.241/27. September 12, 1994. Available online at http://www.unccd.int/Lists/SiteDocumentLibrary/conventionText/conv-eng.pdf. (c) 1994 United Nations. Reprinted with the permission of the United Nations.

PERSPECTIVES

Are Levees the Best Long-Term Solution to Protecting New Orleans from Flooding?

Overview

Hurricane Katrina was the costliest and one of the deadliest hurricanes in the history of the United States. It was the sixth-strongest Atlantic hurricane ever recorded and the third-strongest land-falling U.S. hurricane ever recorded. Katrina occurred late in August during the 2005 Atlantic hurricane season and devastated much of the north-central Gulf Coast of the United States. Most notable in media coverage were its catastrophic effects on the city of New Orleans, Louisiana.

Katrina's sheer size devastated a 100-mile stretch of the Gulf Coast. The storm surge that swept over New Orleans was as high as 27 feet, but that was not the main cause of the damage. The city's levees were fundamentally flawed; they had not been given proper foundations. The soil beneath them was washed away, opening the city to the water of Lake Pontchartrain, a source of water beyond anything that the high storm waters could produce. Much of the city was submerged.

New Orleans is about 45 miles from the Gulf of Mexico and more than twice that distance from the mouth of the Mississippi. It is part of that river's delta, a broad region of bayous and wetlands. Today the main, built-up part of the city is free from marshes as a result of the extensive measures taken to drain or pump away the water. Both natural and built levees run east and west within the city between the Mississippi River and Lake Pontchartrain, which stretches northward for more than 20 miles.

Levees extend along both sides of the lower Mississippi for a total distance of 1,500 miles. Farther up the valley of the river these levees are quite high, as high as 36 feet with base widths of 360 feet, but those around the city area average only 15 feet above the natural ridges on which they were built. Because the differences in elevation between the water level inside the levees and the lowest parts of the city are so big, there is a great need for a thoroughly dependable levee system. In addition to the levees, there is a floodway through which water can be bypassed during a river flood.

Diversion of water is the usual method of minimizing threats to the city. To the west is a large water diversion channel beginning far upstream and continuing down the Atchafalaya basin into the Gulf, affectionately named Old River Control Structure. Half of all the water in the Mississippi, when it is at flood stage, can be bypassed in this way. On the western outskirts of the city, on the main river, is another diversion, the Bonnet Carre Spillway. It can be opened to divert water from the river into Lake Pontchartrain. It is seldom used but is always available. There is a continuing concern about the stability of these protective measures because of the nature of the underlying ground. During a major flood in 1973, for instance, part of the Atchafalaya was undermined and one wall failed.

Because there is so much uncompacted material everywhere in and around the city, compaction of these sediments from time to time is the major cause of subsidence, or sinking. Land sinking, shoreline erosion, and salt water encroachment all are active and add to this problem of maintaining a consistent level of land. At times these forces cause sudden changes to buildings and facilities. If allowance is made for sea level variations, the general picture of subsidence, or sinking, rates is about seven inches a century. Local groundwater withdrawals further aggravate the situation.

About one in 10 homes and the same proportion of commercial buildings, plus one out of every three streets and sidewalks, show signs of subsidence. Typical conditions include buckling of patios and exposure of foundation slabs. Driveways, too, sink to such an extent that it is impossible to drive into carports. Gas and water leaks occur as underground utility lines sag. The problem worsens with development as new nonporous coverings of streets, parking lots, and buildings lead to drying and compaction in the organic soils beneath, hence subsidence.

When the first settlers occupied some high ground on the banks of the Mississippi almost 300 years ago, there was little thought about the problems of growth, but the risks gradually increased as the settlement expanded. Today the city continues to push its frontiers farther and farther into low-lying marshy areas where building is possible only with the best of modern technology. Structures 600 feet tall stand where formerly the ground could not support the weight of one person.

Water levels, when the river is in flood, can be as high as 27 feet above the lowest areas of the city. There seems to be great faith in the stability of the dikes, but those responsible for them are always on alert, especially when strong winds blow.

Angus M. Gunn

Perspective 1: Levees Provide Long-Term Protection

Many who tracked Hurricane Katrina breathed a sigh of relief as the eye of the storm moved to the east of New Orleans. The Big Easy would avoid a direct hit. Still, no one could imagine the devastation that would follow. Hurricane Katrina drowned the New Orleans metropolitan area on August 29, 2005. Nearly 85 percent of the area was under water within 24 hours when the levees, designed to protect the city, were breached in 50 areas. Although the levees were expected to withstand a Category 3 storm, they were not designed for the intense storm surge created by Hurricane Katrina, which railroaded immense storm surges of water into lakes Borgne and Pontchartrain—storm surges of 28 feet and waves as high as 55 feet.

Unprecedented Flooding

Flooding was unprecedented as citizens found themselves amid floodwaters as deep as nine to 22 feet in many areas, particularly in St. Bernard Parish and the Lower Ninth Ward just three miles east of New Orleans. More than 100,000 people were left homeless within hours, and about 1,100 people died. But what if there were no levees at all? The scene would have been apocalyptic and New Orleans would have become a modern-day Atlantis. Unquestionably, a levee system is the best long-term solution to protecting New Orleans. What remains debatable is how strong the levee system needs to be.

Located on the Gulf Coast of the United States, New Orleans is particularly vulnerable to tropical storms and hurricanes. The levees in place during Hurricane Katrina were poorly designed and suffered from municipal neglect over generations. The collapse of the floodwalls along Lake Pontchartrain was responsible for the flooding and the loss of the wetlands in St. Bernard Parish, which had provided natural protection in the past, contributed to the disaster.

For three centuries, landowners, engineers, and the government have tried to control the Mississippi River through floodwalls, and sediments that would normally flow out to southeastern Louisiana were redirected; therefore, the wetlands were unable to flourish and powerless to absorb storm surges. According to the National Oceanic and Atmospheric Administration (NOAA), of the top ten worst hurricanes in United States history, four directly hit New Orleans and its surrounding parishes: the Last Island Hurricane on August 9, 1856; the Cheniere Caminada Hurricane on October 2, 1893; the New Orleans Hurricane of 1915; and Hurricane Katrina on August 28, 2005.

Vulnerable Location

New Orleans was settled upon a natural levee in 1717. A large labor force was necessary to fortify the natural levees and make that land habitable because it was subject to regular flooding from the Mississippi River. Landowners were responsible for creating the earthen levees and used slaves to build them. It became an eternal battle against nature. The first levee collapsed in 1735, resulting in significant damage.

From the time of its settlement through today, at least 109 hurricanes, floods, and fires have ravaged New Orleans. In 1927, nature reclaimed her land. After months of heavy rains expanding the length of the Mississippi, many levees along its route were breached. Seven states were severely flooded, including Louisiana, and the floodwaters, as high as 30 to 100 feet, created lakes where previously there were none. In 1965, during Hurricane Betsy, the levees failed at Florida Avenue and along the Industrial Canal because of the storm surge in Lake Pontchartrain. Again, the New Orleans area was inundated.

Following Hurricane Katrina, investigations were conducted in order to determine the cause of the flooding. Investigators found the levees protecting New Orleans were not built for the most severe hurricanes. The land in southern Louisiana has been slowly sinking and eroding for the past 200 years, though more quickly in the past several decades. It is most evident in New Orleans.

In 1718, the levee system was only 3 feet high; today, the levees are 17 feet high. After Katrina, the Army Corps of Engineers spent $14.5 billion to raise the levees and construct floodgates to protect the city against a 100-year event. However, considering the vulnerability of New Orleans, and the frequency of flooding, which occurs even during a rainy day because the city is below sea level, it should be protected against a 500-year event.

The new levee system in the New Orleans metro area is a far cry from the one in place during Hurricane Katrina—better engineering, computer modeling, and improved construction materials were used in the development. Still, many areas just outside New Orleans along the lake and river are still protected by the old system. Louisiana officials agree that a levee system through St. John, St. James, and Ascension Parishes is necessary. The Army Corps of Engineers favors an environmentally preferred route along the wetlands in St. John Parish. Early estimates suggest it would cost over $700 million.

Still, not all of the levee system needs to consist of man-made levees. Joannes Westerink led a team of University of Notre Dame researchers on the issue. Westerink codeveloped the computer model for storm surges used by the Army Corps of Engineers, the Federal Emergency Management Agency, and Louisiana to assist in the design of the new levee system in New Orleans. According to Westerink, human-made levee systems can work with natural systems to considerably reduce construction costs. The best long-term protection for New Orleans is to use man-made levees to control flooding and natural levees to restore the land in this area.

Liza Lugo

Perspective 2: No New Levees!

The Mississippi River is one of North America's largest rivers in terms of both areal drainage and volume of water. The river floods regularly, which is problematic for human settlements along its banks. It is also a vital waterway for the transport of goods and raw materials. The port city of New Orleans is located on the river's delta, and a great deal of money has been invested in levees to help protect the city from flooding and normalize the flow for transportation. But are levees the best long-term solution to protecting New Orleans from flooding? This essay presents the position that levees are not the best long-term solution, and presents an alternative.

In the United States, the Army Corps of Engineers is responsible for maintaining navigable waterways and is the primary federal agency for dealing with flood control. Historically, the Corps' approach to managing floods has been to build systems of dams and levees designed to regulate the flow of river systems. A levee is a large embankment that runs parallel to a river. Levees are constructed to contain a certain level of water, which is based on the probability of a given magnitude of flood occurring in any given year. Levees are supposed to prevent floodwaters from leaving the channel and spreading onto the surrounding floodplain.

The Problem of Levees

Despite their usefulness for channelizing floods, levees can create problems. The goal of levees in flood-control management is to protect human settlements from inundation. Floods do not necessarily occur at regular intervals, and vary in intensity at any given point along a river. Levees are constructed to be long lasting, and several generations can live in a floodplain that is protected by a levee without experiencing a major flood event.

This can create a false sense of security. Population and associated settlements can grow in a floodplain next to a levee without experiencing a major flood event for years. Eventually, when a flood event occurs that surpasses the levee's designed protection level, it can devastate these communities. Generations of people assume the levee will protect them and build as if they do not live in a floodplain, only to have a major flood cause significant economic damage and, potentially, human casualties.

Further, like all infrastructure, levees wear out without proper maintenance. This upkeep is expensive, and can easily be put off for a later date when elected officials allocate limited tax revenues. When levees fall into disrepair, the floods they were designed to contain can break through weak points and spread into the floodplain.

Lastly, floods provide ecosystem services, natural processes that are beneficial to humans and/or the larger environment. The primary ecosystem service provided by floods is the transport of silt and mineral rich sediment onto the floodplain. When an unregulated river floods, the volume of water in the channel increases along with the river's velocity. This increases the river's ability to erode material from the channel.

As the floodwater spills over the bank of the channel, it loses velocity and its ability to carry the sediment it has eroded. This sediment is then deposited onto the floodplain, replenishing nutrients consumed by vegetation. This is a valuable service for agricultural societies such as ancient Egypt, which was prosperous because of its agriculture, which relied upon the Nile's annual flooding. When a river is channelized by levees, this natural process is interrupted and sediments are transported downstream.

The Mississippi River is heavily regulated, as it floods regularly and is a major transportation artery for the country. All of this regulation has had a major influence on the river. Left unregulated, the Mississippi River meanders and changes course over long periods of time. This process is called avulsion. Because of all the regulation that has occurred, the Mississippi has not been allowed to change course as it normally would.

Today the Mississippi River flows to the Gulf of Mexico in the familiar channel around which New Orleans, Louisiana, has grown. However, if the river were unregulated it would have naturally moved west and entered the Gulf via the Atchafalaya River. New Orleans is no stranger to flooding, and is home to many structural flood control measures, including levees.

Hurricane Katrina's impact on New Orleans brought the debate about levees and flooding to center stage. This catastrophe is a microcosm of all the facets of levees and flooding discussed previously. New Orleans is an important port city, and has been for centuries. Because of its strategic and economic importance, structural control measures have been the primary method of flood management employed there.

Over the city's long history, floods have been an ever-present threat. According to the National Oceanic and Atmospheric Administration, the last known inundation of New Orleans due to spring flooding occurred in 1825. Since then, many levees and diversions have been constructed to mitigate floods. During this time there have been instances when levees have been blown up to prevent high water from flooding the city. Unfortunately for New Orleans, the city also has to deal with hurricanes as a natural hazard. It was Hurricane Katrina that exposed the design flaws in the levee system protecting New Orleans, as there were several failures that exacerbated the disaster. When the levees failed, flooding was made worse, resulting in property damage and loss of life. After the storm, people were allowed back into the parts of the city that flooded to begin reconstruction.

This is the heart of the problem. Rather than rebuilding in areas prone to flooding and constructing a new levee system, New Orleans should vacate those areas that have experienced the greatest damage and relocate them to less vulnerable areas. Additionally, a greater portion of the Mississippi River should be allowed to follow its natural path down the Atchafalaya.

In conclusion, levees are not the best long-term solution to protecting New Orleans from flooding. Instead, the city should take a more comprehensive approach involving the relocation of businesses and homes in areas of the city most prone to flooding and allowing the river to flow more naturally. By taking a portion of the river and the flood-prone sections of the city out of the equation, New Orleans could have greater resilience in the face of high water.

Nathan Eidem

REFERENCES AND FURTHER READING

Abbott, P. L. *Natural Disasters*. New York: McGraw-Hill, 2011.

Freitag, B., S. Bolton, F. Westerlund, and J. Clark. *Floodplain Management: A New Approach for a New Era*. Washington DC: Island Press, 2009.

Heerden, Ivor van. "The Storm That Drowned a City." NOVA. PBS.org. 52:31. November 22, 2005. http://www.pbs.org/wgbh/nova/earth/storm-that-drowned-city.html.

NOAA. "High Flows and Flood History on the Lower Mississippi River." http://www.srh.noaa.gov/lix/?n=ms_flood_history.

Sills, G. L., N. D. Vroman, R. Wahl, and N. Schwanz. "Overview of New Orleans Levee Failures: Lessons Learned and Their Impact on National Levee Design and Assessment." *J. Geotechnical and Geoenvironmental Engineering*, May 2008.

Swenson, D. "Flash Flood: Hurricane Katrina's Inundation of New Orleans, August 29, 2005." Interactive Graphics. *Times-Picayune* (New Orleans, LA), http://www.nola.com/katrina/graphics/flashflood.swf.

Thompson, S. A. *Water Use, Management, and Planning in the United States*. San Diego: Academic Press, 1999.

White, G. F. "Human Adjustment to Floods." Research Paper No. 29, Department of Geography, University of Chicago, 1942.

Wright, J. M., and D. L. Porter. "Floodplain Management and Natural Systems." In *Water Resources Administration in the United States: Policy, Practice, and Emerging Issues*, edited by M. Reuss. East Lansing, MI: American Water Resources Association, 1999.

3 SHARING FRESHWATER: COOPERATION OR CONFLICT?

OVERVIEW

Population growth and development pressures are placing increasing strain on shared water supplies. This chapter explores how competition for freshwater, coupled with rising economic demands and freshwater supply bottlenecks, are recipes for tension among users, not only in the same country but also across political boundaries.

Where two or more countries share water resources such as cross-boundary river basins, waterways, bays, wetlands, seas, lakes, or aquifers, there is a risk of conflict, but also an opportunity for multinational cooperation. This chapter looks at different models of transboundary water-sharing arrangements and a range of approaches and complex issues in the Great Lakes area, the Danube River basin, the Baltic and Arctic Seas, Central America, and other areas of the world.

The specific case studies in this chapter examine examples of collaboration and conflicts related to shared Lake Victoria waters in Africa, the Lempa River Basin in Central America, and the Jordan River Valley in the Middle East. The perspectives section reviews the situation of the shared Colorado River Basin in the United States and explores issues such as the mitigation of interstate conflicts.

Cooperate or Compete?

Over millennia, countries and border regions have had the option to cooperate or to compete for shared water and natural resources. To reduce the potential for conflict in water-sharing arrangements, institutions and water users have to agree on the basic principle of managing water fairly. This is usually achieved through regular political negotiations and collaborative water allocation and distribution arrangements involving country, state, regional, or local stakeholders. However, as some examples in this chapter illustrate, building consensus can be complex, even among countries with long-standing, friendly relationships. In countries or regions already suffering from political, religious, or ethnic conflicts, it is even more complicated.

It is estimated that there are at least 276 transnational rivers, waterways, lake basins, seas, wetlands, and aquifers across the world, which extend across the joint territories of about 148 countries (Chellaney 2015, 37). Water flows freely and visibly across country borders or is hidden in aquifers under state or national jurisdictions.

In the past, when water demand could easily be met by existing sources, there were fewer open conflicts. The problem accelerated with water shortages, climate change, droughts, or excessive water use by one party or the other. With this development, it meant that bordering countries or regions had to agree on or fight for water allocations.

Since the 1950s, many countries have entered into water treaties, agreements, memoranda of understanding, and other formal or nonformal arrangements to guide use of joint waterways and the peaceful sharing of fisheries and watershed or groundwater resources for agriculture, hydroelectric power, and other economic activities. These arrangements have had mixed results. Some are fraught with difficulties; others work well. Most accords are generally weak on conflict resolution mechanisms and binding monitoring and enforcement regulations.

Nonetheless, from a global point of view, international cooperation has helped to stave off many national and multinational armed conflicts over water use. Most joint use agreements do not extend to transboundary underground aquifers, which still lack an internationally ratified legally binding and regulatory framework, although there is a draft resolution on this topic, as this chapter examines.

The most important transnational river basins are well known. These include the Danube and the Rhine in Europe, the Nile, the Congo and the Zambezi in Africa, the Jordan River in the Middle East, the Mekong in Asia, and the Amazon in South America. Shared rivers can crisscross many countries and even joint borders of the same countries several times before meandering their way to seas and oceans.

International rivers support about 40 percent of the world's population, cover almost 50 percent of the earth's surface area, and contain approximately 60 percent of global freshwater resources (Giordano and Wolf 2003, 163). In South America alone, there are 38 transnational river basins in different transboundary combinations among 13 countries. In North America, Central America, and the Caribbean, 40 river basins extend across the boundaries of 13 countries (TFDD 2012). Rivers are essential for providing freshwater for human survival.

In Asia, the Mekong River is a major artery running through Laos, Vietnam, Cambodia, and Thailand. The Nile watershed, cradle of an ancient civilization in Egypt, extends to Egypt, Sudan, Ethiopia, and Uganda. The Danube River encompasses territories of 19 European countries, including parts of Germany, Austria, Slovakia, Hungary, Croatia, Serbia, Romania, Moldova, Bulgaria, and the Ukraine. The Amazon basin area comprises nine countries: Brazil, Peru, Bolivia, Colombia, Ecuador, Venezuela, Guyana, Suriname, and French Guiana. They are governed by an Amazon Cooperation Treaty, which promotes the harmonious and sustainable development of the Amazon basin.

Many countries also share seas—defined as bodies of water wholly or partially bordered by land and/or connected to oceans—around the world. Some seas open up to oceans, e.g., the Baltic, Mediterranean, Caribbean, Arctic, and Black Seas; others are landlocked, e.g., the Caspian and Aral seas. The Mediterranean Sea covers 22 countries from Albania to Tunisia. The Arctic Sea is shared by its five littoral countries: Denmark (Greenland), Canada, the United States, Russia, and Norway, while the Baltic Sea area covers some parts of Finland, Denmark, Sweden, Poland, Russia, and Germany.

Cross-border lakes and wetlands are other examples of shared water resources across the world. In North America, the United States and Canada share the Great Lakes and St. Lawrence Seaway. In Africa, Lake Chad is common to Nigeria, Cameroon, Chad, and Niger, and many African countries share Lake Victoria (see Case Study 1). In South America, the Pantanal, a cross-border wetland, extends across parts of Brazil, Bolivia, and Paraguay.

The Great Lakes/St. Lawrence Seaway: Peaceful and Effective Cooperation

The Great Lakes/St. Lawrence Seaway is a cross-border water ecosystem that is shared peacefully and cooperatively between Canada and the United States. These lakes are so huge, covering an area of 95,000 square miles, that they are sometimes called inland seas. The Great Lakes make up 20 percent of the world's surface freshwater, forming the largest single surface freshwater network on the planet. Only polar ice caps store more freshwater in frozen form.

The Great Lakes are at least 750 miles long. They are separate but interconnected basins: Lake Superior, Lake Michigan, Lake Huron, Lake Erie, and Lake Ontario, which extend from west to east along the U.S.-Canadian border area. Around 30 million people live around the lakes, and 35,000 islands are scattered throughout the entire system (EPA).

The Great Lakes have always been an important backbone of social and economic development in the border area, crucial for transportation, manufacturing, energy generation, recreation, and fishing activities. Their linkage to the Atlantic Ocean through the Saint Lawrence River or Seaway has provided an ocean gateway for travel and the import or export of products to and from lake-connected areas and more distant rivers, lakes, and bays to the south, north, and east and west of the Basin.

Within the United States and Canada, many connecting rivers and canals link the Great Lakes to other waterways such as the Mississippi, Detroit and Niagara Rivers, the Welland Canal, Lake Nipigon, Lake Winnebago, Lake Simcoe, Lake St. Clair, Green Bay, and Georgian Bay.

Benefits and Costs of Sharing

The Great Lakes are a prime example of the advantages of peacefully sharing both the benefits and the costs of an integrated water system. First the benefits: the Seaway makes Canada and the United States completely accessible to oceangoing ships (except for newer, wider container ships) from all over the world. This feature is vital

Photo of the Great Lakes taken from space. The Great Lakes contain 18 percent of the world's freshwater supply. Only the polar ice caps hold more. (EPA)

for handling their imports and exports.

The Seaway is particularly crucial for relatively cheap transportation and trade of bulk cargo such as grain and ores. Due to the moderating influence of water, the lakes generate a favorable microclimate to grow fruits and food crops in the border area, which supplies urban markets throughout North America and beyond. The lakes also provide drinking water for both countries, and for fishing, boating, cruising, and water sports.

Now to the costs: the sharing of the Great Lakes entails administrative and other expenses required to coordinate joint actions of member states and stakeholders, and to implement pollution control, and joint maintenance, monitoring, and enforcement projects. Shared pollution control is especially important, because contaminants enter the Great Lakes from many sources in both countries and become trapped and concentrated within the basin system to the detriment of users in the entire basin area.

Lake basin toxins are comprised of agricultural runoff and chemicals from farms, waste from cities, discharges from industries, and other atmospheric pollutants that fall as precipitation onto the surface of the lakes. A major threat to the lakes' ecosystems—that can only be addressed by joint action—is control of the zebra mussel, a harmful invasive species that once entered the lakes through the ballast water of ships.

The coordination of joint actions of the economic sectors and key public and private stakeholders falls under the 1909 U.S.-Canada Boundary Waters Treaty. The entire Great Lakes system falls within the national jurisdictions of the province of Ontario in Canada and the U.S. states of Minnesota, Wisconsin, Michigan, Illinois, Indiana, Ohio, Pennsylvania, and New York.

Due to the wide breadth of the geographical area covered, a complex, multiagency structure has been put in place for the management and environmental protection of the entire basin. This structure operates according to the principle of Integrated River Basin Management (IRBM), which emphasizes delegation of authority to the lowest administrative level possible.

In 1987, in response to growing pollution problems of the basin system, a Great Lakes Water Quality Agreement was established as an additional provision to manage water quality. At government level, the United States and Canada exercise jurisdiction through federal agencies in both countries (in the United States, it is the Environmental Protection Agency [EPA]) and the oversight of over 100 federal programs of environmental management.

At the local level, each individual lake has its own lake basin management plan. Eight U.S. state governments and the province of Ontario, Canada, are actively involved in lake management, in addition to lake area and local community stakeholders, 40 tribal nations, six urban metropolitan areas, and numerous county and local governments.

To further protect the Great Lakes, coordinate efforts, and reduce duplication, a Great Lakes Interagency Task Force was established in 2004. In the United States the EPA was the lead agency for ecosystem management, charged with providing strategic direction on policies and programs and collaborating with other efforts such as the Council of Great Lakes Governors and the Great Lakes Cities Initiative.

The responsibilities of the Interagency Task Force include the protection of commercial harbors and waterfronts from excessive wave action, continuous repair and maintenance of infrastructure, and regular dredging of salt and sediments in shipping channels. The U.S. Army Corps of Engineers is responsible for dredging Great Lakes harbors and channels on the U.S. side to keep them navigable. It also handles rehabilitation of locks and the collection of user fees within the maritime transportation system to cover costs. To further upgrade the waterway, the U.S. Water Resources Development Act of 2007 authorized the secretary of transportation to repair the Eisenhower and Snell Locks on the St. Lawrence River in upstate New York.

The key to managing this internationally shared seaway has been the strong consensus between Canada and the United States on a comprehensive basin management strategy. Peaceful and friendly relations between the two nations and sound institutional capacity on both sides are pivotal factors to coordinate multiple actors and stakeholders and implement oversight mechanisms. In 2010, both parties launched a Great Lakes Restoration Initiative. Its objectives were to restore the system's environmental soundness by addressing water pollution and invasive species problems, and carry out environmentally safe dredging and the safe recycling and disposal of dredge material and sludge.

Lake Chad: Management and Collaboration Issues

If we compare the management of the Great Lakes system with another globally shared lake basin, Lake Chad in Africa, we notice a sharp contrast in the level of economic development and political stability of the two countries. Lake Chad's littoral countries are faced with pressing challenges typical of most low-income developing countries. These include high levels of impoverishment of lake-area communities; lack of financial resources and poor technical capacity to implement projects; inadequate coordination efforts; and weak institutional, regulatory, and political structures.

Specific social and ecological problems hamper efficient and effective management of Lake Chad and application of the IRBM principles demonstrated in the Canada/U.S. example. The Lake Chad Basin countries are Cameroon, Chad, Niger, and Nigeria, while nonriparian countries are Algeria, Central African Republic, Libya,

A fisherman paddles his canoe on Lake Chad in Koudouboul, Chad, in November 2006. The lake once provided adequate livelihoods for 20 million people in west-central Africa, but lost 90 percent of its original surface area during the last 30 years. (AP/Wide World Photos)

and Sudan. The basin is a highly vulnerable ecosystem, since most countries are located in arid or semiarid areas experiencing lengthy bouts of drought. In addition to poverty and inadequate social development, countries are also faced with political turmoil and the lack of technical capacity to manage and protect shared waters for the benefit of all.

Due to limited funds and the difficulties of getting to consensus, Lake Chad countries do not adequately monitor the lake's pollution levels and hydrologic conditions. There is little reliable scientific information available, minimal sharing of technical knowledge, and low compliance with international environmental and water quality standards. In contrast to the Great Lakes experience, these limitations make it difficult for lake-bordering countries to collectively respond to growing environmental problems of the basin. These include chronically low water levels, impacts of climate variability, ongoing drought, and pollution. Political wrangling, lack of political commitment, and low participation of community stakeholders in decision-making hinder sustained and equitable allocation of water resources.

The Lake Chad Basin Commission was formed in 1964 under an international treaty signed by basin states. It has encountered serious problems in the execution of its mandate. Some challenges are related to the difficulty of monitoring and controlling the excessive removal of water upstream for irrigation, which is impacting downstream communities. Weak fish stock protection regulations also remain a major problem.

Although World Bank reforestation and wetland restoration projects have helped address some issues, the commission's work is plagued by political difficulties, weak coordination, and insufficient funds. As observed in the Great Lakes, the ability to collect user charges or raise funds to pay for lake protection activities is a crucial factor to guarantee sustainability of actions.

Cooperation in the Baltic Sea: A Work in Progress

In another geographical area, the joint management of an international sea such as the Baltic provides us with yet another set of challenges and opportunities for international cooperation. The experiences of the Baltic differs from the Great Lakes and Lake Chad experiences discussed above, but is instructive to illustrate global options of managing shared water. The end of the Cold War provided countries around the Baltic Sea with huge opportunities to manage its use in harmony. Progress has been achieved, but there are ongoing challenges to coordination and achievement of common ground activities.

The Baltic Region and Historical Ties

The global importance of the Baltic region and the need for international cooperation has to be viewed in a historical and geopolitical context. The Baltic Sea is a diverse political and religious region with cultural differences and common interests. The history of settlement in the area dates back to the Bronze and Iron Ages and is related to the development of trade links and cultural ties among Slavic tribes, German Saxons, Scandinavians, Vikings, Lithuanians, Russians, and Prussians.

Kattegat Strait forms part of the connection between the Baltic Sea and the North Sea, trending north-south between Denmark and southern Sweden. (Shutterstock)

In 1974 (before the end of the Cold War), seven Baltic states—Denmark, Sweden, Finland, the then-Soviet Union, Poland, the German Democratic Republic (then East Germany), and the Federal Republic of Germany (then West Germany)—signed the Helsinki Convention on the Protection of the Marine Environment (HELCOM) in Helsinki, Finland, to address the common problem of Baltic Sea pollution.

The HELCOM Convention

The main objective of HELCOM is to protect the Baltic's marine environment through regional cooperation. It was the first treaty to address land-based sources of pollution of the sea as well as marine pollution from oil drilling activities within the basin. The culture of trade among the countries and stakeholders around the Baltic helped foster drafting of this treaty and reduce open conflicts in tackling the emerging problems of the Baltic. After the breakup of the former Soviet Union, separate Baltic states emerged, but again their historical ties provided some impetus to continue with collective action.

Approximately 72 million people live in the wider Baltic basin, of which half are from Poland. Sixteen million inhabitants live directly on the sea's coastline. Intensive industrial development in littoral states is the main source of pollutants in the Baltic Sea. Several thousand tons of phosphorus and millions of tons of nitrogen are dumped annually into the Baltic Sea. This complex ecosystem presents specific challenges to the solving of contamination problems. It is one of the largest brackish water areas in the world due to the intrusion of concentrated salt water from the North Sea. Salt water, heavy metals from municipal and industrial discharges, and other pollutants sink to the bottom of the sea. They deplete oxygen in the lowest layers and produce toxic bacteria.

Although the impact of pollution on Baltic Sea fisheries was known since the 1960s, it was not until 1973 that the countries signed the UN Convention on Fishing and Conservation of the Living Resources of the High Seas. This agreement opened the door for an International Baltic Sea Fishery Commission and other multicountry efforts to recruit independent marine biologists and oceanographers to further study contamination. These measures offer a successful model for joint research activities and peaceful, scientific cooperation among states with very different political ideologies.

Collaboration has continued since the 1970s with a monitoring program to measure metal contaminants and pollution levels in seabird eggs, herring, and marine ecosystems at different research stations throughout the Baltic sea. This program continues to gather reliable data and analyses of water exchanges between the Baltic and the North Sea, promote reduction of organic pollutants, and emphasize the recovery of fish stocks and expansion of international marine science cooperation efforts.

In 1992, the Council of the Baltic Sea States (CBSS) was founded and headquartered in Stockholm, Sweden. This entity represented an additional institutional step to cement peaceful cooperation around the sea, and to create an intergovernmental forum among an increased number of basin countries (Denmark, Germany, Norway, Finland, Sweden, Estonia, Latvia, Lithuania, Poland, Russia, and Iceland).

The Conference of Foreign Ministers, the highest-level body of the council, is the prime institution to coordinate activities. It meets every two years. Expert groups established under the CBSS guidelines have the responsibility to implement programs on economic cooperation, nuclear and radiation security, customs, border crossing, and maritime policy aspects.

The current collaboration arrangements among Baltic states with different political systems and a history of Cold War disagreements provide valuable lessons on regional approaches to joint problems in shared ecosystems. Building upon this experience and using HELCOM as a model, the United Nations Environment Programme (UNEP) established a Regional Seas Programme for the Mediterranean Sea. Sixteen Mediterranean littoral states formally adopted the program in 1976 under four guiding principles: a) scientific cooperation; b) research and monitoring of environmental quality and information sharing; c) institutional and financial arrangements; and d) project planning and development.

Arctic Sea: Options for Cooperation or Conflict

The situation of another international sea, the Arctic Sea, provides a different set of experiences and options for international cooperation. A major backdrop to success or failure of collaboration in the Arctic arena is accelerated ice melt and increasing competition among nations for natural resources on the seabed. The Arctic—much the same way as other international seas or lake basins—has international treaties and regulations to govern the management of common areas.

The Arctic Council is the main coordination institution. Indigenous peoples such as the Inuit (Alaska/Russia) and the Saami (Greenland) participate in council meetings and have opportunities to weigh in on important matters. In addition to the council, 19 countries make up the International Arctic Science Committee, a nongovernmental research institution.

Competition for Resources

Climate change and the growing retreat of the ice have made the need for collaboration and stricter oversight of Arctic Sea regulations more urgent. Ice melt has drastically changed the mental image of the Arctic Sea as an area normally buried under hundreds of feet of snow and ice. This melting is now a magnet for national and investor competition for the exploitation of oil, gas, and mineral resources—and a stimulus to potentially unleash conflicts in the sea.

The 1982 UN Convention on the Law of the Sea (UNCLOS) is the legal framework that governs activities on, beneath, and through the Arctic Sea area. According to UNCLOS regulations, territorial sea boundaries of each Arctic littoral state extend up to 12 miles offshore, exclusive economic zones up to 200 miles offshore, and continental shelf rights up to 350 miles. UNCLOS has also established the International Seabed Authority and associated mechanisms to help resolve seabed conflicts.

Nations with Arctic Sea coastlines, i.e., the United States (Alaska), Canada, Russia, Norway, and Denmark (Greenland), and neighboring countries such as Iceland, Sweden, and Finland have begun to reassess their positions and strategic interests in seabed minerals. It is ironic that the global rise in temperatures—originally unleashed by fossil fuel burning—might eventually free up the Arctic for fossil fuel exploration and potentially release more greenhouse gases through permafrost melt and the impacts of oil and industrial development.

Overfishing

In addition to competition among Arctic states for mineral rights, there are increasing threats of overfishing by Arctic and non-Arctic states. Ocean warming and the migration of new fish populations to a warmer Arctic Sea pose additional risks to marine resources, prompting calls for new fishing regulations. However, final agreements to regulate industrial-scale fishing in Arctic waters are still pending additional research on fishing stocks.

In 2013, eight Arctic Council members signed important agreements to improve oil pollution preparedness and oil spill coordination measures. However, Arctic-specific standards for safety equipment, ship design, construction, shipping, and crew training programs still have to be developed.

Northwest Passage Potentially Open?

Some discussions are already occurring around the possible opening of the so-called Northwest Passage. Within the Arctic Sea, the Northwest Passage is a strait: a three-thousand-mile shipping lane across the top of Eurasia connecting the Atlantic to the Pacific Ocean. It first became ice free for a short period in the summer of 2007, which sparked interest in its viability as a waterway (*Economist* 2012). Both the United States and the European Union have claimed free navigation rights to the passage, while Canada maintains that it is an inland waterway under its territorial jurisdiction.

Use of the passage in summer under conditions of minimal ice cover could potentially shave several days off a Pacific–Atlantic trip through the Panama Canal. The prospect of shorter trade routes through the Arctic has therefore become very appealing to nations involved in international East–West trade such as China, Singapore, Japan, India, and Korea. Of concern is the possibility that increased oil transportation through the area could augment the risk of spills in Arctic waters (in 1989 the *Exxon Valdez* tanker spilled over 250,000 barrels of oil in the Arctic).

Agreements and Disputes

While an agreement between Russia and Norway has peacefully settled a maritime dispute in the Barents Sea between the two parties, other disputes among Arctic states are simmering. Russia continues to claim that parts of the Northern Sea Route (NSR) above Siberia belong to its internal waters, but the United States does not accept this position. Denmark and Canada both claim Hans Island in the Arctic Sea as their territory, and all Arctic nations have staked competing claims to the sea's seabed and mineral resources.

Denmark is claiming mineral rights to the continental shelf through its autonomous territory, Greenland. The latter has changed its laws to grant exploration licenses for radioactive elements such as uranium and thorium, and unveiled plans to extend gold mining and construct an aluminum smelter in the sea area. At the same time, Russia is interested in exploring Arctic oil and gas fields, but many U.S. companies already have drilling leases in U.S. waters off the coast of Alaska.

In 2007 a Russian-led delegation planted a Russian flag below the North Pole in a mineral-rich ridge also claimed by Canada and Denmark. Claims and counter claims by Arctic Sea states as well as the rising interest of other nations in the world appear to be setting the stage for increased rivalry and potential conflicts in this once relatively quiet and ice-blocked area.

The Danube River Basin: A Model of European Collaboration

In addition to international lakes, lake basins, and multicountry seas, there are many river basins with internationally shared water resources that are managed through cooperation agreements. The Danube River in Europe is an example of cross-border collaboration. The Danube is a historically important European river that flows through old and new states of post–Cold War Europe. The largest international river basin in Europe, it inspired a famous waltz called *The Blue Danube*, written by Johann Strauss in the 1800s.

A River Runs through It

The Danube Basin includes at least 83 million inhabitants and 60 large cities. On its approximate 1,800-mile course through Germany, Austria, the Slovak Republic,

The Danube River passes through Bratislava, the capital of Slovakia. Here the river is known as the Dunaj. (European Commission)

Hungary, Croatia, Serbia, Romania, Bulgaria, Slovenia, and the Ukraine, the River Danube drains a catchment area of nine additional countries before emptying into the Black Sea. For centuries, people along the Danube's banks have used its waters to support fishing, navigation, drinking water, trade, agriculture, and industrial activities—and also to dispose of wastewater and other pollutants.

The first Danube Convention was signed by Danube littoral states in 1948 (after the end of World War II) to regulate river navigation, but it only came into force in 1964. A joint institution called the Danube Commission is responsible for overseeing the main areas of institutional action: consultations among member states on economic, technical, and legal issues; joint hydrologic forecasts; navigation regulations and scientific studies; and implementation of regulatory activities.

Challenges on the Danube

In spite of the relatively smooth coordination of joint activities, there are many challenges facing the Danube River Basin and Danube states. The construction of 69 dams on the river over time has provided valuable hydroelectric power, but changed the flow and water quality of the river and increased the volume of suspended sediments. Rising levels of organic pollutants, nutrients, and pesticide loads have reduced dissolved oxygen in the river, added to groundwater contamination, and decreased the biodiversity of the Black Sea, into which the Danube flows. Environmental stress of the entire river system requires joint action to control water pollution and its impact on riverine communities.

In 1994, a new legal framework, the Danube River Protection Convention (DRPC), was signed by 11 of the 13 riparian states. It came into force in 1998, placing greater emphasis on information management, pollution monitoring and reduction, and the implementation of emergency warning and accident prevention systems. It also included a long-term integrated sustainable development strategy for the entire Danube River Basin and the Black Sea.

Experience from the Danube indicates that conventions cannot resolve all conflicts among states sharing water resources. Countries in the Danube River Basin (and in other basins too) have recourse on present cases to the International Court of Justice (ICJ) for dispute mediation, if problems cannot be resolved within the Danube Protection Convention. In 1977, Hungary and Slovakia feuded over the construction and operation of locks on the Danube and resorted to the ICJ to mediate their conflict.

As can be gleaned from the Danube experience, sound institutional frameworks and peaceful conflict resolution frameworks that establish clear rules of the game are vital elements of effective water resource management. Cooperation—rather than conflict—leads to better sharing of ecosystems and the benefits of freshwater use by all river basin communities. Effective collective action means willingness to split the costs of pollution reduction and environmental protection among all members.

Other benefits of multicountry collaboration include more efficient food and energy production, improved trade and transport networks, and the preservation of clean freshwater for future generations. Increased cooperation cements political integration and social and cultural ties of all stakeholders throughout the area. A major plus of harmonious management of cross-boundary waters is that it fosters positive relations among states, and provides a peace dividend for all countries and citizens involved.

Camille Gaskin-Reyes

REFERENCES AND FURTHER READING

Amazon Cooperation Treaty Organization. http://www.otca.info.

American Great Lakes Ports Association. http://www.greatlakesports.org/.

Ansohn, Albrecht, and Boris Pleskovic, eds. *Climate Governance and Development, Berlin Workshop Series*. Washington, DC: World Bank Publications, 2012.

Chellaney, Brahma. *Water, Peace, and War: Confronting the Global Water Crisis*. Lanham, MD: Rowman & Littlefield, 2015.

Council of the Baltic Sea States. www.cbss.org.

Council on Foreign Relations. "The Emerging Arctic: A CFR InfoGuide Presentation." http://www.cfr.org/polar-regions/emerging-arctic/p32620#!/?cid=otr_marketing_use-arctic_Infoguide#!.

The Economist. "Too Much to Fight Over" June 12, 2012. http://www.economist.com/node/21556797.

Environmental Protection Agency (EPA). "The Great Lakes." http://www.epa.gov/greatlakes/basicinfo.html.

Giordano, Meredith A., and Aaron T. Wolf. "Sharing Waters: Post-Rio International Water Management," *Natural Resources Forum* 27 (2003): 163–171.

International River Basin Register: South America. Transboundary Freshwater Dispute Database (TFDD), Department of Geosciences, Oregon State University. 2012. http://www.transboundarywaters.orst.edu/publications/register/tables/IRB_southamerica.html.

Jansky, Libor, Masahiro Murakami, and Nevelina I. Pachova. *The Danube*. Tokyo: United Nations University Press, 2004.

Koch, Wendy. "Denmark Eyes North Pole, but How Much Oil and Gas Await?" *National Geographic,* December 17, 2014. http://news.nationalgeographic.com/news/energy/2014/12/141217/oil-natural-gas-denmark-north-pole-arctic/.

Payoyo, Peter Bautista, ed. *Ocean Governance: Sustainable Development of the Seas*. Tokyo: United Nations University Press, 1994.

Rosenthal, Elisabeth. "Race is On as Ice Melt Reveals Arctic Treasures." *New York Times,* September 18, 2012. http://www.nytimes.com/2012/09/19/science/earth/arctic-resources-exposed-by-warming-set-off-competition.html.

Salkida, Ahmad. "Africa's Vanishing Lake Chad." Africa Renewal Online. April 15, 2012. http://www.un.org/africarenewal/taxonomy/term/491.

United States Arctic Research Commission (USARC). https://www.arctic.gov.

World Bank. "Lessons for Managing Lake Basins for Sustainable Use, Report No. 32877." Environment Department. Washington, DC: World Bank Publications, 2005.

CASE STUDIES

The case studies in this section examine water sharing cooperation or conflict situations in different geographical contexts ranging from Lake Victoria in Africa to the Lempa River Basin in Central America and the Jordan Valley in the Middle East.

Case Study 1: The Impact of Pollution in a Multicountry Lake: A Case Study of Lake Victoria in the Ugandan Context

Unplanned Urban Development and Impacts on the Lake

Rapid population growth, uncontrolled urban growth, deforestation, destruction of wetlands, and lake contamination are among Uganda's main environmental threats. Environmental degradation impacts negatively on human well-being through decreasing agricultural yields, water contamination, and degradation of the population's health status.

This situation is particularly evident in the ballooning peri-urban (peripheral urban communities) environments around Kampala, Uganda's capital. The most water-stressed people are those living in the city's rapidly growing informal settlements and slums. Erratic urban growth is coupled with inadequate sanitation infrastructure, roads, and electricity; the provision of clean water is an almost impossible task.

Compounding the problem is the fact that many unplanned developments occur in low-lying areas that include wetlands. The springs in these wetland areas were once the primary source of water to the city, but development and pollution have eliminated their capacity to function as natural water filtration systems. Informal settlements (along with other developments) have polluted every wetland and freshwater spring surrounding Kampala. These areas are also prone to severe flooding

Kampala now gets the majority of its water from Lake Victoria. As a result, this once-thriving lake is now a receding, highly contested, and polluted resource that requires concerted international cooperation and management efforts. In addition to supplying Kampala with valuable drinking water, the lake supports Africa's largest inland fishery and is also important for two other countries, Kenya and Tanzania.

The Plight of Lake Victoria

Lake Victoria, formed by a shallow depression between two rift valleys, is the largest of Africa's Great Lakes, the second largest freshwater lake in the world, and the source region for the White Nile and the Katonga River. The lake sits in a high plateau dotted by rivers and lakes, and is an international body, bordered by Uganda, Kenya, and Tanzania.

The region experiences a biannual rainy season based on its equatorial latitude and a moderate temperature range due to the plateau's altitude. Much of the agriculture in the region is rain fed. The catchment basin for Lake Victoria stretches into the aforementioned bordering countries as well as Rwanda and Burundi. From Lake Victoria, the White Nile flows north to join the Blue Nile and continue through Egypt towards the Nile delta.

Uganda makes up 45 percent of Lake Victoria's shoreline. The lake is under pressure from human activities. These include overfishing, siltation from erosion caused by deforestation of the watershed, spread of invasive species, industrial pollution, oxygen depletion, and the fallout of climate change. The lake is receding rapidly by about 3 to 4 percent each year as a consequence of recent drought, but excessive water withdrawals from the lake are also a major contributing factor.

Water Vulnerability in the Lake Region

The lake's surrounding communities rely heavily on its waters for drinking water and employment in fisheries. Water for Kampala is drawn from the lake through a large intake pipe. The government has moved the pipe farther and deeper into the lake because of increasing levels of pollution and the drop in water level. Even water treated with chemical additives sometimes contains heavy metals. Although officials have blamed dropping water levels on drought, they have also admitted that there is a problem of pollution (Kagenda 2015). There is also an increasing incidence of water shortages in Kampala.

Two of the greatest threats to Kampala's development are encroachment of informal settlements on wetlands and ongoing pollution of Lake Victoria. Disposal of untreated solid waste, wastewater, and industrial effluents places a high negative burden on human welfare and environmental safety, both in terms of unclean drinking water and heavy metal loads in the food chain.

Source Pollution

One of the leading causes of pollution is heavy metals, particularly from nonpoint sources such as motor vehicles. The use of lead-based gasoline in cars significantly increases chemicals in the water; and the practice of washing cars directly in the lake area also adds to the amount of heavy metals entering the system. Uganda's environmental regulations prohibit this practice, but they are rarely enforced.

Another concern is that heavy metals bio-accumulate, making urban agriculture in polluted soil a significant issue. Urban agriculture is common in Kampala's sprawling settlements and provides some nutrition and revenue for the poor. One study found that vegetables grown in urban and peri-urban areas of Kampala are significantly more contaminated with heavy metals (cadmium and lead) than those grown in rural areas (Owor 2007). This issue poses a serious risk to human health.

The same study discovered that mercury found in the belly fat of Nile perch (common to Lake Victoria and rivers in the area) most likely originates from battery-recycling plants, which deposit waste directly into rivers. Mercury concentrations in fish, an important protein source for local communities, are dangerous.

Water and Sewage Problems

The population of Kampala's urban area dramatically increased from 330,000 in 1970 to 1.2 million in 2012. According to the 2002 census, 60 percent of the total city population lives in slums. This number continues to increase. Much of this growth has occurred in low-lying areas of the city, including the valuable wetlands (protected only in theory). Today, most slum settlements experience drainage problems, which lead to heavy flooding in the rainy season. The most common disease affecting slum dwellers is malaria, but waterborne diseases such as cholera and giardia are also endemic.

Unplanned and rapid urban development does not allow the government time to plan and install the necessary infrastructure. The lack of adequate basic sanitation services contributes to water pollution, as human waste is often dumped in drainage ditches or directly into the lake. Only 7 percent of Kampala's population is served by a sewer system. There is no centralized sewer system that covers the entire city, and most planned sanitation measures are in the form of poor quality septic tanks that frequently leak into groundwater (UN-Habitat 2006).

Low sewer coverage is predicted to continue for decades, as a citywide sewer system is neither affordable nor feasible. Rapid population growth is also resulting in increased garbage generation, which the city is unable to address. Uncollected solid waste clogs storm water drains, exacerbates flooding, and pollutes surface water, which eventually enters the lake. Many wetlands around Kampala had traditionally offered a natural filtration system for effluents before they reach the lake, but this function has been drastically reduced due to wetland draining for farmland and settlements.

Industrial Pollution

Inadequate environmental protection and enforcement have resulted in high levels of pollution of the lake's catchment area, resulting from the Mpererwe landfill and the Kilembe copper mining district. Leachate (liquid runoff) from the landfill is also affecting environmental quality downstream. The Kilembe copper mine in the Lake George district, although abandoned, continues to pollute the surrounding area and Lake Victoria. Heavy metals from this mine drain into Lake George and into the Kagera River, and finally into Lake Victoria.

Spread of Invasive Species

The spread of invasive species, such as water hyacinth in Lake Victoria, poses a dangerous risk to local well-being and people's livelihoods. Water hyacinth is both a pollutant and a catalyst for other pollutants. Over the last 150 years, it has migrated into Lake Victoria and begun to obstruct the shoreline. Water hyacinth has a very rapid growth rate, which makes effective elimination of the weed difficult.

The invasion of the water hyacinth has pushed floating plants such as papyrus and elephant grass out of the area, resulting in reduced biodiversity and the extinction of most of these other plants. Lagoons, previously one of the area's main water filtering systems, are now replaced with floating mats of water hyacinth, which increase nutrient pollution that creates more favorable conditions for algae blooms (Tiwari et al. 2007).

The spread of water hyacinth is related to increased lake pollution. Areas upstream are polluted with fertilizers or human refuse, which increase nutrient loads in the lake and foster an environment for water hyacinth to thrive. Although water hyacinth could potentially be addressed by mechanical or biological means on the lake surface, a better and more permanent solution would be to reduce the pollutants that enter the surrounding tributaries.

Malaria and Schistosomiasis Diseases

There is a proven correlation between water hyacinth and the increase in malaria and schistosomiasis. Hyacinth provides a shelter for mosquito larvae when it is stationary or caught in trees next to the shoreline. Trees on the shoreline were originally planted to discourage hippopotami from residing on the shores near areas of human activity. The trees displaced native hippo grass, which had deterred mosquito larvae formation, since waves from the lake would periodically sweep over the grasses and dislodge the larvae.

Free-floating water hyacinth plants provide an ideal habitat for snails that carry the disease schistosomiasis. These snails latch onto a floating mat of water hyacinth after a storm and essentially "ride" the plant until it becomes trapped in a lagoon or in trees on the shoreline. There, the snail comes into contact with people, often while they are drawing water from the lake or fishing (Ofulla et al. 2010). Areas traditionally free of schistosomiasis have seen dramatic increases through the spread of water hyacinth.

Addressing the Problem

Lake Victoria is an instructive case study of an international-level socioenvironmental set of problems, requiring strong, collective action. Rural-urban migration and rapid population growth in Kampala have created unplanned and unchecked settlements, and led to the pollution and eventual destruction of vital wetlands, once the primary source of water for the city. Flooding and the discharge of industrial effluents and sanitation wastes from the city have compounded Lake Victoria's pollution problems. The lake has become dangerously degraded, compromising the once-vibrant fishing industry and freshwater supplies for the area.

It is imperative for Uganda to find ways to curb lake pollution and enforce environmental regulations. Since Lake Victoria is part of a regionally shared resource, efforts may prove futile without joint agreements and actions, financial resources, and the consistent enforcement of regulations. To be effective, collective activities need to cover many sectors, ranging from urban planning to improved sewage and sanitation infrastructure, to reduction of source pollution and invasive species and protection of the lake's freshwater resources. Without progress in these areas, negative health effects, further water insecurity, and social and environmental vulnerability of the region will continue to occur.

Amy Krakowka Richmond, James Link, and Dylan Malcomb

REFERENCES AND FURTHER READING

Kagenda, Patrick. "Uganda: Kampala Water Scarcity." *Independent* (Kampala), February 1, 2015. http://allafrica.com/stories/201502020045.html.

Kiage, Lawrence M., and Joyce Obuoyo. "The Potential Link between El Nino and Water Hyacinth Blooms in Winam Gulf of Lake Victoria, East Africa: Evidence from Satellite Imagery." *Water Resources Management* 25, no. 14 (2011): 3931–3945. doi:10.1007/s11269-011-9895-x.

Ofulla, A. V. O., D. Karanja, R. Omondi, T. Okurut, A. Matano, T. Jembe, R. Abila, P. Boera, and J. Gichuki. "Relative Abundance of Mosquitoes and Snails Associated with Water Hyacinth and Hippo Grass in the Nyanza Gulf of Lake Victoria." *Lakes & Reservoirs: Research & Management* 15, no. 3 (2010): 255–271. doi:10.1111/j.1440-1770.2010.00434.x.

Ogwok, P., J. H. Muyonga, and M. L. Sserunjogi. "Pesticide Residues and Heavy Metals in Lake Victoria Nile Perch, Lates Niloticus, Belly Flap Oil." *Bulletin of Environmental Contamination and Toxicology* 82, no. 5 (2009): 529–533. doi:10.1007/s00128-009-9668-x.

Opande, George, John Onyango, and Samuel Wagai. "Lake Victoria: The Water Hyacinth (Eichhornia Crassipes [Mart.] Solms), Its Socio-Economic Effects, Control Measures and Resurgence in the Winam Gulf." *Limnologica—Ecology and Management of Inland Waters* 34, nos. 1–2 (2004): 105–109.

Owor, Michael, and Tina Hartwig. "Impact of Tailings from the Kilembe Copper Mining District on Lake George, Uganda." *Environmental Geology* 51, no. 6 (2007): 1065–1075. doi: 10.1007/s00254-006-0398-7.

Oyoo, Richard. 2015. "Deteriorating Water Quality in the Lake Victoria Inner Murchison Bay and Its Impact on the Drinking Water Supply for Kampala, Uganda." http://wldb.ilec.or.jp/data/ilec/WLC13_Papers/others/3.pdf.

Tiwari, Suchi, Savita Dixit, and Neelam Verma. "An Effective Means of Biofiltration of Heavy Metal Contaminated Water Bodies Using Aquatic Weed Eichhornia Crassipes." *Environmental Monitoring and Assessment* 129, nos. 1–3 (2007): 253–256. doi:10.1007/s10661-006-9358-7.

UN-Habitat. *Situation Analysis of Informal Settlements in Kampala by UN-HABITAT*. UN-Habitat: 2006. http://unhabitat.org/books/situation-analysis-of-informal-settlements-in-kampala/.

Case Study 2: The Lempa River Basin in Central America: Water as a Regional Public Good

The Trifiñio Region

The Trifiñio region is a biosphere reserve in Central America, which extends through parts of El Salvador, Guatemala, and Honduras. The reserve covers 7,070 square miles and encompasses the headwaters of the Lempa and the Motagua Rivers, two key watershed basins in Central America. The wider geographical area includes mountainous ecosystems such as the Montecristo and the El Pital cloud forests, which form part of the area's protected biosphere and biological reserves. The Lempa and Motagua Rivers traverse wetlands and forests of the San Diego–La Barra ecosystem, which is noted for its biological diversity and sanctuary function for birds migrating between North and South America.

The Upper Lempa River Basin composes 47 percent of the Trifiñio region. It is the largest river system of the Pacific Ocean watershed and the only river basin in Central America that is shared by three countries: El Salvador (56 percent), Guatemala (14 percent), and Honduras (30 percent). The river system is 208 miles long and runs completely through Central America. An estimated 1 million people live in the watershed area and share the River Lempa's abundant freshwater resources. The watershed population is expected to double by 2026 (Bocalandro and Villa 2009).

Importance of the Lempa River Basin

The Lempa Basin presents an interesting model for trinational cooperation around the concept of water as a regional public good. The water in the Lempa Basin is critical for the socioeconomic life of the watershed communities, food security, and the general health of the ecosystem. Forty-five municipalities in the three countries that share the basin rely on its water resources as a strategic backbone for social and economic survival.

The basin is the cornerstone of agricultural and nonagricultural activities for the riverine communities in the watershed, in particular farming, ranching, trade, and energy production. Irrigation is of major importance. Water from the Lempa irrigates approximately 34,600 acres of cropland. It also provides drinking water for close to a million inhabitants, and water for electricity generation from four hydroelectric plants.

The Trinational Commission

Within the framework of regional collaboration in the Lempa Basin, the national legislatures regulate water allocation in order to promote equitable distribution of water resources. This system is one of few successful models in Latin America for managing cross-border water resources and staving off conflicts. The Trifiñio biosphere presents a positive opportunity for collaboration, operationalized through a trinational commission, which has been set up to implement a strategy called the Trifiñio Blueprint.

The Blueprint represents a strategic effort among the three countries, El Salvador, Honduras, and Guatemala, called Water Without Borders, to manage shared waters and use them as a regional public good. The main benefits of collectively managing resources include sharing the burden of the management costs, stimulating regional innovation, and exchanging lessons learned. Such coordination and collaboration efforts also reduce the risks of natural hazards such as hurricanes, severe storms, earthquakes, and floods.

As part of its regional cooperation and Regional Public Goods (RPG) support program in Latin America and the Caribbean, the Inter-American Development Bank (IDB) has been instrumental in supporting the activities of the three countries in the Lempa Basin and strengthening the Tripartite Commission. A major component of this support was an in-depth assessment of the human capacity of the three countries to jointly manage shared natural resources, and the improvement of common governance and operation of the Lempa River Basin.

The IDB also contributed to the cocreation of technical and legal instruments to manage the trinational water agenda. These measures include an integrated information system for data collection throughout the basin, development of municipal ordnances, and the generation and monitoring of socioeconomic and environmental information. As part of the project's emphasis on conservation awareness, the activities included environmental education programs and environmental monitoring of shared water resources.

In 2007 the riparian municipalities of the three countries jointly established the Trinational Federation of the Upper Lempa Municipalities, which achieved autonomous status. One of the important features of collaboration involved the coordination of budgets among the participating municipalities to finance the implementation of the jointly negotiated Water Agenda.

Concept of an Integrated Water Management System

The comanagement of an integrated water system in Central America falls under the concept of viewing water as a regional public good that should be managed for the benefit of local communities and users and preserved by all stakeholders for future generations. This approach reduces the possibility of depletion of resources caused by short-term interests and generates local accountability and ownership of the process. It has fostered the training of local leaders in 45 municipalities of the countries and strong involvement of local communities, users, and consumers in all sectors of the economy.

The environmental information generated by the project has facilitated decision making and the spread of information on water resource management and resources among cross-border communities. These activities support cooperation efforts, which in turn have reduced potential conflicts among countries and municipalities sharing the basin. They also enable important exchanges on methods and technologies to protect forestry and water resources in the entire watershed area for the benefit of all residents.

Many challenges remain, including those related to agreements on regional zoning aspects, implementation of joint measures to adapt to climate change, and follow up on national and municipal commitments and enforcement of basin agreements. Other matters of joint importance that need to be tackled by all parties include mechanisms to recharge underground water and safeguard potable water for human consumption, watershed reforestation campaigns, and deforestation control regulations. All of these actions require continued efforts to address social, environmental, and natural disaster vulnerabilities in the area and implement efficient and innovative approaches towards sustainability.

Water as a Regional Public Good

So far, the joint collaboration of the three countries under the umbrella concept of "Water Without Borders" has produced initial outcomes. This concept regards water as a regional public good that should be shared by current and future communities in the region. At the institutional level, the collaboration of El Salvador, Honduras, and Guatemala on water management has resulted in the establishment of an institutional mechanism for peaceful dialogue on shared water, multinational consensus, and the strong involvement of local communities and stakeholders. Other concrete outputs have been the establishment of mechanisms to strengthen local and community municipal participation and bring together all stakeholders.

Based on the principles of productive dialogue, social participation, consensus building, and ongoing validation of results, the three involved countries have created a trinational information system for the Upper Lempa Basin to coordinate geospatial data and develop water quality standards. Regulatory measures such as municipal ordnances provide a sound basis for the administration, conservation, and management of water for human consumption and irrigation. The process of developing and adopting the Trinational Water Agenda is expected to guide and regulate future joint actions among the countries.

Creation of Legal and Technical Mechanisms

A further success of joint collaboration embedded in the project was the creation of legal and technical mechanisms and tools to manage water resources and define and implement regional agreements to manage water as a regional public good. In addition to improving coordination of activities, the Trinational Federation fostered institutional integration at the national, state, federation, and municipal levels among all municipalities in the Trifiñio region.

An independent evaluation of the program confirmed that it has energized individual states to comply with trinational agreements on water administration. A key conclusion of the evaluation was that increased participation and ownership of water management issues at the municipal or local level had strengthened tripartite cooperation and filled an institutional void in the national agendas of the three countries.

This new technical and legal basis at the municipal level has had other side benefits. It helped create attitudinal changes in the countries and awareness in the communities of the importance of conserving natural resources. It is evident that increased sensitization of communities and stakeholders is a product of the environmental education and water management awareness programs. The population is more aware of the negative impacts of unsustainable water management practices and of water contamination on its quality of life. Communities have become more motivated to preserve natural resources, particularly the water reserves located within their municipalities as well as the forests that protect the watersheds.

Environmental education programs were carried out in approximately 200 schools in the three countries and incorporated into school curricula at the national level of each participating state. The preparation of environmental education materials and the implementation of actions to promote conservation and sustainable water use included didactic guides on environmental awareness, kits to measure water quality, maps of water quality, and games and coloring books for children. These activities led to increased demand for teacher training and additional material on environmental education.

High Stakeholder Participation

The independent evaluation of the program also revealed that the level of participation in the tripartite program was extraordinary. On the public sector side, it involved 45 municipalities, public federations, and other governmental institutions charged with monitoring water resources. On the community side, it involved nongovernmental organizations, hundreds of grassroots groups and education centers within the river basin, thousands of students and parents, and regional, national, communications, and media institutions. Numerous private sector and workers' representatives were involved in environmental education exchanges and the trinational environmental management system, as well as discussions on protection of water resources and development of territorial ordnances and information systems.

Sustainability

Looking ahead at how the process of Central American regional collaboration can be sustained in the future, it appears that the creation of a favorable collaborative environment is a big recipe for future success. Cooperative attitudes among the three countries are conducive to attracting high levels of public and private investments, and building long-term incentives for water conservation and sustainable ecosystem management practices.

New projects in forestry, health, and environmental health are expected to be developed following the river basin sharing model. There is the potential to expand the concept of watershed basin management as a regional public good to the agricultural and industrial sectors. The validation of shared water basin efforts in the three countries at the national and community levels is also expected to provide a model for other Latin American countries to improve management of their water resources.

Laura Bocalandro

REFERENCES AND FURTHER READING

Bocalandro, Laura, and Rafael Villa. *Regional Public Goods: Promoting Innovative Solutions for Latin America and the Caribbean*. Washington, DC: Inter-American Development Bank Publication, 2009.

Inter-American Development Bank. "Trifiñio Trinational Commission." http://www.iadb.org/en/projects/Project-description-title.1303.html?id=RG-T1157.

Tri-national Trifinio Plan Commission. "Report of the Independent Final Evaluation of the Trifiñio Project." www.aguasinfronteras.net.

Case Study 3: Middle East Water: A Bone of Contention

History teaches us that competition and the relative scarcity of natural resources can lead to political and even armed conflicts. The Middle East is no exception. Throughout history, water has always been a contentious issue in the region; more recent conflicts are heightened by low rainfall levels, water scarcity and drought, population pressures, political strife and the Israeli-Palestinian conflict, national security interests, and geopolitical considerations. Turkey, Syria, Jordan, Israel, the Palestinian Territories, and Lebanon are increasingly tapping water from the Jordan, Euphrates, Tigris, Litani, Orontes, and Yarmouk Rivers, and drawing down on aquifers to meet increasing demand.

This case study zooms into the Israeli-Palestinian water conflict, which is inseparably linked to past and current political and land disputes, including armed conflicts between the two sides. The Israel/Palestine/Jordan area is one of the most heavily contested water sharing regions in the world. As in many water-deficient zones, domain over water is closely linked to economic and political power.

Palestinian, Jordanian, and Israeli actors all lay claim to the area's water supply, whether from the Jordan River or aquifers below the ground. The availability of freshwater in Israel, the West Bank, and Jordan heavily depends on the climate and rainfall regimes, which vary throughout the region, and on water distribution and allocation mechanisms.

The northern and coastal regions of Israel as well as the westernmost edge of Jordan have a Mediterranean climate with long, hot summers and cool, rainy winters.

The West Bank, the southern and eastern regions of Israel, and the central and eastern regions of Jordan have a mostly arid climate with very little rainfall. Due to inadequate rainfall, aquifers and the control of aquifers play a major role in water supply.

Jordan River Basin

Water scarcity is widespread in the Jordan River Basin, of which 40 percent is located in Jordan, 37 percent in Israel, 10 percent in Syria, 9 percent in the West Bank, and 4 percent in Lebanon. The region has one of the lowest per capita water resources in the world. Water has become even more scarce due to pollution of the Jordan River Basin and the surrounding tributaries. While the Upper Jordan River area enjoys adequate water quality, the Lower Jordan River consists mainly of untreated sewage, agricultural return flows, groundwater seepage, and brackish water diverted from Lake Tiberias in the north.

Agreements and Conflicts among Israel, Palestine, and Jordan

The Jordan River Basin is an area of great contention when it comes to water sharing. The river covers a distance of 138 miles from north to south, and flows into the Dead Sea. Five riparian countries border and share the river's water resources: Israel, Jordan, Lebanon, Palestine, and Syria. Current water use in the basin is disproportionate. Israel extracts the most water from the basin, followed by Syria, Jordan, and Lebanon. Although there are bilateral agreements regarding shared water use in the area, the variance in actual withdrawal amounts is related to the region's multiple political and geographic tensions. The challenge is how to transform current water conflict situations to water sharing experiences through the use of more diplomacy and less confrontation.

After the 1967 Six-Day War in the region, Israel gained control of the headwaters of the Jordan River (originating in the Golan Heights), and also of the main underground aquifers of the areas. This is a major source for the Israeli Lake Tiberius (Sea of Galilee), which is Israel's largest freshwater reservoir and the starting point of its National Water Carrier pipeline, indispensable for water distribution throughout Israel.

The 1994 peace treaty between Israel and Jordan linked water with peace; it was the first agreement that incorporated water into security issues within an overall peace framework. It was agreed that Israel would receive annual water units from the Yarmouk River (a tributary of the Jordan River), provided that it transferred water to Jordan in summer.

Other agreements regarding water use in the river basin are between Jordan and Syria and between Israel and the Palestinian Territories. The 1987 agreement between Syria and Jordan defined the Syrian share of the Yarmouk River, limiting Syria to 25 dams in the basin area. Jordan was also slated to receive some desalinated water from saline springs in the Lower Jordan River. Article 40 of the Israeli-Palestinian Oslo II Accords addressed the issue of sharing water in Israel and Palestinian areas.

The Oslo agreement included provisions for both parties to establish the Joint Water Committee, charged with regulating water resources in the West Bank de jure.

This has shifted most of the responsibility over the Palestinian water sector to the Palestinian Authority, while the Israeli Authorities maintain de facto dominion over most of the water resources in the West Bank through West Bank aquifers. Following Oslo, the West Bank was split into three administrative sections. Area C, which makes up approximately 60 percent of the West Bank, is a closed military zone. It includes access points of the West Bank to the Jordan River, further increasing tensions.

The allocation and sharing of aquifers between Israel and the Palestinian Authority remains problematic. The primary groundwater source in the West Bank and Israel is the Mountain Aquifer. This aquifer is broken down into three different subaquifer basins: Eastern, Western, and Northeastern. The Western Aquifer basin is the most productive. The Oslo Accords originally determined water discharges to be drawn down by Israelis and Palestinians, but the breakdown of peace talks, the intractable conflict, and insufficient follow-up on commitments have led to severe overdrawing of aquifers—beyond the guidelines of the Oslo Peace Accords, and well above recharge rates.

Overdrawing of Water

Settlements in the West Bank and irrigation practices use large amounts of underground water. The second main aquifer used by Palestinians and Israelis is the Coastal Aquifer. This aquifer lies on the westernmost side of Israel and all of the Gaza Strip. Gaza's extraction rate provides subpar quantities of water per capita, and most of this water is contaminated with saline water and sewage and unfit for human consumption. Poor water quality in Gaza is associated with high poverty levels, poor health, and high incidences of waterborne diseases. Because tap water is too polluted for personal consumption, most Palestinians in Gaza purchase water from external vendors, including Israeli companies. The few desalinization plants in Gaza have been rendered inoperative in armed conflicts.

Water and Political Conflicts Interwoven

The Israeli-Palestinian conflict is interwoven with complex water-sharing regimes and variations in water use. This prolonged conflict—heightened by water scarcity and regional political instability in the entire Middle East—complicates current water problems and future prospects of appropriate and equitable water allocation and sustainable water management in the region.

The growth of Israeli settlements and population pressures in the West Bank, coupled with difficulties for Palestinians to obtain permits to drill new wells or rehabilitate existing ones, are contributing to further water tensions. A closed military zone known as Area C prevents access to the Jordan River, deepening water scarcity in the West Bank. All these factors hamper the resolution of chronic water shortages.

According to the World Bank, Palestinian domestic water availability averages 13 gallons per capita per day with some networks providing as little as one-fifth of that amount. According to B'Tselem, the Israeli Information Center for Human Rights in the Occupied Territories, total consumption per capita per day in the West Bank is 19 gallons for those connected to a water grid, and 5 to 13 gallons for those not connected to a water grid. In Israel, water consumption is approximately 148 gallons per capita per day. The World Health Organization (WHO) recommends a minimal amount of 26 gallons per capita per day.

Work of Nongovernmental Organizations (NGOs)

There are countless NGOs and international aid groups that are working to alleviate some of the water scarcity and water distribution problems and monitor and respond to ongoing water problems. An example of such an NGO is the Palestinian Hydrology Group, which has been studying the effect of the conflict on water availability over the past two years.

In addition, Friends of the Earth Middle East (FOEME) works in environmental conflict resolution and has put forth numerous publications and policy recommendations on the water situation. FOEME also works on projects related to sustainable water use and infrastructure. Many NGOs participate in a regional Emergency Water and Sanitation-Health Committee (EWASH) with the intention of improving the efficiency of the delivery of water-related aid by coordinating the activities of local village groups.

The future of sustainable and more equitable water use in Israel, Palestine, and Jordan, and the ability of this region to manage water stress, drought, and climate change impacts will depend heavily on whether a solution to the Israeli-Palestinian conflict is reached. Concrete steps toward a solution to water problems require a detailed review of the situation and agreement on measures to improve water sharing and conflict management practices. Although sustained solutions to the political and water-related issues facing the region may not yet be in sight, increased municipal and regional coordination efforts, such as those of the Joint Water Committee, can play an important role in the interim.

The future of sustainable water use in the area will also depend on the sharing of innovative technologies and more sustainable water harvesting methods in areas experiencing the most conflicts and water shortages. Israel has invested heavily in recycling techniques and desalinization plants, capable of producing almost 70 percent of domestic water consumption. However, such plants are expensive and technology-intensive, and Palestinians have little access to such technology.

The future of water sharing looks bleak in a region facing drought and desertification. At this juncture, the prospects for achieving workable Israeli-Palestinian water-sharing arrangements and joint water cooperation frameworks at the local and national level are subject to the prospects of political compromises, peace agreements, and the resolution of wider geopolitical issues.

Anour Esa

REFERENCES AND FURTHER READING

AQUASTAT. "Jordan Basin." Food and Agriculture Organization of the United States. http://www.fao.org/nr/water/aquastat/basins/jordan/index.stm.

B'Tselem. "Water Crisis: Discriminatory Water Supply." March 10, 2014. http://www.btselem.org/water/discrimination_in_water_supply.

El-Fadel, M., R. Quba'a, N. El-Hougeiri, Z. Hashisho, and D. Jamali. "The Israeli Palestinian Mountain Aquifer: A Case Study in Ground Water Conflict Resolution." *Journal of Natural Resources and Life Sciences Education* 30 (2001): 50–61. https://www.agronomy.org/files/jnrlse/issues/2001/e00-23.pdf.

Hassan, Marwan A., Graham Mcintyre, Brian Klinkenberg, Abed Al-Rahman Tamimi, Richard K. Paisely, Mousa Diabat, and Khaled Shahin. "Palestinian Water I: Resources, Allocation and Perception." *Geography Compass* (2010): 123.

Schein, Jonah. "The Role of NGOs in Addressing Water Access in Israel and the Palestinian Authority." *Sustainable Development Law & Policy* 5, no. 1 (Winter 2005). http://digitalcommons.wcl.american.edu/cgi/viewcontent.cgi?article=1405&context=sdlp.

United Nations Economic and Social Commission for Western Asia (ESCWA). "Country Paper on Water Resource Statistical Records in Palestine."

Water Resources Action Project (WRAP). "A Comparative Study of Water Data across Israel, West Bank, and Jordan." December 2013. http://www.wrapdc.org/wp-content/uploads/2013/12/WRAP-Rainfall-Trends-Study-FINAL.pdf.

Weibel, Catherine, and Sajy Elmughanni. "A Fresh Solution to Gaza's Water Crisis." UNICEF Middle East and North Africa. January 14, 2014. http://www.unicef.org/mena/media_8765.html.

World Bank. "Average Precipitation in Depth." http://data.worldbank.org/indicator/AG.LND.PRCP.MM.

ANNOTATED DOCUMENT

2008 DRAFT ARTICLES OF THE UNITED NATIONS ON THE LAW OF TRANSBOUNDARY AQUIFERS

Background

In 1947, after the end of World War II, the International Law Commission (ILC) was established by the General Assembly of the United Nations to initiate legal studies and make recommendations to foster the development of international law and its codification.

On August 5, 2008, the International Law Commission (ILC) adopted nineteen draft articles as the first step towards an international framework convention on transboundary aquifers. These articles applied to single transboundary aquifers and to transboundary aquifer systems comprising a series of two or more hydraulically connected aquifers.

In 2008, a two-step approach was recommended to the General Assembly of the United Nations: a) that they take note of the draft articles, adopt them, and recommend that members make arrangements to manage their transboundary aquifers on the principles of the draft articles; and b) that a convention be elaborated on and approved at a later stage on the basis of the draft articles. This latter step has not yet occurred.

Summary

The 19 draft articles on transboundary aquifers cover the following main areas: a) rules for the utilization of transboundary aquifers; b) activities that are likely to have an impact upon those aquifers; and c) measures for the protection, preservation, and management of shared aquifers. The following summarizes the content of a number of selected articles. The excerpt at the end of this section presents the resolution adopted in 2011 by the United States on the draft articles, in which it urges member states to take the guidelines into account but defers a final, binding, legal convention to a later date.

Comments on the Draft Articles

The draft articles highlight the importance of groundwater (aquifer) resources in general, and transboundary aquifers in particular—underscoring the threats to their sustainable use. The draft articles attempt to strike a balance between the rights of water users to extract water resources and the obligations of these users and regulators to protect and manage aquifer resources sustainably.

No exact global inventory of transboundary aquifers exists, but there are estimated to be about 250 worldwide. Insufficient attention has been paid to their depletion rates. In addition, the importance of aquifers is not adequately taken into account in international agreements and economic development plans of individual countries or groups of countries sharing aquifers and other water resources.

The draft articles emphasize the significance of groundwater reserves for potable water supplies, industry, irrigation, and other uses, and draw attention to the problem of the increasing overabstraction and pollution of aquifers.

Prior to this draft document on transboundary aquifers, there was no unified legal or regulatory instrument for countries and international organizations that mentioned the issues of joint management of shared groundwater or attempted to organize or regulate sharing arrangements among nations.

However, it is important to note that in these draft articles, the "polluter-pays" principle is not included. Without this principle, it is difficult to hold polluters of groundwater reserves accountable for their actions. There is also no legal obligation in the draft articles for states to undertake environmental impact analyses before carrying out planned activities.

Even though the draft articles have not been ratified, they are an important step towards international coordination and the regulation of increasingly depleted resources—in much the same way as internationally shared watercourses, seas, and oceans above the surface.

Some articles recognize the need to protect aquifers against harm resulting from pollution coming from fertilizers, pesticides, or industry in aquifer recharge zones. They establish the principle of the sovereignty of aquifer states to make decisions, but also highlight the principle of equitable and reasonable utilization of aquifers. In other words, decisions on aquifer use should be made in such a way that all users receive proportionate benefits from their share of joint resources without overstepping the boundaries of sustainable drawdown of aquifer waters.

Other articles include the obligation that states should not cause significant damage in their use of transboundary aquifers, and should avoid excessive water abstraction as well as pollution or saline intrusion of aquifers. The obligation to not cause harm is important, since a relatively small amount of pollutants can cause greater harm to an aquifer than to a surface water source.

States are thus obliged to cooperate with each other on aquifer use and exchange scientific data and information on the status of joint aquifers, including volume, quality, and extent of its resources. This information is needed to foster appropriate water management and avoid unsustainable depletion rates, which would occur if each border nation or region tried to extract as much as possible before the water runs dry without regard to the rights of others sharing the resource.

To facilitate harmonious sharing of water, some articles include calls for the future establishment of joint aquifer utilization agreements dedicated to the protection, preservation, and management of aquifer waters. However, these agreements should take into account the present and future needs for water and the presence of alternative water sources; proactive approaches to prevent, reduce, or control pollution; effective functioning and maintenance of aquifers; and protection of ecosystems and water recharge/discharge zones.

Numerous articles zero in on the joint monitoring of aquifers and water resources. They stress the need for joint planning of activities of all parties sharing the aquifer, including the requirement to notify neighboring countries of planned aquifer programs that may affect or draw down upon joint resources. They also address the need to provide technical assistance to developing programs and action plans to help them manage their transboundary aquifers and cope with emergency situations and times of armed conflicts.

A two-step process was followed in the presentation of the draft articles. First, they would need to be adopted by the General Assembly, and second, countries would then formally approve and ratify a binding framework convention in a second stage.

A ratification of the convention and codification of rights and obligations would provide legally binding principles to regulate the use and protection of transboundary aquifers. This second step is yet to be implemented in spite of the urgent situation of aquifers.

It is hoped that a negotiating conference would be convened in the near future to set the stage for specific multilateral, bilateral, or regional treaties to jointly manage transboundary aquifers and coordinate common-ground actions among nations.

Camille Gaskin-Reyes

Excerpts of the Resolution adopted by the General Assembly on December 9, 2011, on the Report of the Sixth Committee (A/66/477)] 66/104 on Transboundary Aquifers

The General Assembly,

Recalling its resolution 63/124 of 11 December 2008, in which it took note of the draft articles on the law of trans-boundary aquifers formulated by the International Law Commission,

Noting the major importance of the subject of the law of trans-boundary aquifers in the relations of States and the need for reasonable and proper management of trans-boundary aquifers, a vitally important natural resource, through international cooperation,

Emphasizing the continuing importance of the codification and progressive development of international law, as referred to in Article 13, paragraph 1 (a), of the Charter of the United Nations,

Taking note of the comments of Governments and the discussions in the Sixth Committee at its sixty-third and sixty-sixth sessions on this topic,

1. Further encourages
the States concerned to make appropriate bilateral or regional arrangements for the proper management of their trans-boundary aquifers, taking into account the provisions of the draft articles annexed to its resolution 63/124;

2. Encourages
the International Hydrological Program of the United Nations Educational, Scientific and Cultural Organization, whose contribution was noted in resolution 63/124, to offer further scientific and technical assistance to the States concerned;

3. Decides
to include in the provisional agenda of its sixty-eighth session the item entitled "The law of trans-boundary aquifers" and, in the light of written comments of Governments, as well as views expressed in the debates of the Sixth Committee held at its sixty-third and sixty-sixth sessions, to continue to examine, inter alia, the question of the final form that might be given to the draft articles."

82nd plenary meeting/9 December 2011

Excerpts of Selected Draft Articles

Article 3
Sovereignty of Aquifer States

Each aquifer State has sovereignty over the portion of a trans-boundary aquifer or aquifer system located within its territory. It shall exercise its sovereignty in accordance with international law and the present draft articles;

Article 4
Equitable and Reasonable Utilization

Aquifer States shall utilize trans-boundary aquifers or aquifer systems according to the principle of equitable and reasonable utilization, maximizing the long-term benefits derived from the use of water contained therein;

Article 6
Obligation not to cause significant harm

1. Aquifer States shall, in utilizing trans-boundary aquifers or aquifer systems in their territories, take all appropriate measures to prevent the causing of significant harm to other aquifer States or other States in whose territory a discharge zone is located.

2. Aquifer States shall, in undertaking activities other than utilization of a trans-boundary aquifer or aquifer system that have, or are likely to have, an impact upon that trans-boundary aquifer or aquifer system, take all appropriate measures to prevent the causing of significant harm through that aquifer or aquifer system to other aquifer States or other States in whose territory a discharge zone is located.

3. Where significant harm nevertheless is caused to another aquifer State or a State in whose territory a discharge zone is located, the aquifer State whose activities cause such harm shall take, in consultation with the affected State, all appropriate response measures to eliminate or mitigate such harm, having due regard for the provisions of draft articles 4 and 5.

Article 7
General obligation to cooperate

1. Aquifer States shall cooperate on the basis of sovereign equality, territorial integrity, sustainable development, mutual benefit, and good faith in order to attain equitable and reasonable utilization and appropriate protection of their trans-boundary aquifers or aquifer systems.

2. For the purpose of paragraph 1, aquifer States should establish joint mechanisms of cooperation.

Article 8
Regular exchange of data and information

1. Pursuant to draft article 7, aquifer States shall, on a regular basis, exchange readily available data and information on the condition of their trans-boundary aquifers or aquifer systems, in particular of a geological, hydrogeological, hydrological, meteorological, and ecological nature and related to the hydrochemistry of the aquifers or aquifer systems, as well as related forecasts.

2. Where knowledge about the nature and extent of a trans-boundary aquifer or aquifer system is inadequate, aquifer States concerned shall employ their best efforts to collect and generate more complete data and information relating to such aquifer or aquifer system, taking into account current practices and

standards. They shall take such action individually or jointly and, where appropriate, together with or through international organizations.

3. If an aquifer State is requested by another aquifer State to provide data and information relating to an aquifer or aquifer system that are not readily available, it shall employ its best efforts to comply with the request. The requested State may condition its compliance upon payment by the requesting State of the reasonable costs of collecting and, where appropriate, processing such data or information.

4. Aquifer States shall, where appropriate, employ their best efforts to collect and process data and information in a manner that facilitates their utilization by the other aquifer States to which such data and information are communicated.

Article 9
Bilateral and regional agreements and arrangements

For the purpose of managing a particular trans-boundary aquifer or aquifer system, aquifer States are encouraged to enter into bilateral or regional agreements or arrangements among themselves. Such agreements or arrangements may be entered into with respect to an entire aquifer or aquifer system or any part thereof or a particular project, programme, or utilization except insofar as an agreement or arrangement adversely affects, to a significant extent, the utilization, by one or more other aquifer States, of the water in that aquifer or aquifer system, without their express consent.

Article 10
Protection and preservation of ecosystems

Aquifer States shall take all appropriate measures to protect and preserve ecosystems within, or dependent upon, their trans-boundary aquifers or aquifer systems, including measures to ensure that the quality and quantity of water retained in an aquifer or aquifer system, as well as that released through its discharge zones, are sufficient to protect and preserve such ecosystems.

Source: United Nations Draft Resolution: The Law of Transboundary Aquifers. November 3, 2011. A/C.6/66/L.24. Available online at http://www.uncsd2012.org/content/documents/348text%20resolution%20the%209%20of%20November%2011.pdf. (c) 2011 United Nations. Reprinted with the permission of the United Nations.

PERSPECTIVES

How Should the Colorado River Basin Be Managed to Mitigate Interstate Conflict?

Overview

The Colorado River is the gem of the West. Its once-powerful flows carved a 1,450-mile path through the landscape of seven U.S. states, eventually crossing the

Mexican border and depositing into a thriving delta ecosystem that supports a diversity of human and animal life. The Colorado River is not just a thing of beauty; it is also a lifeline for those who inhabit the arid environment of the region. Because of the high demand for water from the river, the once-raging force of nature that carved the Grand Canyon is now a tamed flow regulated by human policies and structures. While efforts to normalize river flow may have worked in the past, changing circumstances in the basin may make "normal" impossible in the coming decades.

Climate change and population growth, both of which are putting increased pressure on local resources, are the main culprits in increasing uncertainty in the Colorado River region. According to the Intergovernmental Panel on Climate Change's Fourth Assessment Report, changes in the amount of water flowing through the Colorado basin are likely to be so dramatic by 2020 that even current water needs will not be met. Despite the looming threat of water shortage, population growth in the American Southwest continues to outpace the national average, and demand for an ever-dwindling water resource continues to grow.

While it is important to recognize the impacts of climate change and population growth, many of the problems in the Colorado River Basin are political rather than physical. Water distribution between states in the basin was originally outlined in the 1922 Colorado River Compact, which divided the river into an upper basin and a lower basin. States in the upper basin were allocated water based on a percentage of the estimated total flow—often considered to be 15 Million Acre-Feet or MAF—while lower basin states were allocated specific quantities based on need (see table below). The compact is one of the most important components of a series of legal cases, laws, and treaties that are collectively known as the "Law of the River."

There are several fundamental problems with the Law of the River that result in conflict between states within the basin. First, the 1922 compact, which provides the baseline measurements for how water is allocated between states, may significantly

Table 3.1: State Allocations of Water as Stipulated in the 1922 Colorado River Compact

State	Basin	Allocation
Wyoming	Upper	14% of total flow
Colorado	Upper	51.75% of total flow
New Mexico	Upper	11.25% of total flow
Utah	Upper	23% of total flow
Arizona	Lower	2.8 MAF
California	Lower	4.4 MAF
Nevada	Lower	.3 MAF
Mexico	Lower	1.5 MAF

Source: M. Troy Burnett, *Natural Resource Conflicts: From Blood Diamonds to Rainforest Destruction*. Santa Barbara: ABC-CLIO, forthcoming 2016.

overestimate the volume of water in the river. Studies show that the early 20th century was an unusually wet period in U.S. history, and, despite the fact that 17.5 MAF are allocated between the seven U.S. states and Mexico, the actual average flow of the river may be closer to 14.7 MAF (Kenney 2014, 108–109). If this is indeed the case, then the 1922 compact allocates more water than physically exists in the river. This has been a point of debate over the last several decades, and will undoubtedly need to be addressed as demand continues to increase.

Second, as illustrated above, the Law of the River is complex, often contradictory, and largely inequitable in how it allocates water. One inequality in particular could pose potential problems for peaceful water management. Water added to the Colorado River from tributaries in the upper basin (Wyoming, Colorado, New Mexico, and Utah) is included in the total amount of allocated water, but water added from tributaries in the lower basin (Arizona, California, and Nevada) is disregarded. In fact, one of the only places where the Law of the River treats all basin states equally is when it outlines that the upper and the lower basin states are each responsible for 50 percent of the water that the United States promises to Mexico each year. Because of the design of the water allocations written into the Law of the River, the states in the upper basin are disproportionately affected during years of shortage. Known as the "upper basin climate squeeze," this issue is likely to be a major source of conflict in the coming decades.

Finally, conflict may arise from the lack of leadership in the basin. In part, this is because established laws position states in the upper basin and lower basin in conflict against each other. Because of fear over losing their own rights to water and losing favor with their constituents, political leaders lack incentives to work in a unified way. In addition, U.S. basin states have no effective platform for communication. The International Boundary and Water Commission, which governs resource relations between the United States and Mexico, has little authority in the United States, and no other political body exists in the American West to help manage the river holistically. Many have called for support from a basin-wide organization; however, states are unlikely to willingly relinquish their control over resources.

In conclusion, water managers in the Colorado River basin face a number of challenges in the years ahead. The Law of the River, which has comprised the key governing documents for the river to date, may be too convoluted to address the challenges of water shortage that basin states are likely to face. As climate change and population growth continue to widen the gap between supply and demand, it will be vital for states to find ways to mitigate conflict before shortages occur.

Leeann Sullivan

Perspective 1: The Difficult Task of Managing the Colorado Basin with Multiple Stakeholders

The Colorado River is the lifeblood of the American Southwest. Beginning in the magnificent Colorado Rockies, the river meanders and flows through seven states and 1,400 hundred miles of dramatic landscapes and diverse ecosystems. More than 30 million people (close to the population of Canada), the cities of Denver, Los Angeles,

Phoenix, and Las Vegas, as well as 20 Native American nations and Northern Mexico, rely on its waters for their livelihoods. Though, as a matter of fact, 80 percent of the river's annual flow is consumed by 3 million acres of farmland. Along with a complex and heavily regulated system of dams, canals, and pipelines, numerous treaties, decrees, court decisions, laws, and compacts exist to manage all the often-contentious demands. The primary agreement is the 1922 Colorado River Compact; created by the seven basin states with general provisions for water allotments.

Water conflicts arise when there are multiple, competing interests for the river's resources. Farmers, cities, ecologists, recreation seekers, and even spiritualists demand their share. In the case of the Colorado River and its tributaries, the dilemma is as stark as it is straightforward: more water is promised to river users than is available. Compounding the issue are other factors: the recent ten-plus year drought in the American West; the exponential population growth in Nevada, Arizona, Colorado, Utah, and California; climate change and the resultant increase in temperature and further desiccation of the desert environment; Mexico's increasing pressure and demands for more water and higher quality and less toxic flow; as well as calls by environmentalists to preserve habitat and restore flow into the river's delta in the northern gulf of California.

Evaporation and over-usage have dramatically reduced the primary reservoirs of the river (Lake Powell, Lake Mead, and Lake Havasu), exposing long-buried channels and mesas. Conflicts and disputes, no stranger to the basin, seem destined to continue and, according to many, worsen. The question is how best to manage the basin to mitigate the existing and future conflicts, including conservation of the river's unique ecological heritage.

Environmental Impacts and Conservation Strategies

The river basin, from the headwaters to the delta, supports an immense variety of fish and wildlife, including unique, endemic species of trout, warm water fish, hundreds of bird species, and a wide array of mammals, from bears to coyotes to bats. With the development of the basin and the numerous demands placed on the system, much of this ecological heritage is threatened. Altered and reduced stream flow have damaged and destroyed habitat, and nonnative trees like tamarisk have taken over the native communities, further stressing the life cycle of native birds and wildlife. In the lower basin, less than 10 percent of native riverside habitat remains. It has become increasingly obvious that the existing management of the river system is not ecologically sustainable, and indeed, if the management strategies don't change, the environmental damage will worsen as future demand continues to outstrip supply.

While balancing the multiple needs in the basin presents many challenges, viable solutions can be found when all the stakeholders work creatively and cooperatively. In general, to achieve a sustainable, healthy river ecosystem there are two approaches: 1) protection of the minimally impacted and existing functional habitats, and 2) restoration of degraded systems. Both approaches require a customized mix of strategies at varying geographic scales. Existing water supply and delivery systems need to be managed flexibly and efficiently.

Sustained protection of healthy habitat has benefited from successful state programs, such as Colorado's in-stream flow water rights legislation or Utah's native fishery water-leasing program. Both are useful models that can be tailored to the needs and issues of the other basin states so as to strike a balance between environmental protection and human use.

Restoration relies primarily on increasing river flow and, in particular, needs strategic thinking and flexible approaches. Strategies and tools vary with location; examples include: infrastructure repair and updating; retrofitting of existing dams to improve flow rates and storage capability; voluntary, compensated water-rights transactions to discourage waste and misuse; water banking and downstream water exchanges; increased agricultural efficiency and switching to less water-intensive crops; and more aggressive water conservation strategies in the basin's largest cities.

Stakeholders and environmentalists must prioritize and dedicate resources to areas and habitats of key concern, such as the Upper Colorado stem from Granby to its confluence with the Eagle River. Full restoration will also require an aggressive effort to remove the invasive tamarisks, followed by native species revegetation, the removal of inoperable, outdated barriers to fish movement, and other site-specific, in-channel, and riparian habitat improvements.

Conclusion

Water management, planning, and conflict mitigation for the Colorado River Basin is going to require innovation in thinking, flexibility, and the continued evolution of the existing Law of the River. The days when users knew with certainty how much water the river could reliably provide are gone. Risk and uncertainty are the new norm for water planning as climate change ushers in an era of irregular and unpredictable flow rates. As the basin continues to grow, avoiding conflicts between the multiple, interstate stakeholders while promoting ecological sustainability, stakeholders will require an aggressive combination of water efficiency, reuse, voluntary sharing and trading between agriculture and municipalities, and technological innovation that includes smaller-scale, structural projects.

Mark Troy Burnett

Perspective 2: Colorado Basin Management—A Simmering Conflict

As resource depletion in the Colorado River Basin worsens, conflict may be inevitable. Inequality in how water is allocated to each state dates back to the middle of the 20th century. Despite the threat of climate change, overpopulation, and conflict between states, little has been done to substantially address these policy failures. In order to mitigate conflict in the Colorado River Basin, two major changes must be made: first, the method for allocating water between basin states must be standardized so that resource shortages impact both halves of the basin equally; second, the upper and lower basin commissions must aid in the establishment of a basin-wide forum—composed of policy makers, business leaders, scientists, and community groups—that can provide leadership to water managers. By focusing attention in the short term on these two goals, water managers can head off potential conflict in the decades to come.

Flaw in the Law

Before residents of the Colorado River Basin can hope to find technical solutions to water shortages, policymakers must look to correct some of the fundamental flaws of the Law of the River. Most importantly, they must address the inequalities in how resources from the river are divided between states. By shifting all seven states to a percentage-based allocation system, resource scarcity would be distributed throughout the West instead of disproportionately impacting residents in upper basin states like Colorado and Utah.

Current law dictates that lower basin states can make a call upon their northern neighbors if they do not receive their promised allocation, essentially providing them with legal favor over the entire basin's resources. By making all allocations percentage-based, water managers eliminate the need for this costly and time-consuming litigation, and distribute the hardships of scarcity among a lot of people instead of a few.

There are a number of benefits to reallocating resources based on percentage of total flow for all states. First, it allows for a basin-wide reduction of water withdrawals during drought. As climate change impacts the water cycle and reduces precipitation and snow melt in the Colorado basin, it is likely that drought events will increase in both frequency and scope. Water managers must have agreeable options for dealing with the resulting water shortages. Second, a percentage-based allocation system may be easier for states to reach agreement on than a system in which quantitative allocations must be set for each state, which could result in years of legal dispute. There are still, however, inherent challenges in reallocating water. Uncertainty about what baseline to use may mean that percentages are based on predictions about the river rather than physical hydrology. In addition, river infrastructure such as dams and reservoirs would need to be flexible in order to handle dramatic changes in the river's flow.

Renovations of current river infrastructure may be costly and politically contentious.

Alongside a shift in how water is allocated, relationships between basin states could benefit substantially from a basin-wide forum with equal input from both the upper and lower states. By creating a regional management system focused specifically on the Colorado River, state water managers could benefit through input from technical experts, policy experts, and local residents who rely on and understand the river.

Most importantly, the forum could play a significant role in helping to mitigate conflict in the face of rapid development and climate uncertainty. As the primary governing body for the river, representatives from the forum could approve or deny projects that impact flow (such as the construction of new dams or diversions) based on the best interests of the whole region. This could mean more peaceful relationships and, importantly, a more ecologically resilient basin.

Regional Management Body Needed

A regional management body would have benefits but also challenges. The forum could be a useful tool in facilitating data exchange between states, which would make management more representative of the hydrologic reality in the basin. It could serve as a platform for voices from nongovernment entities, including community groups, business leaders, and representatives from local Indian tribes who rely on resources from the river for sustenance. In addition, many river basins that currently use forums or joint institutions for management rely on technical, policy, or economic committees to aid in management.

This specialization, housed under a secretariat that manages and monitors day-to-day interactions in the basin, has in some cases been shown to improve the health of the river and the communities that rely on it. Despite the success of these management systems in other regions, it is possible that a basin-wide forum could be seen as a threat to state sovereignty over resources in the Colorado basin. If the forum were seen as a threat, states would be less likely to support, fund, or comply with mandates from the organization. Additionally, some critics of basin-wide forums argue that the Colorado River would benefit most from less governmental "red tape" and that a forum of this sort may result in more gridlock over how to manage resources.

Whatever approach water managers take toward the Colorado basin in the coming decades, it will be important that they shift away from the status quo and recognize the inherent uncertainties of managing a complex basin like the Colorado. Two relatively simple changes can be made now to mitigate conflict as the gap between supply and demand widens. First, a basin-wide percentage-based allocation system would evenly distribute the negative impacts of water shortage throughout the region rather than disproportionately putting burden on upper basin states. Second, the development of a basin-wide forum could provide leadership and technical expertise to water managers and allow them to manage the river as it exists rather than continue to rely on outdated and fundamentally flawed policies. These changes have vast potential benefits along with certain challenges, but may be necessary to the successful management of the river in an uncertain resource future.

Leeann Sullivan

REFERENCES AND FURTHER READING

American Rivers and Western Resource Advocates. "The Hardest Working River in the West: Common Sense Solutions for a Reliable Water Future for the Colorado River Basin." American Rivers Web site, 2014. http://www.americanrivers.org/newsroom/resources/the-hardest-working-river-in-the-west-common-sense-solutions-for-a-reliable-water-future-for-the-colorado-river-basin/.

Colorado River Governance Initiative. "The 'Upper Basin Voluntary Demand Cap' as a Means of Mitigating Legal Uncertainty in the Colorado River Basin: Modeling Results." Boulder: University of Colorado at Boulder, 2013.

Gleick, Peter. "Water Planning and Management under Climate Change." The World's Water. Washington, DC: Island Press, 1999.

Jones, P. Andrew, and Thomas V. Cech. *Colorado Water Law for Non-Lawyers*. Boulder: University Press of Colorado, 2009.

Kenney, Douglas, Sarah Bates, Anne Bensard, and John Beggren. "The Colorado River and the Inevitability of Institutional Change." *Public Land and Resource Law Review* 32 (2011): 103–152.

Milly, P. C. D., et al. "Stationarity Is Dead—Whither Water Management?" *Science* 319 (February 1, 2008).

Schlager, Edella, and Tanya Heikkila. "Water Scarcity, Conflict Resolution, and Adaptive Governance in Federal Transboundary River Basins." In *Federal Rivers: Managing Water in Multi-Layered Political Systems*, edited by Dustin E. Garrick, George R. M. Anderson, Daniel Connell, and Jamie Pittock, 57–72. Northampton: Edward Elgar Publishing, 2014.

Summitt, April R. *Contested Waters: An Environmental History of the Colorado River*. Boulder: University Press of Colorado, 2013.

4 OCEAN AND COASTAL ENVIRONMENTS

OVERVIEW

Since time immemorial, humans have looked out at the oceans and pondered their origins, often creating legends to explain these vast expanses of water. People have always relied on ocean environments to provide food and support social and economic activities, as well as their trade and transportation networks.

We call our planet the water planet for good reasons. Photographs of planet Earth from space reveal a huge ball of ocean blue veiled by swirling clouds. These pictures provide a strong visual reminder that oceans cover 75 percent of earth's surface area and are the largest ecosystem on the planet. They provide half the world's supply of oxygen through the photosynthesis of aquatic organisms, and supply protein from fish and marine products for about 3 billion people.

This chapter looks at the importance of oceanic and island environments and the impacts of climate change and human economic activity on marine resources and ecosystems. These changes include ocean pollution, increased stress on island ecosystems, depletion of fish stocks and marine mammals, and damage to coastal and coral reef environments.

The chapter also examines the international framework of laws governing activities on the high seas and the problem of monitoring and enforcing regulations. The case studies in the chapter review increasing threats to the Galapagos Islands, lessons learned from the *Exxon Valdez* oil spill in Alaska, and the problem of ocean pollution through plastic. The perspectives section zooms into the challenges of oil drilling in the Gulf of Mexico.

Advances in Oceanography

A branch of science called oceanography studies the oceans, structure, and properties of seabed environments and marine life, and the utilization of marine resources.

For centuries, explorers, navigators, and oceanographers have been curious not only about what lay above the oceans but also what lay below their depths. Over time, different methods were developed to measure ocean depths and map the ocean floor's contours.

In the 1900s, the improvement of technology allowed the calculation of ocean depths through bouncing sound waves off the ocean bottom. In 1930, Charles William Beebe invented an important device called the bathysphere. It was a large sphere—suspended through a steel cable from a ship—that descended to 3,000 feet below the ocean (Newton 2003, 224).

Other inventors and researchers eventually devised a wide range of submersibles, undersea laboratories, scuba equipment, motorized underwater pods, and sophisticated cameras, which permitted people to dive or travel below the surface of the ocean to study marine ecosystems.

These methods provide valuable information on life below the oceans and continue to document the response mechanisms of marine ecosystems to emerging challenges such as ocean warming, acidification, pollution, and resource depletion. Each successive human attempt to explore and research oceans and ocean flows has yielded more sophisticated methodologies to gather reliable data on wind and ocean currents, continental shelves and ocean trenches, marine ecosystems, water quality, and temperature changes.

Modern tools of oceanography and renowned research institutions such the Woods Hole Oceanographic Institution in Massachusetts (founded in 1930) and the National Geographic Society have provided further impetus to surveys of the sea floor and the monitoring of ocean pollution, salinity, temperature trends, and acidification.

The groundbreaking oceanic research of Dr. Sylvia Earle, the National Geographic Society's explorer-in-residence, who led more than 100 underwater expeditions, including the first team of women aquanauts in 1970, has raised awareness on the beauty of the oceans and emerging threats. Dr. Earle also founded "Mission Blue," an ocean conservation advocacy and research group to educate the public on threats to marine resources, press for more protected areas or reserves (only 4 percent of oceans is protected), and alert humankind on the impacts of human pressures on the oceans for human survival.

The National Oceanic and Atmospheric Administration (NOAA) of the United States—with roots in 1807—also carries out cutting-edge scientific research on oceans, including sea level rise, tidal currents, marine fisheries, and ecosystems, seabeds, coastal restoration and storm resilience, and ocean warming, acidification, and contamination trends.

A NOAA expedition ship, on an exploration trip between 2011 and 2013, discovered about 570 methane gas plumes in the Atlantic Ocean between Cape Hatteras in North Carolina and Georges Bank in Massachusetts. Scientists estimate there may be at least 30,000 such gas plumes in the world's oceans that may have resulted from warmer ocean waters melting ice on the seabed, and uncorking natural gas deposits. There are two main concerns: 1) that this methane, a greenhouse gas, could be released into the atmosphere, and 2) that when dissolved in seawater, it could contribute to further acidification of the ocean and damage to marine life (NOAA 2014).

This steady accumulation of scientific knowledge has also produced valuable information about the mechanics of oceans and ocean currents. In past centuries, people believed that ocean currents moved by supernatural forces, and conjured up myths and legends to explain these huge volumes of moving water. Through modern research, we understand that ocean currents form when wind currents push ocean water in their same direction horizontally, or when winds blow from the tropics to other zones, whipping up water in their wake. We also know that changes or disruptions to such currents can be extremely disruptive to global climatological patterns.

Ocean Currents

In the 15th century, in the age of early transocean navigation, human knowledge of wind and ocean currents was an important factor that enabled sailing ships to explore distant and unknown areas. Maps and charts of oceans and ocean currents, even if incomplete or incorrect, were priceless, and treated as state secrets, since they held the keys to claiming new lands. Although early navigators, explorers, and pirates might not have fully understood the science of ocean currents, their familiarity with wind flows and reflows was critical for riding waves to new places—and sometimes for bringing them back home.

There are five main patterns of ocean currents that extend over the north and south Atlantic Ocean, the north and south Pacific Ocean, and the Indian Ocean. These currents vary in speed, width, temperature, and depth of water across ocean areas. The spinning of the earth causes ocean currents to flow in a clockwise direction in the Northern Hemisphere, and a counterclockwise direction in the Southern Hemisphere.

Oceans absorb roughly 90 percent of the earth's heat. Since the late 1990s, research has shown that higher temperatures and cyclical changes in tropical ocean currents—combined with strong trade winds—have taken surface warmth deeper into the oceans (NOAA 2014).

El Niño/La Niña Disruptions

Anomalies in the flow and temperature of ocean currents also change weather patterns on the land. El Niño and La Niña are examples of such changes, which occur through the periodic reversals of winds and ocean currents across the equatorial Pacific Ocean. Normally, trade winds and surface waters in the tropical Pacific flow westwards from the Americas towards the Asian continent. As the ocean warms on its way, it generates heat and energy, bringing rain to Asian countries. When these trade winds stop moving, an El Niño phenomenon occurs, as the flow of currents piles up and traps water against the Pacific coasts—causing volumes of heated, stacked water to flow back eastward to the Americas. El Niño is said to be one of the major factors in the 2015 record heat registered in the United States (Bromwich December 23, 2015).

El Niño creates huge impacts. The return of warm water waves to the coasts of the Americas overrides the regular cold currents (Humboldt Current) on the Pacific Ocean and often displaces cold waters and their rich nutrients, the mainstay of the fishing industry. El Niño typically unleashes heavy rains and damage to the fishing industries of coastal Pacific nations, while at the same time causing imbalances such as droughts, and disruptions of monsoon and other patterns in different parts of the west Pacific.

The typhoon season in the tropical west Pacific has shown a tendency to become more pronounced due to warmer oceans and the El Niño effect. El Niño also disturbs the paths of other winds (the jet streams), which can cause Pacific rainstorms on the California coast. Its impacts usually abate after eighteen months, after which a reversal could occur, i.e., heavy rains in the west, drought in the east. This reverse phenomenon is called La Niña. Together, El Niño and La Niña cause severe, periodic disruptions to global patterns on both sides of the Pacific Ocean with significant economic impacts (Bromwich December 16, 2015).

Emerging Challenges

Currently, the most serious threats to oceans include climate change, the ongoing loss of biodiversity, global mismanagement and depletion of marine resources, and disregard for international marine laws. Much in the same way that human activities have impacted the supply and quality of freshwater on the earth's surface, they have affected ocean and coastal ecosystems through pollution, biodiversity loss, and the over-harvesting of marine resources.

The degradation of mangroves, coral reefs, barrier islands, and estuaries—which provide vital protection for coastal and marine environments from storm surges and coastal erosion—has reduced the capacity of coastlines to withstand intense weather events. Estuaries (the mouths of rivers emptying into the sea) are specific microecosystems of ocean coastline environments. In these areas, freshwater from rivers flows into salty oceans; and these mixing bowls of fresh and saltwater create specific ecosystems of bays and delta areas.

Silt carried down by rivers to estuaries provides a rich environment for plants, animals, waterfowl, and marine life that live in and adapt to this particular ecosystem. Estuaries such as the Chesapeake Bay and the San Francisco Bay demonstrate these features, but are under pressure due to many factors: dredging and filling in for buildings and fish farming; sewage discharge and industrial pollution; contamination from pesticides and fertilizers; reduced oxygen levels; and loss of habitat, fish, avian, and wildlife resources.

Warming Waters

Oceans are showing the effects of climate change and warmer temperatures. Ocean chlorophyll—essential to produce oxygen—has fallen by around 10 percent in the last decade in the North Pacific, Indian, and North Atlantic Oceans. Higher temperatures and ocean acidification (as a result of absorption of carbon dioxide in the oceans) are damaging coral reefs, which host intricate webs of living ecosystems and function as lines of defense against storms. Warmer temperatures are disrupting marine life as well, causing fish species to migrate to cooler waters.

The North Atlantic Ocean has been warming more quickly than any other ocean. Higher temperatures affect organisms above the surface, and also those that live below.

In the North Atlantic, overfishing has created increased imbalances in the age, composition, and quantity of fish stocks. Changes in ocean currents and temperatures also alter the location of plankton or fish stocks, reduce their numbers, or lead to the outright loss of fish populations. Plankton stocks are particularly impacted by shifts in ocean circulation.

Fish are good barometers of temperature change. Since they are unable to regulate their temperature independently of the surrounding water, they usually swim to other areas to find waters in their ideal temperature range. Southern fish species such as red mullet, anchovy, sardine, and poor cod are now found in the North Sea due to the warmer temperatures there. When fish change their latitude or swimming depth, predators usually try to follow in a migratory response, affecting the entire chain.

Depletion of Fish Stocks and Marine Resources

Due to the increasing consumption of fish protein for human diets and predatory harvesting practices, two-thirds of all fish stocks on the high seas are already overexploited (*Economist* 2014). North Sea cod is already overfished. During the 1930s, British fishermen harvested roughly 300,000 tons of cod annually. By 1999, the catch went down to 80,000 tons; currently, cod stocks in the Atlantic are at historically low levels (Johansen 2015). Cod populations also tend to decline with the rise in ocean temperature because cod do not breed well in warmer water.

Other species are under stress. Millions of sharks are killed yearly for their fins; stocks of big species such as tuna, marlin, and swordfish have fallen since the 1950s. Other species such as flounder, grouper, hake, halibut, and Chilean sea bass are also threatened. Salmon stocks have also disappeared from many of their usual breeding areas around the world. Wild Atlantic salmon has been halved over the last two decades due to global warming and infections spread to natural stocks by hatchery-bred fish. In 2002, a marine parasite associated with warmer stream waters began spreading among Alaska's Yukon River Chinook salmon stocks. In 2008, a parasite in Norway's fish farms infected about 10 percent of the country's rivers, devastating wild salmon stocks (*Economist* 2009).

The Tragedy of the Commons

Declining levels of fish stocks have led to a global race among nations to harvest what remains in common high seas areas, a phenomenon often called the "tragedy of the commons." This process occurs when collective global assets such as oceans become depleted through short-term action of individuals or groups acting in their private interest rather than through long-term considerations for the preservation of the resources for all. The ironic tragedy of this situation is that those nations or private enterprises which damage collective resources do not necessarily shoulder the cost of replenishing them.

Marine animals and organisms continue to be under attack. History teaches us that human depletion of marine animals is no new occurrence. From the 16th to the 19th centuries, hunters relentlessly killed fur seals for their furs—from Australia to the

South Atlantic, and from the Pacific to the Indian Ocean. Alone in the first two decades of the 19th century, human predators killed 6 million fur seals around the world, leading to their decline (Ponting 2008, 160).

In the North Atlantic, after the demise of the fur seal, hunters turned to the harp seal. The harp sealing industry in Newfoundland peaked in the 1850s with the killing of about 600,000 seals per year—then declined sharply after near-extinction of these seals. Hunters also pursued elephant seals on the islands around Antarctica and off the west coast of North America, prizing them not for their fur but for their oil. A million elephant seals were killed in the 19th century (Ponting 2008, 160).

Human predators hounded the walrus too for its oil and skin, and for ivory from its tusks. Millions of walruses were killed before the industry collapsed at the end of the 19th century. These unsustainable hunting practices built in their own demise: by the turn of the 19th century the fur seal and walrus industries had failed due to the elimination of the animals.

Hunters then turned to whaling, following the same pattern: pursue, overexploit, or decimate before moving on to another marine species. The new targets became the right whale, the fin whale, and the toothed sperm whale. The right whales were particularly slow movers in the ocean and therefore easy prey for hunters, who pursued them with harpoons, dragged them to shore, and processed them for their byproducts. There are only about 300 surviving North Atlantic right whales, which are currently affected by a decline in plankton, their main food source.

Before electricity, whale blubber provided oil for lighting, and later became important for food products, cosmetics, and lubrication of industrial machinery. The popularity of whalebone for women's corsets, hooped skirts, umbrellas, whips, fishing rods, and dagger and cutlery handles endangered whales even further.

Technical innovations from the 18th century onward introduced large steam-powered ships and brick ovens to render blubber on board. Ships could thus remain at sea for extended periods and process more whales. The indiscriminate slaughter of both young calves and mature whales precipitated further decline of whale stocks. However, in the 20th century, more efficient fat-processing methods gave whaling a new boost. Whaling resurged in the 1930s and the 1950s due to European demand for whale meat, eventually reducing the species even more drastically.

Whaling Quotas

In 1946, the International Whaling Commission intervened to set quotas for whaling nations and to ban the killing of gray and white whales entirely. Nonetheless, whaling stocks continued to decline due to ongoing impacts of previous slaughter of stocks. By the early 1960s, most countries were forced to reduce whaling due to depletion of whales and worldwide condemnation, although a few nations continued the practice. In 1970 the United States banned all whale products; and in 1972 it prohibited the slaughter of all whales within U.S. waters under the Marine Mammal Protection Act. In 1986, most commercial whaling halted, with the exception of Norway, Iceland, and Japan, and some permitted exceptions for indigenous groups (see case study on whaling in this chapter).

Decline of Fish Stocks and the Threat of Extinction

While whaling was declining, the harvesting of fish and shellfish species was intensifying. Human activity on the high seas turned to intensified fisheries. From the 1950s onward, fishing nations and private fishing companies began to master modern technology and the use of sophisticated equipment to extract ever-increasing quantities of fish from the ocean on a grand scale. It became easy to locate and harvest fish shoals with radar. This new technology plus the lack of regulations on the oceans stimulated the growth of predatory fishing practices. Large scale fishing fleets could travel to the furthest corners of the oceans, using high-tech equipment such as radar, dragnets with steel plates, and heavy rollers to comb the ocean floor for fish.

Fleets on the high seas became floating factories that could harvest, process, and package the catch in one place. It is estimated that daily about 15,400 square miles of seabed are scraped by bottom trawling methods, leaving virtually fish-deficient zones in their wake (Ziegler 2014). However, due to the lack of oversight over high seas fishing, it is difficult to give an exact volume. Given large scale trawling and new technologies, fish, which once seemed a virtually inexhaustible resource in the ocean since the first man or woman ever cast a net, have started to decline at alarming rates.

The relentless chase for fish means that about 30 percent of the world's fish stocks are overexploited or already depleted—up from 10 percent in 1974 (*Economist* 2009). Unless governments of the world and large fishing concerns come together to severely restrict or stop trawling of the sea bottom, it is unlikely that the severe degradation of world fish stocks can be reversed.

Ninety percent of large predatory fish such as tuna, swordfish, and sharks have disappeared, and many smaller fish species are also in decline (*Economist* 2009). Salmon have disappeared from about 40 percent of their historical range in the Pacific Northwest, with Chinook, sockeye and coho salmon facing the risk of depletion (Klein 2014, 441). The Bluefin tuna (prized for sashimi), has become one of the most desirable fish in the world, commanding top prices, but is becoming the most endangered large-fish species.

Annual cod landings in the Gulf of Maine have also steadily declined, from 70,000 metric tons in 1861 to 54,000 metric tons in 1880, down to about 20,000 tons in the 1920s and only a few thousand in recent years. This trend once led to a ban on cod fishing in the area for a period to allow fish stocks to recover (Bolster 2015).

To preserve fish stocks and regulate fishing in European waters, European Union states have established a system of quotas and licenses. Iceland has been particularly successful in setting up individual transferable quotas allocated to fishing boats for each species caught over a three-year period.

Nonetheless, fish stocks in Europe are still overexploited. Growing demand for fish as a protein product for people, increased global trade in fish products (even for animal food production), and the worldwide popularity of deep-sea sports fishing continue to stress fisheries and marine ecosystems. When fish species drop below certain levels, there is a chain of impacts on predators in the food chain, but also on baby birds, for example, since they depend on fish for their survival.

Ocean Pollution

In addition to overfishing, ocean pollution is a serious problem. After World War II, in the 1950s, some parts of the ocean became dumping grounds for obsolete or surplus munitions (Agence France-Presse 2014). During the Cold War, some developed nations conducted nuclear tests on ocean atolls of the Marshall Islands in the Pacific and displaced Marshall residents to other islands. Ironically, some of these people or their later generations are currently facing seawater encroachment problems and the likelihood of having to move again due to climate change.

After the 1960s, the main sources of ocean pollution became pesticides and fertilizers from farms, effluents from port and industries, garbage and sewage from the land, solid waste from cruise and container ships, and oil spills from tankers or ocean drilling platforms.

The construction of offshore and deep-sea oil drilling platforms and the increase of oil tanker traffic on the high seas and in regional seas such as the Arctic and the Gulf of Mexico are major factors in the growing frequency of marine oil accidents in oceans—ranging from small leaks to megaspills. These spills cause damage to ocean ecosystems; destroy fish stocks, marine organisms, and seabirds; and contaminate beaches, estuaries, wetlands, mangroves, and barrier islands.

Waste flowing out of a pipe into the ocean contaminates marine ecosystems. (Charles Wagner, Jr./Dreamstime.com)

Major spills have included the *Exxon Valdez* spill in Alaska in 1989, the Guimares oil spill in the Philippines in 2006, and the Gulf of Mexico BP spill in 2010. The *Exxon Valdez* oil spill off Prince William Sound released millions of gallons of fuel oil after the tanker struck a reef, causing one of the largest human-made marine disasters (see case study). The Guimares spill discharged more than 130,000 gallons of fuel oil and impacted fishing grounds, marine sanctuaries, and mangrove forests in the area around the Sulu Sea and the Guimares Strait in the Philippines.

The *Deepwater Horizon* or British Petroleum (BP) oil spill in 2010 was a major example of ocean contamination and damage caused by oil drilling gone awry on an oceanic platform. Due to a rig explosion, an underwater oil well gushed for about three months, and discharged about 5 million barrels of crude oil into the Gulf of Mexico. This incident led to the deaths of eleven oil platform workers and created an oil slick over 68,000 square miles of ocean. It severely affected the tourism and fishing industries along the Gulf Coast in Texas, Louisiana, Mississippi, Alabama, and Florida.

Even though BP used a range of containment, dispersal, and oil removal strategies to cope with the disaster on land and sea, the marine food web and ecosystems in the area were severely disrupted. Huge quantities of dolphin, tuna, amberjack fish, and sea birds perished; oil sludge, tar balls, and deposits contaminated beaches, marsh sands, estuaries, and coastal and land organisms. Under the principle of "the polluter pays,"

Fire boat response crews battle the blazing remnants of the offshore oil rig *Deepwater Horizon* on April 21, 2010. (U.S. Coast Guard)

BP paid billions of dollars in penalties to the U.S. Department of Justice and was required to compensate Gulf Coast residents for disruption of their livelihoods.

Invasive Marine Species

An additional problem affecting the balance and stability of marine ecosystems is the spread of invasive marine species in the oceans. It usually occurs when ocean vessels pass from one marine area to another. Ships take on ballast water at ports of departure and discharge it at ports along ocean routes; this practice transports invasive marine species from one area to another and contributes to the displacement of native species and transformation of ecosystems.

Coral Reef Destruction

Coral reefs are yet another cause for concern in marine environments. They have declined worldwide by 40 percent (Zimmer 2015). Corals are complex and diverse marine environments, living colonies of tiny animals that contain microscopic algae within their structure. They have become severely endangered through ocean pollution and climate change effects, i.e., higher ocean temperatures and acidification (caused through ocean absorption of carbon dioxide). These factors stunt the reproduction and growth of coral organisms, a process called bleaching.

Coral reefs occupy less than 0.1 percent of the world's ocean surface, but host at least 25 percent of all marine species. Tropical reefs usually grow horizontally about 1.2 inches per year and vertically up to 9.8 inches annually. Reef structures build up in layers through deposits of calcium carbonate coral skeletons, which form the foundation for living organisms on the upper layers. They need sunlight to grow and are mostly located in relatively shallow ocean areas, usually at 30 degrees north or south of the Equator (Coral Reef Alliance 2014; NOAA Ocean Service Education 2008).

The Great Barrier Reef in Australia is the world's largest barrier reef. It is located along the continental shelf and stretches for approximately 1,300 miles. Coral damage is already being reported on the Great Barrier Reef; it is estimated that 10 percent of all coral reefs around the world are dead and about 60 percent are endangered. Threatened coral reefs are found off the coast of Central America in Honduras (the Bay Islands), the Bahamas, Bermuda, Florida, the Maldives, Indonesia, and the Philippines (Ailing 2010, 107).

In addition to bleaching, there are other factors causing coral damage: excessive accumulation of bacteria, algae, sediments, and pesticides; sewage discharge from the land; and ocean deposits of microplastics (smaller than one-fifth of an inch). Around 88 percent of oceans contain these microplastics. When living coral organisms ingest them, they wrap themselves around internal tissues and disrupt digestion. Coral reef destruction also occurs through the use of dynamite for fishing, destructive diving and snorkeling practices, expansion of algae blooms, and construction of pathways through reefs for tourism, boating, and cruising activities.

Ocean Warming and Acidification

Warmer and more acidic ocean environments threaten shellfish production. In the Pacific Northwest, the oyster industry lost more than $110 million in 2014, an extremely warm year on record (*Barbados Advocate* 2015). Rising temperatures and sedimentation from land runoff are increasing the production of harmful micro algae and excessive blooms of bacteria, weeds, jellyfish, and red tides in oceans as well as the deterioration of sea grasses, kelp forests, and coastal wetlands. The discharge of large pieces of nonbiodegradable plastic into oceans is also harmful to turtles, sharks, seabirds, and other marine species. A case study in this chapter looks at the problem of swirling plastic in the Pacific Ocean, called the Great Pacific Garbage Patch.

Impacts of Expanding Aquaculture

Due to the decline of natural fish species through overfishing—caused by rising demand for fish protein since the 1950s—there has been a global explosion of aquaculture or fish farming in coastal areas. In the mid-1970s, the technology of domesticating and farming Atlantic salmon in captivity succeeded in changing the flow of salmon's natural migration and breeding patterns in the wild. In the 1980s and 1990s, European countries produced over 250,000 tons of farmed salmon per year; in 1998, farmed salmon exceeded wild salmon catch for the first time (Matsen 2010, 97).

Over 65 percent of the world's salmon is currently harvested in marine farms and offshore pens in countries such as the United States, Canada, Norway, Japan, New Zealand, and Chile (Matsen 2010, 97). There is increasing commercial interest to build on the salmon industry and expand offshore fish farming activities to other species such as cod, tuna, mahimahi, and other high-value fish. Since the 1990s, an increasing number of countries have also expanded fish farming to include trout, crawfish, oysters, and shrimp.

The development of the fish farming or aquaculture industry has positive aspects such as providing protein for people and creating jobs. Large aquaculture facilities around the world raise and feed fish and shellfish in contained coastal pens until harvesting, and contribute to the economic output and exports of many countries. Likewise, small family and communal fish farms are vital for family and women's incomes and provide an important source of protein for poor rural families in developing countries.

However, there are serious environmental concerns regarding large, industrial fish-farming operations. Impoundments may discharge wastes, antibiotics, uneaten fish food, and chemicals into coastal waters, estuaries, and oceans. In Ecuador, the marine shrimp farming industry—developed in the early 1960s—produces millions of dollars per year and creates thousands of jobs, but generates significant environmental impacts as well.

Farmed fish and shrimp by-products can damage intricate marine ecosystems and may also contain heavy metal and mercury that could enter the human body through the food chain. Harmful bacteria or plankton also spread through fish farms and can damage farmed fish. In addition, when farmed species escape from pens, they compete with wild fish for food or weaken natural stocks through interbreeding.

The rapid expansion of fish and shrimp farming industries has destroyed many mangrove forests along African and South American coasts and around the Bay of Bengal. These mangroves and coastal wetlands have for centuries provided a buffer against storms and storm surges. They also function as natural nurseries for baby shrimp and other organisms, they process effluents coming from the land, and they are naturally adapted to saline environments through their ability to filter salt through roots and leaves. Mangrove destruction has increased erosion and weakened shoreline defenses. It has been suggested that mangrove loss was a major factor for the high levels of devastation caused by a 1999 supercyclone that struck Orissa in eastern India, causing 20,000 deaths and huge land and crop losses in the area (Black 2004, 93).

There are also social ramifications associated with the expansion of large-scale commercial shrimp and fish farms in former mangrove forests: mangrove loss affects the livelihoods of coastal inhabitants, who depend on mangrove forests for the small-scale harvesting of wild shrimp, fish, and bivalves. Mangrove loss also increases the social and environmental vulnerability of low-income groups, who are affected by loss of income, greater exposure to storms, and more frequent displacement of coastal communities.

Current proposals to locate large-scale commercial aquaculture operations further away from shorelines and more directly in the ocean raise concerns about potential impacts of extending fish farms into deep-sea habitats of large fish species and ocean sanctuaries. Such ventures may complicate the challenges of making marine aquaculture sustainable and financially viable, as well as minimizing ocean pollution and the spread of diseases to natural species.

Melting Arctic Sea and Seabed Mining

The melting of Arctic sea ice and glaciers has produced another worry: the race to prospect and mine seabed mineral resources. Deep seas contain valuable metals such as silver, gold, copper, manganese, cobalt, and zinc. Russia, the United States, Japan, China, India, South Africa, and some European countries are spearheading Arctic exploration efforts to mine new mineral sources, a development which could have potential environmental impacts on marine ecosystems.

At least 19 exploratory contracts for ocean seabed mining in the high seas, covering an area of about half a million square miles, have been issued (up from zero in 2000) (Zimmer 2015 and Ziegler 2014). While international rules and regulations on deep-sea mining are contemplated in the United Nations Law of the Sea, effective monitoring of these rules in global waters is a contentious issue. The International Seabed Authority is charged with regulating deep-sea mining operations outside each nation's 200–nautical mile limit; however, increased globalization and growing competition for ocean mining make it imperative to update prevailing regulations and enforcement mechanisms.

Run on Medicinal Products from the Oceans

In addition to the race for minerals, there is a run on products from seabed and marine organisms that could be harvested for pharmaceutical, medical, and dietary reasons. The number of patents for such products has been rising 12 percent per year. Many companies are selling anticancer products derived from marine organisms, estimated to top $1 billion in sales per year (Urbina 2015); others are marketing the health benefits of krill and extolling the value of kelp and other types of seaweed as a source of human food, biofuel, and animal feed (Goodyear 2015). Overharvesting of kelp and krill is already having a negative impact on certain fish species and ocean mammals that rely on them for food.

Camille Gaskin-Reyes

REFERENCES AND FURTHER READING

Agence France-Presse. "Bikini Atoll Nuclear Test: 60 Years Later and Islands Still Unlivable." *Guardian*, March 1, 2014. http://www.theguardian.com/world/2014/mar/02/bikini-atoll-nuclear-test-60-years.

Ailing, Abigail. "Coral Reefs in Crisis." In *Oceans: The Threats to Our Seas and What You Can Do To Turn the Tide*, edited by Jon Bowermaster, 107–116. New York: Public Affairs, 2010.

Barbados Advocate. "Oceans Become More Acidic, Threatening U.S. Shellfish Industry." *Barbados Advocate*, February 25, 2015. www.barbadosadvocate.com.

Black, Maggie. *No Nonsense Guide to Water*. Oxford: New Internationalist Publications, 2004.

Bolster, W. Jeffrey. "Where Have All the Cod Gone?" *New York Times*, January 1, 2015. http://www.nytimes.com/2015/01/02/opinion/where-have-all-the-cod-gone.html.

Borger, Julian. "Marshall Islands Sues Nine Nuclear Powers Over Failure to Disarm." *Guardian*, April 24, 2014. http://www.theguardian.com/world/2014/apr/24/marshall-islands-sues-nine-nuclear-powers-failure-disarm.

Bowermaster, Jon, ed. *Oceans: The Threats to Our Seas and What You Can Do to Turn the Tide*. New York: Participant Media, 2010.

Bromwich, Jonah. "A Fitting End for the Hottest Year on Record." *New York Times*, December 23, 2015. http://www.nytimes.com/2015/12/24/science/climate-change-record-warm-year.html.

Bromwich, Jonah. "Understanding El Niño." *New York Times*, December 16, 2015. http://www.nytimes.com/2015/12/17/science/understanding-el-nino.html.

Coral Reef Alliance. "How Coral Reefs Grow." 2014. http://coral.org/coral-reefs-101/coral-reef-ecology/how-coral-reefs-grow/.

The Economist. "Governing the Oceans: The Tragedy of the High Seas." February, 22, 2014. http://www.economist.com/news/leaders/21596942-new-management-needed-planets-most-important-common-resource-tragedy-high www.economist.com/news/leaders/21596942-new-management-needed-planets-most-important-common-resource-tragedy-high.

The Economist. "Troubled Waters: A Special Report on the Seas." January 3, 2009.

Goodyear, Dana. "A New Leaf: Seaweed Could be a Miracle Food—If We Can Figure Out How to Make it Taste Good." *New Yorker*, November 2, 2015. http://www.newyorker.com/magazine/2015/11/02/a-new-leaf.

Johansen, Bruce E. "Marine Life, Fisheries, and Global Warming." In *World Geography: Understanding a Changing World*, Santa Barbara: ABC-CLIO, 2015.

Klein, Naomi. *This Changes Everything: Capitalism vs. the Climate*. New York: Simon & Schuster, 2014.

Kunzig, Robert. *The Restless Sea: Exploring the World Beneath the Waves*. New York: Norton & Company, 1999.

Matsen, Brad. "Barging Down to the Sea: Farming the Ocean without Regret." In *Oceans: The Threats to Our Seas and What You Can Do to Turn the Tide*, edited by Jon Bowermaster, 93–101. New York: Public Affairs, 2010.

Mother Jones. "The 15 Biggest Oil Spills." September/October 2010. http://www.motherjones.com/environment/2010/09/bp-ocean-15-biggest-oil-spills.

National Oceanic and Atmospheric Administration (NOAA). August 26, 2014. http://oceanexplorer.noaa.gov/about/what-we-do/oer-updates/2014/seeps-082614.html.

Newton, David E. *Encyclopedia of Water*. Westport, CT: Greenwood Press, 2003.

NOAA: Ocean Service Education. "Corals." 2014. http://oceanservice.noaa.gov/education/kits/corals/coral04_reefs.html.

NOAA, Ocean Facts. What is a Thermocline? http://oceanservice.noaa.gov/facts/thermocline.html

Novaresio, Paolo. *The Explorers: From the Ancient World to the Present*. New York: U.S. Media Holdings, 1996.

Pearce, Fred. *With Speed and Violence: Why Scientists Fear Tipping Points in Climate Change*. Boston: Beacon Press, 2007.

Ponting, Clive. *A New Green History of the World: The Environment and the Collapse of Civilizations*. New York: Penguin Group, 2008.

Prager, Ellen J. and Sylvia A. Earle, *The Oceans*. New York: McGraw-Hill, 2000.

Rirdan, Daniel. *The Blueprint: Averting Global Collapse*. Louisville, KY: Corinno Press, 2012.

Urbina, Ian. "Protecting the Untamed Seas." *New York Times*, July 31, 2015. http://www.nytimes.com/2015/08/02/sunday-review/protecting-the-untamed-seas.html.

Ziegler, Dominic. "A Sea of Expectations." In *The Economist: The World in 2014*, edited by Daniel Franklin. London: Yvonne Osman, 2014.

Zimmer, Carl. "Ocean Life Faces Mass Extinction, Broad Study Says," *New York Times*, January 15, 2015. http://www.nytimes.com/2015/01/16/science/earth/study-raises-alarm-for-health-of-ocean-life.html?login=email&mtrref=www.google.com&assetType=nyt_now.

CASE STUDIES

The following case studies address some threats to oceans, coastal environments, and island ecosystems as a result of human activities in the oil sector, tourism and fisheries, and contamination of ocean waters. They discuss the fallout of a major oil tanker accident in Prince William Sound in Alaska, the risks of unregulated tourism and overfishing to a unique island environment such as the Galapagos in Ecuador, and the serious problem of ocean contamination with small and large particles of plastic.

Case Study 1: The *Exxon Valdez* Spill

On March 24, 1989, the *Exxon Valdez*, a supertanker carrying crude oil pumped from Alaska's north shore, ran aground in Prince William Sound, Alaska. Almost 11 million gallons of oil were spilled, causing an environmental disaster, which damaged the pristine wilderness of the sound and killed thousands of birds and marine wildlife. While this disaster was a media nightmare for the oil industry, it eventually led to some positive changes in the regulation of the oil industry.

Other oil spills had occurred at sea prior to the *Exxon Valdez* accident. In fact, the large-scale shipment of oil on supertankers since the 1960s raised the risks that serious accidents were inevitable. One of the worst was also one of the first. On March 18, 1967, the *Torrey Canyon* ran aground off the coast of Cornwall in the United Kingdom. The tanker spilled 919,000 barrels of crude oil into the sea. Much of the oil dispersed before reaching land. Other accidents followed. In 1970, the tanker *Othello* lost 438,000 barrels of oil in Tralhavet Bay, Sweden. Yet another large spill occurred on

March 16, 1978, off the Breton Coast of France when the *Amoco Cadiz* ran aground and 1.6 million barrels gushed out of its tanks. The resulting oil slick eventually covered 125 miles of French coastline.

The first large oil spill in U.S. waters occurred in December 1976, when the *Argo Merchant* ran aground off Nantucket, Massachusetts. More than 180,000 barrels spilled into the Atlantic Ocean. The slick it caused was 100 miles long and 60 miles wide. Fortunately, the winds and ocean currents took most of the oil out to sea, where it was eventually dispersed.

Most of these large spills took place overseas and were in proximity to relatively well-developed coastal areas. Although they were reported by the media at the time, the American public took little notice. Even the threat in 1976 to the eastern seaboard of the United States was not regarded with great concern, since the region had suffered from widespread pollution for over a century.

Loss of Paradise?

The *Exxon Valdez* spill in the Prince William Sound differed from other spills because of the pristine location in which it occurred. It was a picturesque bay, with mountains and a temperate rainforest surrounding it. It was the home of the largest annual migration of seabirds, the largest populations of sea otters and bald eagles, and the place where millions of salmon spawned. The public reacted with anger to news of the spill due to the sense of loss that a pure and sacred environment had been spoiled by the careless actions of humans.

Alaska had first become the focus of environmental concerns when oil companies proposed to drill in the Alaskan wilderness in the late 1960s. Despite opposition to the construction of the Alaskan pipeline from environmentalists and Native Americans, among others, construction began in 1974 and was completed in 1977. Specially trained teams of workers, with equipment, were located along the pipeline to deal quickly and efficiently with spills. The pipeline's terminal was at Valdez in Prince William Sound. There, supertankers would load oil and take it to refineries in the lower 48 states. Equipment and personnel to deal with spills from the tankers were also located in the sound.

During the 1980s, safety standards declined dramatically. The crews assigned to tankers were reduced by half to an average of only 20. The specially trained crews to handle spills were assigned other duties, and much of the equipment was dispersed or needed repair. Even the U.S. Coast Guard station at Valdez was cut back. A less-powerful unit radar was installed, leaving the Coast Guard unable to track tankers as they exited Prince William Sound.

Just after 9:00 p.m. on March 23, 1989, the *Exxon Valdez* cast off to make a five-day trip to Long Beach, California. It carried 53 million gallons of crude oil. The *Exxon Valdez*'s captain was Joseph Hazelwood. Hazelwood was regarded as an accomplished captain, but had a history of alcohol abuse. He later admitted consuming alcohol before sailing that night. After conning the ship into the outward-bound lane for ships, Hazelwood turned control over to Third Mate Gregory Cousins. Cousins was not certified to command the tanker, however. Hazelwood went to his cabin, but was called back when the ship's radar detected ice in the water ahead. Hazelwood decided to turn left, into the lane used by inbound tankers. He ordered the tanker to turn, then left Cousins alone on the bridge.

Cousins steered the *Exxon Valdez* across the inbound lane, but failed to make a turn back towards the open sea. Instead, the huge tanker sailed directly for Bligh Reef, a well-known and marked hazard. Just before midnight, the lookout reported seeing the flashing red buoy that marked Bligh Reef on the right-hand side of the ship, instead of the left, where it should have been. Cousins immediately tried to turn the tanker and notified Hazelwood. The *Exxon Valdez* was two-tenths of a mile long and needed at least six-tenths of a mile to make a turn; the distance to the reef was far less.

At 12:04 a.m. on March 24, the *Exxon Valdez* struck Bligh Reef. A later investigation suggested that Hazelwood might have tried to rock the tanker off the reef, causing more damage. Crude oil began spilling into Prince William Sound as Hazelwood reported to the Coast Guard that the ship had "fetched up hard aground" and was "evidently leaking some oil." Hazelwood left the tanker and the Coast Guard was only able to test his blood-alcohol level 10 hours later. At that time, his blood-alcohol content was 0.061, higher than the permissible 0.04 limit to operate a ship. Hazelwood was later found not guilty of operating a vessel under the influence of alcohol. Nonetheless, he was found guilty of a misdemeanor for negligent discharge of oil.

Prince William Sound was an unfortunate place for a massive spill on many levels, not only because of the threat to its natural beauty, but also because of the strong currents active in the sound. The currents quickly moved the oil throughout the sound and up onto the shore, making for a difficult and costly cleanup. To make matters worse, the response to the spill from local authorities was uncoordinated and slow. Even with the help of the U.S. Coast Guard, it took 14 hours for the first barge with booms to make its way from Valdez harbor to the spill. Containment of the oil by booms was the first line of defense. The booms were intended to float on the water and keep the oil from spreading. However, because the booms were intended for the calm waters of protected harbors and not the rougher waters of the sound, they proved ineffective.

By the next morning, another Exxon tanker had taken the remaining oil off the *Exxon Valdez*, but by then a significant amount of oil had already discharged. One-fifth of the cargo, 11 million gallons, formed a slick that spread and moved to the southwest. Containment quickly became impossible, and efforts turned towards cleanup. By the fourth day, the oil had traveled 40 miles. A week later, the slick had gone another 100 miles.

Clean-Up

Methods of cleaning up the oil ranged from the most basic to the most advanced as the media quickly focused its attention on the spill. The Coast Guard and Exxon (today ExxonMobil) were under intense public pressure to ensure that as little harm as possible occurred to the environment. Many different techniques were tried to clean up the oil. Exxon put a layer of napalm-like substance on the oil and ignited it, but the oil became too dispersed for this method to be effective. Two small skimmers were used to scoop up the oil on the top of the water, but their containers soon became full with no place to unload the oil.

Three days after the spill, the Coast Guard finally authorized Exxon to use chemical dispersants to break up the oil. The effort was unsuccessful because too little dispersants were available and because the current was not strong enough to mix the dispersants with the oil. Experts were also concerned that the dispersants would increase the damage to marine life. Bioremediation was also used on some beaches. This method encourages the growth of bacteria that break down oil. However, this also proved dangerous as no one knew the effects of the bacteria on other life forms.

After several days, the crude oil mixed with ocean water to form a tar-like substance. Some of this material drifted out to sea, where it eventually sank. The rest washed onto the beaches of Prince William Sound. Exxon hired local citizens to help clean up the beaches. Television viewers saw people with what appeared to be paper towels rubbing down rocks and beaches. Other beaches were treated with hot pressurized water. However, even this effort proved to have negative effects, as the hot water destroyed certain forms of marine life. Although many beaches were cleaned, much of the oil was forced into the soil, where it remained a hazard.

Aftermath of the Spill

The disastrous effect on wildlife and the surrounding ecosystem was the most apparent problem at the time. Thousands of birds and otters were killed and the salmon and shellfish populations were severely damaged. Only several hundred were saved by cleaning and special care, although the long-term effects on these animals remain unknown.

The *Exxon Valdez* oil spill also had a great impact on the oil industry. Today, better safeguards have been instituted to minimize the effects of an oil spill. Greater safety measures have been taken in the Arctic National Wildlife Refuge, where most of the oil is produced. Environmentalists and their supporters in Congress have resisted any further opening of wilderness lands in Alaska to oil exploration, while emergency teams have been reinstated along the Alaskan pipeline. Unfortunately, these changes could not help the damage already done to Prince William Sound. The recovery of its ecosystem will continue to take many years; while such species as the bald eagle have recovered, many others, such as the salmon, remain below 1989 levels.

After the spill, Exxon was the defendant in a number of lawsuits brought by Native Americans, environmentalists, and the state of Alaska. While Exxon was quick to voluntarily pay for cleanup and offer millions of dollars in settlements to residents, the company was still fined $4.5 billion in punitive damages. ExxonMobil is still in the process of appealing this ruling. Some of the money already collected from the company has gone to programs to monitor the long-term effects of the spill and for research on how to prevent and treat future spills.

Legislation also resulted from the *Exxon Valdez* oil spill. In 1990, Congress passed the Oil Pollution Act, which stated that all oil-carrying vessels holding more than 5,000 tons and operating in U.S. waters are required to have double hulls by 2015. The law also raised corporate liability for spills and established a federal cleanup fund from a new tax on oil. The state of Alaska passed legislation requiring ships' captains to be tested for alcohol before being allowed to sail. Furthermore, tankers may not change channels and are accompanied by two tugs until they are clear of Prince William Sound.

Tim J. Watts

REFERENCES AND FURTHER READING

Davidson, Art. *In the Wake of the* Exxon Valdez: *The Devastating Impact of the Alaska Oil Spill.* San Francisco: Sierra Club Books, 1990.

Keeble, John. *Out of the Channel: The* Exxon Valdez *Oil Spill in Prince William Sound.* New York: HarperCollins, 1991.

Case Study 2: Threats to the Galapagos Islands

This case study examines threats to the biodiversity and unique ecosystem of the Galapagos Islands, such as overfishing, inadequate regulations, unsustainable tourism, and the introduction of invasive plant and animal species.

The volcanic islands of the Galapagos are famous for a wealth of unique plants and animals found nowhere else in the world. They inspired Darwin's theory of evolution, and remain an important living laboratory for scientists from all around the world. These islands have been designated as a UNESCO World Heritage Site, managed by the Ecuadorian government as Galapagos National Park (GNP).

Many external factors threaten this valuable ecosystem. There are four major risks: illegal exploitation of fish species and overfishing; introduction of invasive plants and animals; environmental impacts rising from the demands of more than 160,000 tourists each year; and inadequate regulations to address the threats. These factors threaten the islands' ecosystems and the people who depend on them for their food and livelihoods (WWF 2016).

Unsustainable Tourism

Tourism provides financial support to Galapagos National Park. The park's natural attraction and uniqueness is a huge draw for recreational and educational travel, which may advance social and economic objectives in the islands but also cause negative environmental impacts. According to the International Galapagos Tour Operators Association (IGTOA), tourism threatens the health of the islands through ocean and land contamination from boat engines and oil spills, increased stress on freshwater supply, and the introduction of invasive plants and animals from the Ecuadorean mainland and other places (IGTOA 2015).

Tourism must be kept sustainable, if the Galapagos is to maintain its environmental integrity and preserve its ecosystems. Sustainable ecotourism design and implementation can help mitigate or reduce the negative impact of traditional tourism (Powell and Ham 2008, 467). In the case of the Galapagos, however, this would require stricter regulations: limits to the number of tourists, monitoring and measuring the impact of tourists, and enforcement of the law. Given the high prices that operators demand from tourists and the lucrative nature of Galapagos tourism, it is difficult to change short-term thinking and the current focus on economic gains.

The administration of the islands may therefore have to rethink its tourism model. However, insufficient attention is placed on the carrying capacity of the ecosystem as a whole. The classic way to tour the islands is by ship. Visitors usually go on a tour for one or two weeks on ships that range from a dozen to over a hundred passengers. Tours also include excursions to various parts of the islands, where visitors can take short hikes on designated trails, and also snorkel and dive (Tourtellot 2015).

In 2012, local authorities, scientists, and tour operators made attempts to work out the timing of excursions and limits on the size of visitors. The GNP instituted regulations, which required cruises to operate on a 15-day/14-night schedule to avoid the disruption of wildlife cycles. During its 15-day timeframe, for example, a boat may not visit the same site twice, with the exception of the Charles Darwin Research Station on Santa Cruz Island (IGTOA 2015). Nonetheless, there are concerns about the enforcement of the prescribed flows.

Due to their rich biodiversity, the Galapagos Islands were the first site to be inscribed on the UNESCO World Heritage List in 1978. Some decades later, uncontrolled tourism in the Galapagos was among the 15 issues identified by the World Heritage Committee when it recommended that the Galapagos be placed on the list of World Heritage Sites in Danger (UNESCO 2015). During UNESCO's 38th Session in Doha in 2014, the World Heritage Committee urged Ecuador to step up efforts to put in place environmental protection infrastructure, and to focus on the requirements to rigorously apply international biosecurity standards for cargo ships (UNESCO).

Introduction of Invasive Species

According to the Charles Darwin Foundation, visitors to the Galapagos Islands are leaving footprints containing invasive species from their previous destinations. Invasive plants and nonlocal animals have had devastating impacts on the fragile ecosystems in the archipelago. The most destructive animals in the Galapagos are goats, which were brought to the islands in the 1850s by whalers, needing to stock alternative sources of meat for their survival. Goats adapted to survive in both the dry lowlands and the humid highlands of the islands. They consume food that giant tortoises need to live on (IGTOA 2015).

Goats overgrazed the islands, causing erosion and threatening the survival of rare plants and trees, and competing with native fauna for food (Marris 2009). The Charles Darwin Research Station started its first systematic program in 1965 to rid Santa Fe Island of goats. Ten years later, the last goat was culled on Santa Fe, which allowed that island's plant life to recover its lost richness and native iguanas to thrive again (IGTOA 2015).

Other alien species currently threatening the Galapagos ecosystems are certain types of rats, cats, cattle, dogs, donkeys, horses, and pigs. More than 700 species of invasive plants are recorded on the Galapagos. The quinine tree, guava, and elephant grass are just a few examples. Unknown numbers of nonnative invertebrates such as fire ants also cause negative impacts (SPREP).

To control the introduction of new species, the Galapagos Inspection and Quarantine System (SICGAL) was established in 2000. SICGAL's objective is to prevent new species and organisms from being introduced into the Galapagos. It monitors ports of entry and agricultural zones on the inhabited islands, utilizes protocols to fumigate incoming planes and boats, provides training to inspectors and technicians, and disseminates lists of permitted and prohibited products (IGTOA 2015).

According to the nonprofit Galapagos Conservancy, one of the most pressing challenges remains the control of invasive species. Feasibility studies are being conducted on the use of biological controls to deal with introduced ants, wasps, and mosquitos that can potentially carry the West Nile virus. Methods are also being developed to control parasitic flies that endanger their local host birds. There are ongoing attempts to eradicate fire ants from the larger islands and priority small islands. Eradicating introduced species and keeping new ones from arriving appears to be a never-ending task and a challenge for regulators.

Illegal Fisheries and Overfishing

The Galapagos archipelago has a rich marine ecosystem, nurtured by a confluence of ocean currents. This ecosystem supports all terrestrial and marine life on the islands, but a major threat remains illegal fisheries and overfishing.

When locals cannot find work in tourism, a common option for them is to join the fishing industry. Sea cucumbers and shark fins are targets for exports to Asia, popular for their alleged aphrodisiac qualities. Due to the alarming decrease of sea cucumbers in the early 1990s, the Galapagos National Park Service banned all sea cucumber fishing in the islands until a few years ago. The drastic decline in their numbers has reduced some fishing activity, although illegal fishing still continues.

Seeking to provide a livelihood to residents of the Galapagos, the majority of whom fish for a living, the government allowed locals to collect a maximum of 500,000 sea cucumbers during a 45-day period in 2015. After the harvest, a five-year recovery period was started, where no sea cucumbers could be fished (Alvear and Lewis 2015). However, environmentalists predict that the sea cucumber population in the Galapagos is unlikely to recuperate after the previous periods of such intensive harvesting. Ships from other countries routinely enter the marine reserve illegally in search of sea cucumber catches (IGTOA 2015).

Overfishing has also weakened the marine ecosystem. Illegal fishing has destroyed coral reefs in the archipelago, many of which had persisted for hundreds of years. Sharks are a popular illegal catch. While some sharks in Ecuador are caught and sold for meat, most are targeted for their fins, which are exported to Asia.

Industrial shark harvesting began in the early 1950s. Although the islands were designated a marine reserve in 1986, this decree did not confer the protected area status that would have prohibited shark fishing. After concerns from scientists, the government passed a law in 1993 that required all sharks to be landed with their fins intact. The result was that Ecuadorian fishermen were not allowed to target sharks specifically. However, it was very common to continue the export of fins from an "incidental catch" (Carr 2013).

The above-mentioned threats continue to plague the fragile environment of the Galapagos. There are potential solutions, such as stricter regulations, environmental education, assistance to the inhabitants of the islands to seek economic alternatives, and programs for tourist operators and tourists to understand the need for conservation of this extraordinary ecosystem.

More importantly, the solutions are in the hands of the Ecuadorean people and the authorities to preserve this valuable resource that played an important role in the genesis of the theory of evolution. Further action requires greater attention to enforcement mechanisms and alternative, more sustainable fishing models to provide a more effective path to ecosystem management in the Galapagos.

Andrea Arzaba

REFERENCES AND FURTHER READING

Alvear, Cecilia, and George Lewis. "Nuevo Conflicto Sobre Pepinos de Mar en Galapagos." Galapagos Digital, July 29, 2015. http://www.galapagosdigital.com/espanol/2015/07/29/nuevo-conflicto-sobre-pepinos-de-mar-en-galapagos/.

Carr, Lindsey A. et al. "Illegal Shark Fishing in the Galapagos Marine Reserve," *Marine Policy* 39 (May 2013): 317–321.

Charles Darwin Foundation. "Invasive Plants." http://www.darwinfoundation.org/en/science-research/invasive-species/invasive-plants/.

Galapagos Conservancy. "Invasive Species." 2016. http://www.galapagos.org/conservation/conservation/conservationchallenges/invasive-species/.

Galapagos Conservancy. "Poaching: Marine and Terrestrial Species." 2016. http://www.galapagos.org/conservation/conservation/conservationchallenges/poaching-and-illegal-fishing/.

IGTOA. "Challenges Facing the Galapagos Islands." 2016. http://www.igtoa.org/travel_guide/challenges.

Marris, Emma. "Goodbye Galapagos Goats." *Nature*. January 27, 2009. http://www.nature.com/news/2009/090127/full/news.2009.61.html.

Powell, R. B., and S. H. Ham. "Can Ecotourism Interpretation Really Lead to Pro-conservation Knowledge, Attitudes, and Behaviour? Evidence from the Galapagos Islands." *Journal of Sustainable Tourism* 16, no. 4 (2008): 467–489.

South Pacific Regional Environment Programme (SPREP). "Invasive Species in the Pacific: A Technical Review and Draft Regional Strategy." See http://sprep.org/Roundtable/documents/InvasivesSppRegional.pdf and http://www.issg.org/database/reference/invasive_strategy_and_species.pdf.

Tourtellot, Jonathan. "Galapagos Tourism Backfires." *National Geographic*, January 5, 2015. http://voices.nationalgeographic.com/2015/01/05/galapagos-tourism-backfires/.

UNESCO. "Ecuador Intensifies Actions to Reduce Risks from Ship Stranding in Galapagos Islands." February 6, 2015. http://whc.unesco.org/en/news/1230/.

UNESCO. The 38th Session of the World Heritage Committee at the Qatar National Convention Centre, Doha. June 15–25, 2014. http://whc.unesco.org/en/sessions/38COM/.

World Wildlife Fund. "The Galapagos." 2016. http://www.worldwildlife.org/places/the-galapagos.

Case Study 3: Plastic Contamination of the Ocean: The Great Pacific Garbage Patch

This case study focuses on the serious problem of ocean contamination with plastic. Plastic debris in coastal areas and the high seas ranges from large-scale plastic items to micro-particles dumped into the sea or inadvertently swept out from the land. These articles litter the sea and entangle marine organisms, and endanger marine organisms when ingested.

Plastic Debris Is a Multifaceted Problem

The problem of plastic debris in the Pacific and other oceans of the world is multifaceted. First, there is the wider context of source pollution to consider, i.e., the origin of most of the contamination is related to improper solid waste management on the land. Then, there is the problem of plastic debris arising from human activities on the high seas. These include items from cruise ships and passenger ferries, container ships, oil tankers, buoys, fishing equipment, and debris from oil drilling and mining industries.

Further, there is the issue of inadequate regulations and legal enforcement, on land and sea, to reduce or prevent dumping. Finally, there is the problem of how to tackle removal of plastic debris already in the oceans and the sea floor, since it affects marine organisms and the human food chain.

These problems are pervasive; there is really no part of any ocean on the planet that is not affected. We live in a plastic-filled world on land, and increasingly we are creating a plastic world in the oceans. In the most remote islands, on distant coastlines, and in vast expanses of the oceans, there are large quantities of floating plastic debris or articles washed up on the shore, brought there by ocean currents. It is estimated that by 2025 the amount of plastic entering the oceans will amount to about ten plastic bags per foot of coastline on the planet (Schwartz 2015).

A study carried out from 2007 to 2013 estimated that there are approximately 269,000 tons of mostly nonbiodegradable plastic items in the world's oceans, amounting to over 5 trillion individual pieces of plastic of all sizes with already-dangerous consequences for marine life through entanglement or ingestion of tiny, toxic microplastics. The latter are eventually passed up the food chain to humans (Casey 2014).

Some of the huge quantities of plastic in the oceans are concentrated in special areas or gyres, which are giant, rotating high-pressure zones—currently filled with swirling plastic debris—in five subtropical areas of the world. These areas now cover about 45 percent of the ocean's surface (Casey 2010). A well-known gyre, called the Great Pacific Garbage Patch, is located in the northern Pacific Ocean. This floating plastic debris is the visible symbol of humankind's disposable-oriented, consumer society on the land.

The Rise of Plastic

In the 1950s, plastic was invented. It is a petroleum-based substance that at the time seemed like a miracle product: it was a cheap, versatile, convenient, useful-yet-durable item with many benefits and uses, i.e., for medicine; engineering; computer science; games and sports items, including the Frisbee; bottles; disposable containers for food and household products; and a range of everyday items too numerous to mention.

However, as the use of plastic became more ubiquitous, the problem of plastic pollution became more widespread. With the spectacular rise of the bottled water and beverage industries, and society's rising demand for practical, disposable consumer products, plastic garbage mounds started making their way to landfills and dumping areas on land, and from there into rivers, canals, and waterways, onto boats and ships, and further into seas and oceans.

It is estimated that the average American tosses away about 185 pounds of plastic per year, in the form of food containers, plastic wraps and bags, toys and baby products (pacifiers, disposable diapers, rubber ducks, etc.), packaging material, PVC pipes, furniture, and large household and sports items—even kayaks (Casey 2010, 80). The growth of the bottled water industry and prepackaged or frozen food (including takeout and fast food), and the expansion of sports drinks and the beverage and pharmaceutical industries played a key role in the increase of plastic containers and packaging.

In recent years, however, due to changing consumer tastes and environmental awareness, there has been a growing switch from plastic to paper and cardboard packaging as well as the use of newer biodegradable starch and corn-based materials. Nonetheless, industries churn out more than 60 billion tons of plastic every year in many areas, particularly beverage containers (Casey 2010, 78). Americans consume about 70 million bottles of water per day. Worldwide consumption of bottled water reached over 41 billion gallons in 2004 (de Rothschild 2010, 87).

Plastic in the Ocean

It is estimated that 80 percent of marine plastic in the oceans has its origin in the land (de Rothschild 2010, 86). Of all the marine debris floating in the ocean or on the seabed, about 90 percent is made up of plastic, with the remainder consisting of timber, glass, and other materials that are medical-, chemical-, sewage-, and industry-related by-products (de Rothschild 2010, 84).

Scientists and researchers have major concerns about plastics, namely the nonbiodegradable nature of the material and its molecular structure, which attracts pollutants. Further, there are concerns about marine animals dying when they ingest or become entangled in larger plastic objects, as well as the toxic impacts of the ingestion of microparticles of plastic (called nurdles) on marine organisms and on humans within the food chain.

These nurdles are lentil-sized (or even smaller) plastic pellets; they can also be even tinier specks of dust or particles smaller than grains of rice, which become airborne during the plastic manufacturing process. They blow around like dust, which humans breathe in and become ill. These microparticles remain on the land or are blown or washed into storm drains, creek harbors, estuaries, and oceans.

Along their path to the sea, nurdles attract waste chemicals or persistent organic pollutants (POPs), which latch onto the plastic's molecular structure. They also accumulate on beaches throughout the world, finely mixed in with sand. Microplastics make up about 10 percent of all plastic marine debris and eventually kill live organisms such as coral reefs by attacking their digestive systems when ingested (Casey 2010, 78).

More than 260 ocean species are already impacted by plastic debris such as rope, netting and packaging products, either through entanglement or ingestion (Thompson et al. 2011, 9). Of the 120 marine mammals listed on the International Union for the Conservation of Nature's (IUCN) endangered list, 54 percent are known to have been entangled in or to have eaten plastic. Scientists estimate that over 90 percent of dead birds washed ashore on Brazil's southeast coast in 2011 contained plastic in their gut (Thompson et al. 2011, 9).

Solutions

The challenge remains how to best address the plastic debris problem and how to build consensus around the world to reduce plastic contamination of the oceans. Some attention has already been focused on the problem of the floating swirls of plastic in the gyres, and many creative ideas have been generated. However, so far there has been no concerted global action to deal with the problem in an integrated way, and even less attention to addressing the source of the problem: pollution on land.

Approaches to addressing plastic garbage have to be as multifaceted and complex as the problem itself. The solution requires huge shifts in the unsustainable production and consumption patterns of modern society and the reduction of plastic in manufacturing and daily use, as well as the shift towards more biodegradable, recycled, or recyclable products throughout industries and services currently using plastic products.

Solutions should also encompass the establishment and enforcement of improved solid waste-management policies on land, stricter international regulations on ocean dumping, the setting up of marine protected areas, fish and marine mammal refuges, the protection and management of vulnerable coastal and marine ecosystems, global agreements to clean up gyres, and creative solutions to re-use or recycle floating plastic, once recovered.

To raise awareness of the global plastic problem, a crew of concerned citizens used plastic bottles to construct a 60-foot oceangoing boat called *Plastiki* (a play on *Kon-Tiki*) to sail from California to Australia, to draw attention to the Great Plastic Garbage Patch. The *Plastiki* experience sent a message about the excesses of plastic in daily life and challenged the world's nations and citizens to come up with lasting solutions for this global, human-made problem—beyond recycling of plastic bottles for fleece jackets and carpets. Making and sailing a boat out of plastic bottles across a global ocean was also a statement about the creativity of the human spirit, meant to inspire others to understand and respect the web of life and the value of our global common resources.

Camille Gaskin-Reyes

REFERENCES AND FURTHER READING

Casey, Michael. "World's Oceans 'Plagued' by 269,000 Tons of Plastic Pollution." CBS News, December 11, 2014. http://www.cbsnews.com/news/worlds-oceans-plagued-by-269000-tons-of-plastic-pollution/.

Casey, Susan. "Plastic Ocean: Our Oceans Are Turning into Plastic. . .Are We?" In *Oceans: The Threats to Our Seas and What You Can Do to Turn the Tide*, edited by Jon Bowermaster, 71–82. New York: Public Affairs, 2010.

de Rothschild, David. "Message in a Bottle: Adventures Aboard the *Plastiki*," in *Oceans: The Threats to Our Seas and What You Can Do to Turn the Tide*, edited by Jon Bowermaster, 83–91. New York: Public Affairs, 2010.

Schwartz, John. "Study Finds Rising Levels of Plastics in Oceans." *New York Times*, February 13, 2015.

Scientific and Technical Advisory Panel (STAP). *Marine Debris as a Global Environmental Problem: Introducing a Solutions Based Framework Focused on Plastic*. A STAP Information Document. Washington, DC: Global Environment Facility, 2011.

ANNOTATED DOCUMENT

United Nations Convention on the Law of the Sea: Part V: Exclusive Economic Zone

Background

The United Nations Convention on the Law of the Sea (UNCLOS) was signed in 1982 but only came into force in 1994 after ratification by 60 countries. It represents a unified legal framework on paper to govern peaceful use of the world's oceans, protect marine ecosystems, foster research, and promote equitable utilization of resources. As previously discussed, the reality is that, in spite of the comprehensive nature of the convention, its monitoring and enforcement mechanisms are weak in the wide open areas of the high seas. The high seas belong to everyone, but no one is in charge.

Summary

The convention includes a preamble, 17 parts, and 10 annexes. The preamble outlines the rationale and intent of the Law and the establishment of general principles: the need for international cooperation to use and protect a global public good such as the oceans; and the imperative to maintain peace, justice, and progress on the high seas. The preamble also establishes the rights of individual states to use ocean areas close to their shorelines and the rights of landlocked nations to transit through other countries to access the world's oceans.

The 17 parts cover a range of areas including protection and preservation of the marine environment; marine scientific research and development and transfer of marine technology; action plans and cooperation mechanisms on regional seas; the rights, responsibilities, and obligations of nations to use the oceans; and the definition of territorial limits. They set forth the rights and responsibilities of governmental, merchant, and war ships; the rights and obligations of ships passing close to coastal nations; and the rights and special needs of developing countries.

This section focuses on the specific articles that define the limits of countries' territorial seas and contiguous zones, known as exclusive economic zones (EEZs). It presents comments on the main concepts and rules governing activities of countries in territorial waters and the EEZs. It also outlines transportation and shipping regulations, criminal matters, and the settlement of conflicts in territorial waters.

Innocent Passage

Innocent passage is an important concept within UNCLOS. While a sovereign state is obliged to grant a foreign ship passage through its 12-mile territorial limit, the passage of that foreign ship must be innocent passage. Innocent passage means that any ship passing through territorial waters of another nation does so in good faith, and may not engage in weapons testing, spying, smuggling, pollution, fishing, or unauthorized scientific research. Where territorial waters include waterways used by international ships for trade, i.e., the Straits of Gibraltar, Malacca, or Hormuz, foreign ships transiting these channels have navigational rights, but they have to obey the innocent passage rule.

Territorial Limits and the EEZ

According to the provisions of the Law, the territorial waters of a country usually extend to 12 nautical miles beyond its coast. Outside the territorial limits of 12 nautical miles, every coastal country has the right to establish an EEZ up to 200 nautical miles from its shoreline. Within the EEZ, a state may use and regulate fish resources, build installations, generate energy from waves, and regulate scientific research of foreign ships. The Law also addresses the obligation to protect fish stocks within a country's exclusive economic zone, and outlines conflict resolution mechanisms and the attribution of rights and jurisdictional responsibilities in the EEZ.

Rights of Island Nations and Special Situations

The convention also covers the rights of island nations, and the special situations of archipelagic states and nations located on enclosed or semienclosed seas. It outlines principles to encourage the cooperation of states bordering enclosed or semienclosed seas, and ways to address the situation of landlocked states without ocean access or their own EEZ. It establishes the limits of the continental shelf, the rights of coastal states over the continental shelf, and regulations regarding the construction of artificial islands and installations in the EEZ as well as oil-drilling activities on the continental shelf.

The Law clarifies the legal status of archipelagic states (which may not have wide expanses in their individual EEZ) and their rights to the use of the seabed and ocean subsoil. It also upholds the right of innocent passage through archipelagic sea-lanes and the obligations of ships during such passage. It reiterates the rights of landlocked states to uninhibited passage through archipelagos and other areas to achieve access to oceans.

Regulation of Transportation of Dangerous Substances

In addition, many of the convention's articles regulate the transportation of dangerous substances in the EEZ and the high seas, and the charges to be levied on foreign ships operating commercially. They outline the rules regarding freedom of high seas routes for all countries and the obligations of all nations to maintain peaceful use of the high seas. Some of the articles establish traffic safety separation rules in international straits and the use of navigational and safety aids in shipping. The Law defines specific rules for ships flagged under individual nationalities as well as for ships travelling under the United Nations flag, in addition to provisions made to uphold traditional fishing rights and rules on compensation when ships damage submarine cables or pipelines.

Criminal Matters

On criminal matters, the Law defines criminal and civil jurisdictions regarding many areas. These include the collision of ships or other navigation incidents within the EEZ; the transport of slaves and piracy; noncompliance of warships with laws of coastal states; illicit traffic in narcotic drugs; and unauthorized high sea broadcasting. It also outlines the obligation of ships to render assistance and cooperate with each other in the repression of piracy in the EEZ and the open seas.

Control of Pollution

The Law outlines provisions for prevention, reduction, and control of marine pollution, the conservation of living marine resources, and scientific research and technology transfer. While promoting conservation, the Law also affirms the right of nations to fish on the high seas and their global responsibilities to preserve humankind's collective fish resources. It also establishes the right to peaceful and sustainable sharing of seabed and mineral resources that lie beyond territorial waters, and outlines nations' liability for damage to marine ecosystems.

Settlement of Disputes

While the Law places the responsibility for adhering to the rules squarely on countries and ocean users, it recognizes that conflicts may occur. The default mechanism for dispute settlement on ocean matters is arbitration. A state may choose among the following: The International Tribunal for the Law of the Sea; the International Court of Justice in The Hague; ad hoc arbitration; or Special Arbitral Tribunals on special topics. These are binding arbitration and conciliation procedures. The convention establishes consequences for the failure of states to submit to arbitration or abide by Tribunal decisions.

Camille Gaskin-Reyes

Selected Excerpts from the Law of the Sea Regarding the EEZ

Article 55: Specific Legal Regime of the Exclusive Economic Zone

The exclusive economic zone is an area beyond and adjacent to the territorial sea, subject to the specific legal regime established in this Part, under which the rights and jurisdiction of the coastal State and the rights and freedoms of other States are governed by the relevant provisions of this Convention.

Article 56: Rights, Jurisdiction and Duties of the Coastal State in the Exclusive Economic Zone

1. In the exclusive economic zone, the coastal State has:

(a) sovereign rights for the purpose of exploring and exploiting, conserving and managing the natural resources, whether living or non-living, of the waters superjacent to the seabed and of the seabed and its subsoil, and with regard to other activities for the economic exploitation and exploration of the zone, such as the production of energy from the water, currents, and winds;

(b) jurisdiction as provided for in the relevant provisions of this Convention with regard to:

(i) the establishment and use of artificial islands and installations;
(ii) marine scientific research;
(iii) the protection and preservation of the marine environment;

(c) other rights and duties provided for in this Convention.

2. In exercising its rights and performing its duties under this Convention in the exclusive economic zone, the coastal State shall have due regard to the rights and duties of other States and shall act in a manner compatible with the provisions of this Convention.

3. The rights set out in this article with respect to the seabed and subsoil shall be exercised in accordance with Part VI.

Article 57: Breadth of the Exclusive Economic Zone

The exclusive economic zone shall not extend beyond 200 nautical miles from the baselines from which the breadth of the territorial sea is measured.

Article 58: Rights and Duties of Other States in the Exclusive Economic Zone

1. In the exclusive economic zone, all States, whether coastal or land-locked, enjoy, subject to the relevant provisions of this Convention, the freedoms referred to in Article 87 of navigation and over-flight and of the laying of submarine cables and pipelines, and other internationally lawful uses of the sea related to these freedoms, such as those associated with the operation of ships, aircraft, and submarine cables and pipelines, and compatible with the other provisions of this Convention.

2. Articles 88 to 115 and other pertinent rules of international law apply to the exclusive economic zone in so far as they are not incompatible with this Part.

3. In exercising their rights and performing their duties under this Convention in the exclusive economic zone, States shall have due regard to the rights and duties of the coastal State and shall comply with the laws and regulations adopted by the coastal State in accordance with the provisions of this Convention and other rules of international law in so far as they are not incompatible with this Part.

Article 59: Basis for the Resolution of Conflicts Regarding the Attribution of Rights and Jurisdiction in the Exclusive Economic Zone

In cases where this Convention does not attribute rights or jurisdiction to the coastal State or to other States within the exclusive economic zone, and a conflict arises between the interests of the coastal State and any other State or States, the conflict should be resolved on the basis of equity and in the light of all the relevant circumstances, taking into account the respective importance of the interests involved to the parties as well as to the international community as a whole.

Article 61: Conservation of the Living Resources

1. The coastal State shall determine the allowable catch of the living resources in its exclusive economic zone.

2. The coastal State, taking into account the best scientific evidence available to it, shall ensure through proper conservation and management measures that the maintenance of the living resources in the exclusive economic zone is not endangered by over-exploitation. As appropriate, the coastal State and competent international organizations, whether sub-regional, regional, or global, shall cooperate to this end.

3. Such measures shall also be designed to maintain or restore populations of harvested species at levels which can produce the maximum sustainable yield, as qualified by relevant environmental and economic factors, including the economic needs of coastal fishing communities and the special requirements of developing States, and taking into account fishing patterns, the interdependence of stocks, and any generally recommended international minimum standards, whether sub-regional, regional, or global.

4. In taking such measures the coastal State shall take into consideration the effects on species associated with or dependent upon harvested species with a view to maintaining or restoring populations of such associated or dependent

species above levels at which their reproduction may become seriously threatened.

5. Available scientific information, catch and fishing effort statistics, and other data relevant to the conservation of fish stocks shall be contributed and exchanged on a regular basis through competent international organizations, whether sub-regional, regional, or global, where appropriate and with participation by all States concerned, including States whose nationals are allowed to fish in the exclusive economic zone.

Section 3. Development of Resources of the Area

Article 150: Policies Relating to Activities in the Area

Activities in the Area shall, as specifically provided for in this Part, be carried out in such a manner as to foster healthy development of the world economy and balanced growth of international trade, and to promote international cooperation for the overall development of all countries, especially developing States, and with a view to ensuring:

(a) development of the resources of the area;

(b) orderly, safe, and rational management of the resources of the Area, including the efficient conduct of activities in the Area, and, in accordance with sound principles of conservation, the avoidance of unnecessary waste;

(c) expansion of opportunities for participation in such activities consistent in particular with articles 144 and 148;

(d) participation in revenues by the Authority and the transfer of technology to the Enterprise and developing States as provided for in this Convention;

(e) increased availability of the minerals derived from the Area as needed in conjunction with minerals derived from other sources, to ensure supplies to consumers of such minerals;

(f) promotion of just and stable prices remunerative to producers and fair to consumers for minerals derived both from the Area and from other sources, and the promotion of long-term equilibrium between supply and demand;

(g) enhancement of opportunities for all States Parties, irrespective of their social and economic systems or geographical location, to participate in the development of the resources of the Area and the prevention of monopolization of activities in the Area;

(h) protection of developing countries from adverse effects on their economies or on their export earnings resulting from a reduction in the price of an affected mineral, or in the volume of exports of that mineral, to the extent that such reduction is caused by activities in the Area, as provided in article 151;

(i) development of the common heritage for the benefit of mankind as a whole; and

(j) conditions of access to markets for the imports of minerals produced from the resources of the Area and for imports of commodities produced from such minerals shall not be more favorable than the most favorable applied to imports from other sources.

Source: United Nations Convention on the Law of the Sea: Part V: Exclusive Economic Zone. November 16, 1994. United Nations Treaty Series, Vol 1833, No. 31363. http://www.un.org/depts/los/convention_agreements/texts/unclos/unclos_e.pdf. (c) 1994 United Nations. Reprinted with the permission of the United Nations.

PERSPECTIVES

Will the Drilling of Oil in the Gulf of Mexico Be a Source of Regional Conflict between Domestic Stakeholders in the United States?

Overview

The Gulf of Mexico laps the coast of the southeastern United States. At 615,000 square miles, it is the ninth largest body of water in the world. The gulf is rich in two very lucrative commodities: seafood and oil. With fishermen, oilmen, and the tourist industry, environmentalists, and local and state governments all having different interests at stake in the same place, it has become a hotbed for conflict in recent years as oil has come to dominate, delivering both dollars and destruction to the region.

The roots of offshore oil in the gulf can be traced to 1938, when the first freestanding drilling platform there was constructed. Standing one mile off the shore of Louisiana in 14 feet of water, Superior-Pure State No. 1 successfully began pumping oil out of a subsea well until a hurricane destroyed the platform (Cavnar, 2010). This accident served as a warning sign of the risks of natural disasters posed by operating in the subtropical climate of the gulf. Undeterred, the company Brown & Root rebuilt their platform, which eventually produced a total of 4 million barrels of oil from 10 wells.

The real watershed moment for offshore drilling in the Gulf of Mexico came in 1947, when Brown & Root built a platform 10 miles offshore in 20 feet of water—the first time that a platform was built out of sight of land. This bold feat of engineering marked the start of a new era for oil and gas production in America, one that would make offshore resources an integral part of the nation's energy supply.

In 2003, gulf federal offshore oil production accounted for 25 percent of the nation's total, although that number fell to 17 percent by 2015 (U.S. Energy Information Administration) largely due to the shale gas boom in the Bakken Shale in North Dakota. As of June 2015, there were 2,388 active platforms in the gulf; 2,321 of those are in depths of 0 to 200 meters, while the remaining 67 are located in what is called deep-water, i.e., 4,000 to 7,000 feet (Bureau of Safety and Environmental Enforcement 2015). This is where the real frontier of oil and gas drilling lies, for as shallower wells dry up, oil companies are forced to design new technologies to extract more challenging, deeper, and farther offshore wells.

The proliferation of drilling platforms and wells on the continental shelf in the gulf has not come without conflict. In April 2010, the explosion of the *Deepwater Horizon* drilling rig, which resulted in 4.9 million barrels of oil spilling into the gulf before the well was capped, focused national attention on the hazards of deepwater drilling and spurred the Department of the Interior to enact a six-month moratorium on all deepwater offshore drilling. Yet even before what one journalist deemed "the world's biggest accidental oil spill" (Chediak 2010), local fishermen, environmentalists, and state governments tussled with the oil industry and the federal government over resource exploitation in the sensitive marine ecosystem.

Fisheries

The Gulf of Mexico is one of the most productive fisheries in the world. In 2011, fishermen captured 1.8 billion pounds of finfish and shellfish, which raked in $818 million (NOAA 2011). The fish, shrimp, and oysters from its waters are sold in New Orleans' famous restaurants and to consumers all around the country. Concerns over the health of the fisheries and the quality of the seafood, however, have fomented since the early days of offshore oil. In 1970, Louisiana oyster fishermen and shrimp fishermen brought lawsuits against Chevron after one of the company's wells blew out and caught fire. The *Deepwater Horizon* disaster wrecked many fishermen's intakes that year, with British Petroleum having to pay out billions of dollars in compensation.

Environmentalists

Environmentalists have been some of the harshest critics of the offshore oil industry in the Gulf of Mexico for decades. Except for the abandoned oil platforms, called "Idle Iron," that sometimes turn into flourishing coral reefs, environmentalists and oil companies find little to agree on. *Deepwater Horizon* catalyzed greens into even more strident criticism of the industry. The oil slick from the disaster polluted large areas of the gulf's wetlands, which total about half of the country's entire wetlands area (UNEP 2010). Pelicans, turtles, dolphins, and many other marine species are still recovering. Yet given government and corporate plans to expand drilling in the gulf, environmentalists have refocused their efforts on preventing any new offshore drilling areas from opening up in the Atlantic and the Arctic, where President Barack Obama ended a longstanding ban on offshore drilling in March 2010 (Neuman 2010).

State Governments

Almost all drilling in the Gulf of Mexico takes place in federal waters, as state waters only extend three miles out from their coastlines. This difference in jurisdiction has created tensions between the states and the federal government, especially when the state has traditionally borne the burden of drilling's negative effects without seeing much in the way of direct revenues.

That began to change in 2006, when Congress passed the Gulf of Mexico Energy Security Act. This splits oil and gas revenues between the federal government and the four oil-producing Gulf States of Louisiana, Alabama, Mississippi, and Texas. Under the act, the states receive 37.5 percent of the revenues from new leases (and beginning in 2017, leases issued since 2006) issued in their offshore areas beginning in 2017 (Bureau of Ocean Energy Management). Environmentalists were dismayed at the act's passage, however, for they worry that states will now be more pro-drilling—a clear sign that in the gulf, nothing can please everyone.

Conflicting Horizons Ahead

Although the gulf has many stakeholders, it is tough for fishermen, environmentalists, and state representatives to get their voices heard. That their views largely fall on deaf ears is due to the sheer economic weight and inertia of the oil industry. No industry can compete with the dollars raked in by oil and gas: $75 billion in revenue in 2007 and $9 billion in royalties to the U.S. government (Hayward 2012).

Further, three of the top 10 proven remaining reserves of oil in the U.S. are located in the gulf, meaning that the province will stay in Big Oil's sights for many years to come. Shell is also planning to drill the first offshore wells in the Alaskan Arctic this summer after years of false starts. With the offshore frontier expanding in the U.S., it appears likely that the same fights that have riven the gulf since 1947 will soon be expanding to other coastal areas in America.

Mia Bennett

Perspective 1: The Simmering Conflict in the Gulf of Mexico

The Gulf of Mexico is sometimes referred to as the United States' Third Coast, making it seem like a thoroughly American body of water. Yet the gulf also bears another moniker: the Mediterranean of the Americas. That's because the 615,000-square mile ocean basin washes the shores of Mexico and Cuba, too, making it an international body of water. A huge amount of oil drilling has taken place in the gulf since 1947, largely in America's exclusive economic zone but also with drilling in Mexico's offshore areas and exploration in Cuba's waters, too.

Since the drilling industry took off in 1947, it has avoided setting off any major regional conflicts in the gulf. The relatively peaceful waters draw a contrast with other bodies of water around the world like the South China Sea. China's expansion of its offshore oil fleet complete with ships, rigs, and a dedicated coast guard (Yep and Ma 2014) have angered neighboring countries like the Philippines and Vietnam that also have territorial claims to the same area. Fortunately, the gulf does not seem to suffer from the aggressions and saber rattling that characterize the South China Sea.

Still, the trilateral relationship between the United States, Mexico, and Cuba has been fraught with tension at various points over the past century, most notably between the United States and Cuba during the Cold War. Relations between the United States and Cuba have recently improved thanks to a rapprochement between Washington, D.C. and Havana. Yet a decades-old border conflict still persists. In the gulf, there are two submerged areas of continental shelf that extend past the 200-nautical mile limit as defined by the United Nations Convention on the Law of the Sea. While the United States and Mexico delimited the so-called "Western Gap" by treaty in June 2000, the "Eastern Gap," where the exclusive economic zones of the United States, Cuba, and Mexico overlap, still has not been resolved.

Warmer ties between the United States and Cuba, however, may open the door to border delimitation and ultimately the expansion of U.S. oil and gas companies' activities to Cuba's offshore. In December 2014, the White House expressed, "The United States is prepared to invite the governments of Cuba and Mexico to discuss shared maritime boundaries in the Gulf of Mexico" (Office of the Press Secretary 2014). Of course, if Cuba's industry takes off, this could come back to haunt the United States. Some in Florida have expressed concern about the potential for oil and gas drilling to occur just 50 miles away from their southern coastline (Omestad 2009).

Generally speaking, however, the three countries' desire to exploit their oil and gas reserves could eventually facilitate the resolution of the Eastern Gap territorial conflict. Rather than cause regional conflict, then, oil drilling, or at least dreams of it, could reduce it. This is precisely what happened in the Western Gap, where U.S. oil corporations and pro-oil Congressmen spoke in favor of ratifying the boundary delimitation treaty between the United States and Mexico.

They were considerably more eager to delimit the border than their Mexico counterparts for a number of reasons. First, U.S. offshore drilling in the gulf had been expanding southward over the past several decades and was now approaching the unexploited Western Gap. Second, unlike Mexico, the United States had already advanced significantly ahead in terms of the technology needed to drill this far offshore, meaning that its companies would be able to benefit right away from border delimitation and begin drilling.

Unsurprisingly, U.S.-based petroleum institutes and councils issued a statement during congressional deliberations underscoring that ratification would provide territorial stability that would allow sound commercial planning to move forward. Former Senator Frank Murkowski (R-Alaska), then chairman of the Senate Committee on Energy and Natural Resources and himself a representative of a big oil producing state, urged that ratification would facilitate "the orderly acquisition and development of oil and gas leases along the U.S. side of the international line" (Duff).

But even south of that international line, oil and gas development may benefit from the border delimitation and an additional agreement that has since been passed. In 2013, the U.S. Transboundary Hydrocarbons Agreement was enacted, establishing a framework for the development of transboundary reservoirs. It is important to remember that even though borders neatly divide land, reservoirs underneath them in the seabed can span those arbitrary lines, which is why further regulations are necessary.

The transboundary agreement has ended the moratorium on oil and gas development in the Western Gap and has made an estimated 172 million barrels of oil and 304 billion cubic feet of natural gas accessible (Department of the Interior 2013). U.S.

offshore oil and gas companies and Mexico's Petróleos Mexicanos (Pemex), the country's state-owned oil and gas company, will now be able to partner to develop these deepwater resources. Mexican law had also previously prohibited foreign investment in its energy industry, meaning that the agreement was a breakthrough in a number of ways.

While international conflict in the gulf may be at a minimum, conflict within each country still exists between its multiple stakeholders. Farther south, in 2013, the government of Honduras leased oil and gas exploration rights to a British multinational firm, BG Group, in a 35,000 square kilometer block in the Caribbean. Indigenous peoples in Honduras are not as organized as they are in nearby Belize and Costa Rica, where national-scale alliances have fought the oil companies. As a result, indigenous Hondurans were not well equipped to put up a united front against the oil lobby (Cuffe 2014).

The ease with which information travels might one day enable transnational networks of fishermen, environmentalists, and other stakeholders opposed to Big Oil. This type of regional conflict would transcend typical interstate conflict along international borders, dividing the gulf by transnational interest groups rather than national interests. In each country, fishermen and environmentalists are relatively small fish in a big pond, but if they were to band together, the oil industry might have to take them more seriously in the gulf.

Mia Bennett

Perspective 2: Environmental Devastation as the Source of Conflict in the Gulf of Mexico

In the wake of the catastrophic British Petroleum (BP) oil spill in the Gulf of Mexico, environmentalists and concerned citizens from the various countries in the gulf have risen up to demand a proper accounting of the negative impacts of drilling in the area. Many are distraught and dismayed by the environmental risks being taken in the gulf and are prepared to fight against the industry and its supporters.

Evidence of profound environmental risk-taking in the gulf is better documented than ever before. For example, three subsequent incidents highlight the willingness of the oil and gas industry to engage in environmental risk-taking associated with offshore drilling. In its marking of the two-year anniversary of the Gulf of Mexico disaster in 2012, the *Washington Post* noted that, "In the North Sea, French oil giant Total is still battling to regain control of a natural gas well that has been leaking for nearly four weeks. Meanwhile, Brazil has confiscated the passports of 11 Chevron employees and five employees of drilling contractor Transocean as they await trial on criminal charges related to an offshore oil spill there. And in December [2011], about 40,000 barrels of crude oil leaked out of a five-year-old loading line between a floating storage vessel and an oil tanker in a Royal Dutch Shell field off the coast of Nigeria" (Mufson 2012).

In the aftermath of what is now widely considered the world's worst offshore drilling-related environmental disaster, the economic losses associated with the BP oil spill remain incalculable. According to *ProPublica*, "Nearly two years after the disaster, perceptions are still shifting about how much damage it has done. As oil washed ashore on hundreds of miles of coastline and the Gulf's tourism and fishing industries faltered.

Yet even now (2012), tar balls are still turning up on beaches, residents complain of health problems due to the spill or chemical dispersants used to clean it up, and an increasing number of dolphin deaths have raised biologists' concerns. Environmental scientists say it will be years before the true extent of the disaster can be known . . ." (Lustgarten March 3, 2012). Like environmental disasters of the past, the costs associated with the environmental harm that continues to result from the BP Gulf of Mexico disaster may never be fully known.

What is better known, however, is that market-oriented legal and regulatory incentives for energy companies engaging in these activities to do so in an environmentally sensitive manner tend to be ineffective (*Tampa Bay Times* 2012). One reason for the lack of legal and regulatory efficacy is that the financial penalties associated with noncompliance are comparatively minimal in comparison to the financial rewards derived from offshore drilling activity.

As reported by the *New York Times*, "Two years after analysts questioned whether the extraordinary cost and loss of confidence might drive BP out of business, it has come roaring back. It collected more than $375 billion in 2011, pocketing $26 billion in profits" (Lustgarten April 19, 2012). Of that $375 billion dollars, it is estimated that BP has, over the two years since the Gulf disaster, paid roughly $37 billion dollars in fines, settlement costs, and environmental reparations. Thus, as most observers have noted, the fines and penalties associated with not observing federal and international laws, regulations, agreements, and/or treaties that are designed to protect the environment amount to nothing more than a financial slap on the wrist.

In the face of this economic disparity, *New York Times* columnist Abrahm Lustgarten concludes that "What the gulf spill has taught us is that no matter how bad the disaster (and the environmental impact), the potential consequences have never been large enough to dissuade BP from placing profits ahead of prudence. That might change if a real person was forced to take responsibility—or if the government brought down one of the biggest hammers in its arsenal and banned the company from future federal oil leases and permits altogether. Fines just don't matter" (Lustgarten April 19, 2012).

Making this turnaround, however, has come at great cost to our environment. By some estimates, the environmental damage done by the BP Gulf of Mexico oil spill and the costs associated with that damage alone could take years to fully understand. What is known is that a wide swath of the Gulf of Mexico's ecosystem has been destroyed by the BP spill—and along with it, the biological diversity that once supported thriving coastal industries such as fishing, shrimping, oyster harvesting, and tourism. The environmental impact of the BP spill on the economies of the Gulf Coast states of Louisiana, Mississippi, Alabama, and Florida, as well as that on the environment of the Gulf of Mexico itself, remains unprecedented.

The NOAA report documents that the damage done to the extensive coral reefs that extend as far as seven miles from the site of the spill, a once-thriving colony of biodiversity, is now nothing more than a seven-mile-long stretch of "brown muck." It was, according to Erik Cordes, a biologist from Temple University, "like a graveyard of corals" (Magner 2012). While all seemed perfectly normal on the surface two years after the spill, ". . . beneath the surface—in some cases only a few feet below the sand and surf—lurk[ed] a host of hazards: thick tar mats, petrochemicals absorbed by marine life, and particles of oil broken down by chemical dispersants that [had] settled to the ocean bottom" (Magner 2012). Across the Gulf of Mexico and emanating from the point of the BP oil spill are vast stretches of what marine biologists refer to as "dead zones" in which life no longer exists.

Far less understood, or, for that matter, quantifiable in economic terms, is the intrinsic value derived by members of society interacting with the oceanic environment itself. Whether that value is one that is derived from engaging in recreation or sport, or is of a spiritual nature, human engagement with the environment has value beyond its commoditization as a tangible resource. It is a value that does derive an economic return, even if that return is quantifiably illusive. Given all of the information provided here, we must ask ourselves: is conflict inevitable as the poorly regulated energy industry wantonly and dangerously drills for oil in the Gulf of Mexico with little regard for the environmental risk? The answer appears to be an unequivocal yes.

Albert C. Hine

REFERENCES AND FURTHER READING

Bureau of Ocean Energy Management. "Gulf of Mexico Energy Security Act (GOMESA)." http://www.boem.gov/Revenue-Sharing/.

Bureau of Safety and Environmental Enforcement. "Offshore Statistics by Water Depth." 2015. http://www.data.bsee.gov/homepg/data_center/leasing/WaterDepth/WaterDepth.asp.

Cato, James C. *Gulf of Mexico: Origin, Waters, and Biota*. Vol. 2, *Ocean and Costal Economy*. College Station: Texas A&M University Press, 2009.

Cavnar, B. *Disaster on the Horizon: High Stakes, High Risks, and the Story behind the Deepwater Well Blowout*. White River Junction, VT: Chelsea Green Publishing, 2010.

Chediak, Mark. "Transocean Profit Falls on Declining Demand for Rigs." *Bloomberg*, August 4, 2010. http://www.bloomberg.com/news/articles/2010-08-04/transocean-s-profit-declines-on-costs-tied-to-gulf-spill-lower-rig-demand.

Cuffe, Sandra. "Drilling the Caribbean: Indigenous Communities Speak Out against Oil and Gas Exploration in Honduras." *Upside Down World,* November 24, 2014. http://upsidedownworld.org/main/international-archives-60/5128-drilling-the-caribbean-indigenous-communities-speak-out-against-oil-and-gas-exploration-in-honduras.

Duff, J. "U.S. Ratifies Maritime Boundary Treaty with Mexico." Mississippi-Alabama Sea Grant Legal Program.

Freudenburg, William R., and Robert Gramling. *Blowout in the Gulf: The BP Oil Spill Disaster and the Future of Energy in America*. Cambridge, MA: MIT Press, 2011.

Hayward, Steven. "Who Makes the Most from Oil and Gas Leases on Public Land?" *Powerline,* February 7, 2012. http://www.powerlineblog.com/archives/2012/02/who-makes-the-most-from-oil-and-gas-leases-on-public-land.php.

Lustgarten, Abrahm. "BP Settlement Leaves Most Complex Claims Unresolved." ProPublica, March 3, 2012. https://www.propublica.org/article/bp-settlement-leaves-most-complex-claims-unresolved.

Lustgarten, Abrahm. "A Stain That Won't Wash Away." *New York Times*, April 19, 2012. http://www.nytimes.com/2012/04/20/opinion/a-stain-that-wont-wash-away.html.

Magner, Mike. "Two Years after BP Oil Spill, Marine Life in the Gulf of Mexico Still Reels." *National Journal*, April 21, 2012.

Mufson, Steven. "Two Years After BP Oil Spill, Offshore Drilling Still Poses Risks." *Washington Post*, April 19, 2012. https://www.washingtonpost.com/business/economy/two-years-after-bp-oil-spilloffshore-drilling-still-poses-risks/2012/04/19/gIQAHOkDUT_story.html.

Neuman, Scott. "Obama Ends Ban on East Coast Offshore Drilling." NPR, March 31, 2010. http://www.npr.org/templates/story/story.php?storyId=125378223.

NOAA Office of Science and Technology. "Regional Summary: Gulf of Mexico Region." https://www.st.nmfs.noaa.gov/Assets/economics/documents/feus/2011/FEUS2011%20-%20Gulf%20of%20Mexico.pdf.

Office of the Press Secretary. "Fact Sheet: Charting a New Course on Cuba." December 17, 2014. https://www.whitehouse.gov/the-press-office/2014/12/17/fact-sheet-charting-new-course-cuba.

Omestad, Thomas. "Cuba Plans New Offshore Drilling in Search for Big Oil Finds in the Gulf of Mexico." *U.S. News & World Report*. February 3, 2009. http://www.usnews.com/news/energy/articles/2009/02/03/cuba-plans-new-offshore-drilling-in-search-for-big-oil-finds-in-the-gulf-of-mexico.

Tampa Bay Times. "BP Criminal Fine Sends Strong Message." November 15, 2012. http://www.tampabay.com/opinion/editorials/bp-criminal-fine-sends-strong-message/1261767.

U.S. Department of the Interior Press Release. "Secretary Jewell Applauds Passage of U.S.–Mexico Transboundary Hydrocarbons Agreement." December 23, 2013. https://www.doi.gov/news/pressreleases/secretary-jewell-applauds-passage-of-us-mexico-transboundary-hydrocarbons-agreement.

U.S. Energy Information Administration. "Gulf of Mexico Fact Sheet." http://www.eia.gov/special/gulf_of_mexico/.

UNEP Global Environmental Alert Service. "The Gulf of Mexico Oil Spill: The World's Largest Accidental Oil Spill." 2010. http://na.unep.net/geas/getUNEPPageWithArticleIDScript.php?article_id=65.

Yep, Eric, and Wayne Ma. "China Expands Offshore Oil Fleet for Contested Waters." *Wall Street Journal*, August 1, 2014. http://www.wsj.com/articles/china-expands-offshore-oil-fleet-for-contested-waters-1406941177.

5 WATER GOVERNANCE AND CONFLICTS

OVERVIEW

This chapter examines the importance of effective management of water resources at all levels (international, country, regional, and local), and the significance of building and maintaining competent institutions and sound water management legal, regulatory, and enforcement frameworks. Water problems throughout the world are not just a matter of supply and demand or the competition of users for dwindling resources; they also factor in good governance of these resources, robust institutions and regulations, and whether stakeholders take collaborative or conflictive approaches to resolve problems.

The case studies in this chapter explore the drying up of an Iranian lake, illegal trafficking of people on the high seas, and whaling conflicts. The perspectives section discusses the emergence of conflicts over water rights and sharing mechanisms.

What Is Governance?

Governance is about management: it is a process of policy formulation, organization, regulation, decision-making, financial administration, and institutional action to achieve objectives. Good governance in managing water resources involves stakeholder consultation and collaboration among and within nations. This translates into better interaction and coordination among regional, state, or local authorities, and involvement of the business sector, public institutions, and local communities to solve water problems. In fact, some water problems stem from mismanagement or poor institutions. When stakeholders lack the problem-solving or conflict-resolution skills to address emerging water issues, problems can easily develop into crises.

Sound water management means enabling access of vulnerable groups to clean water, and managing water more efficiently to reduce wastage and improve conservation. Too often, poor institutional capacity or political differences among nations of the world inhibit their ability to respond to global problems such as climate change, water scarcity, water pollution, and water-related border conflicts. Within states, authorities often lack commitment to resolve national differences in water endowment, water rights, water pricing, distribution and allocation mechanisms, and people's ability to pay for water.

No Magic Bullet to Resolve Water Issues

Water governance is complicated, because there are many water sources, and because the management of these different sources doesn't occur in one place. Water is found in many different places and spaces: underground aquifers, wells, rivers, lakes, seas, streams, marshes, frozen ice, tundra, ice packs, oceans, coastal waters, bays, and estuaries. There is an international context to water, and national and local contexts to take into consideration as well. There are problems with freshwater, whether on the surface or in aquifers, and water issues in coastal environments, island ecosystems, and oceans. There is no one solution to the problem, and no one institution capable of managing the entire sector. To complicate matters, water is not evenly distributed across the world, across borders, within individual countries, or even within groups in the same village or city.

Hence, it is difficult to coordinate water distribution and sharing arrangements under one umbrella, given the thousands of jurisdictions, consumers, managers, stakeholders, and sectors using, managing, and competing for the resource. This makes it hard to grapple with problems from a holistic point of view, and particularly to monitor compliance with international norms and standards, including use of globally shared waters. Conflicts continue to pop up at different times, at local, community, regional, state, or global levels.

To manage water at the national level, many countries have put in place frameworks for allocating resources and mediating disputes. But these mechanisms have not always been effective, judging from the ongoing depletion and degradation of surface water and aquifers in the world. Some water users have developed a "winner takes all" approach. Rising water shortages and water demand, drought, contamination, and depletion of water supplies have intensified competition and led to the growing commodification of water. Clean water commands high prices, affecting the ability of poor people to pay for it.

International Water Management

On the international front, most countries that share water basins, lakes, seas, bays, and waterways try to manage joint resources through a range of instruments: treaties, conventions, commissions, working groups, research institutions, legal structures, monitoring tools, and enforcement mechanisms for violation of agreements.

The effectiveness of these arrangements to implement, monitor, and enforce agreements on water sharing and allocation rights depends on the political will of those signing the agreements. Without commitment and enforcement, agreements are pieces of paper akin to barking dogs without a bite. Some countries lack agreements on paper. Others may have them, but if they have not ratified them in their respective parliaments, the agreements are not legally binding.

As previously discussed, the United Nations Convention on the Law of the Sea (UNCLOS), established codes of conduct and governance rules for activities in international waters. These include regulations for protection of water reserves in common areas, legal prohibitions on the transport of specific toxic substances, and prevention of pollution or overfishing on the high seas.

Nonetheless, the cooperation of all states and global enforcement of the rules is problematic, mainly because there are common areas on the high seas that are up for grabs by all countries. Two of the case studies in this chapter examine the issue of adhering to or interpreting international laws.

Robust water governance requires water administrators and users to understand water supply problems early on, and be able to balance competing water interests to avoid outright conflicts and crises further down the road. Effective arrangements (both formal and informal) are needed to guarantee access of all citizens to water, consistency of supply, water quality, and the conservation and sustainable use of shared resources.

Water Sharing Arrangements

Peaceful and collaborative water sharing agreements mean that all parties involved restrain from excessively tapping shared resources or unilaterally moving into the territories of others to take over their water sources. However, since water is becoming more scarce and costly, sharing has become more complicated and contentious. Countries or communities may not always agree on water sharing and allocation rights.

Regions also experience conflicts, for example, when upstream users of a shared river basin siphon off river flows from downstream users by building dams and irrigation canals to divert the water. In situations where rich farmers are able to sink more wells into aquifers than poor farmers can, shared community resources are likely to be unfairly depleted. In the case of transnational water basins, if one country pollutes or depletes a joint water resource, it infringes upon the rights of the other country to clean water or consistent water supply. It is easy to see how disagreements can emerge in these situations.

Dublin Principles

The Dublin Guiding Principles arose from the findings of an International Conference on Water and the Environment (ICWE) in 1992, and set forth guidelines for water governance and management of conflicts among users of oceans, inland bodies of water, and groundwater reserves. Four key principles emerged. The first principle is that freshwater is a finite and vulnerable resource, essential for human life, and water quality should be maintained at certain minimum standards. The second is that water management should be as participatory as possible and involve all users and stakeholders. The third is that women's roles in the provision, management, and safeguarding of water need to be acknowledged. The fourth principle is that water is not an infinite or cost-free resource. This means that even if water falls from the sky or can be accessed from a river, lake, or aquifer, it can be depleted, and there are costs to transport and access it at the individual, household, or business level. Due to increasing distance of water sources, water storage and distribution systems have to be built or extended. Since source water may be polluted, there is also the cost of cleaning it prior to human consumption (WMO 1992).

Water: Human Right or Commodity?

United Nations declarations, the constitutions of some countries around the world, and the value systems of most indigenous or traditional communities maintain that water is a human right and must be accessible to all people for their survival. This belief holds that humans, whether poor or rich, have inherent rights to water as a global good. Some nations believe that water is a public good, and have enshrined this concept within their constitutions. Other countries make good faith efforts to equitably allocate water to poor communities, or build in upfront subsidies or support mechanisms to enable people with low-incomes to obtain water fairly and easily.

However, there is another view of water, namely, that water has an economic value as a commodity and, like any other commodity, can be traded, marketed, sold, or privatized for profits. According to this approach, private sector companies can play a major role in accessing, managing, distributing, or selling water. Private sector predominance is more widespread in countries that have privatized their water supply sources, water utilities, or management and distribution systems. However, water privatization has been fraught with political and social conflicts, particularly in developing countries. A case study on Cochabamba, Bolivia, in chapter six of this book examines this point.

Tackling Water Issues Globally

In the case of nations sharing world oceans, UNCLOS prescribes expected behaviors on the high seas, but there is no international police force that patrols the high seas or can prevent individual sovereign states from violating UN norms. It is up to individual nations and ocean users to abide by the rules. However, there are international courts of law that enforce regulations once claims are made against other parties.

Worldwide problems such as climate change, ocean pollution, overfishing in global maritime areas, overexploitation of seabed minerals, and oil spills can best be resolved by globally coordinated responses. Cross-border water resources also require sharing and collaboration agreements among two or more countries; at the local level, community water shortages or water pollution need municipal, local, and community responses. Getting to international agreements on globally accessible water (40 percent of all oceans) involves political negotiations at the highest levels to hammer out consensus. Many countries have difficulties setting aside sovereign interests for the common good.

According to one of the Dublin principles, effective action on water calls for decentralized and integrated approaches to engage civil society, the public sector, the private sector, and nongovernmental groups. The extent to which collaboration of all players works depends on their capacity to agree on approaches and implement them. The players and stakeholders in the water sector are numerous. They may range from public sector ministries in health and water, water authorities, international or regional water basin commissions, to private farmers, urban and rural consumers, industries, local governments, citizens, and water rights advocacy groups.

Stakeholder Involvement

Stakeholders have the right to access information to better understand the problems at hand and coordinate with others in resolving the issues. People have the right to know if their local water sources are polluted, if aquifers are running dry, if there are outbreaks of waterborne diseases, or if politicians or local governments have made decisions to endanger or protect their water supply. Vulnerable, low-income communities often have limited access to reliable information and warning about precipitation levels, threats of floods, and the probability of natural disaster risks, droughts, or water shortages.

In addition, there is often little communication and coordination among various ministries (energy, industry, agriculture, etc.) or entities in charge of water supply, allocation, and distribution among communities. Overlapping of functions and executive powers among government institutions, as well as competition for the use of water in industrial, social, or agricultural sectors can also hamper the achievement of outcomes.

Within this context, the interrelationships between the water sector and other sectors of the economy merit a closer look. Growth in agriculture, irrigation, and industry, as well as the impacts of climate change on water resources, affect countries and groups differently. Some countries are very small islands with limited means to implement key water management and climate change adaptation actions.

To face up to pressing water quality and quantity challenges at the local farm or community level, areas need not just robust local governments but also the active involvement of communities and local groups coping with or contributing to water problems. Local participation is crucial for the functioning of joint river basin management arrangements and the planning and execution of broad-based solutions for communities across borders.

Local consultation helps ensure more equitable use of watershed resources and reduces the influence of vested interests. For local water governance to work, the participation of water users in onsite water management or consultation committees is important. It is crucial to include the business sectors in getting to appropriate solutions, since they use large amounts of water for agriculture, mining, energy, manufacturing, and services, which may impact water quality.

Joint Problem Solving

In the case of watercourses that meander across borders, the governance of joint resources can be complicated. Synergy is possible through multicountry governing systems that jointly plan, administer, and enforce common rules and mete out penalties for violations. Failure to manage cross-border water disagreements increases the potential for conflict among neighbors. Local communities on both sides of borders often face common problems of drought, floods, aquifer depletion, or pollution of common ground and surface water sources. They can overcome many of these challenges working together. Experience shows that it is difficult to address shared water resource problems on an individual basis.

Building institutional capacity at state, government, and international levels is an important tool with which to tackle current and emerging water crises. At the global level, there are many mechanisms to support governance efforts such as the Regional Seas Programme of the United Nations Environment Programme (UNEP), the Regional Centers for Marine Science and Technology, and multicountry transoceanic tsunami warning networks.

Expanded Use of Oceans

If we trace the development of ocean governance mechanisms, we see that it parallels the expanded use of oceans for trade and communications after the end of World War II. This postwar period of peace ushered in the rise of globalized communications, the laying of transoceanic undersea cables, increased trade and maritime transportation routes across international waters, and global activities in scientific marine research.

Along with this expansion came the need for coordination and the recognition that global cooperation agreements were necessary to address emerging problems, i.e., ocean pollution; regulation of shipping and navigation routes; trafficking of people and transport of hazardous materials; and overfishing and predatory seabed mining. Increasing problems such as climate change, rising sea levels, ocean acidification, and the destruction of coral reefs, islands, and coastal environments, make an even stronger case for more effective ocean cooperation agreements and conflict resolution mechanisms.

General and Regional Seas Conventions

A global landmark convention is the aforementioned United Nations Convention on the Law of the Sea (UNCLOS) but there are other conventions that are regionally focused, i.e., the Regional Seas Programme of the United Nations Environment Programme (UNEP), which contains nine Regional Action Plans (RAPs) for the protection of marine and coastal areas shared by several countries.

These RAPs cover the Red Sea and Gulf of Aden; the Persian Gulf; the Caribbean Sea; East Asian Seas; the South-East Pacific region; the West and Central African coasts; the South Pacific Sea; and eastern Africa. The action plans cover more than 120 states, which have signed agreements to protect and enhance the marine environment within their 200-mile Exclusive Economic Zones. Anything beyond an EEZ is considered open sea and has to be covered by UNCLOS norms.

Multistate cooperation and scientific agreements exist for the Arctic Sea, the Baltic Sea, the Mediterranean Sea, and Antarctica. In addition, there are specialized international conventions for specific areas such as oceanography, marine research, fisheries, maritime safety, piracy, and ecosystem conservation. The compliance of nations with these agreements varies. The ability to manage shared areas is only as good as the political will, competency, and commitment of signatory countries. Many agreements remain on paper for years before becoming ratified by their respective parliaments under agreed protocols. Some do not even get ratified.

When states sign such international or regional conventions, they are in effect promising to play their part to comply with their roles and responsibilities. However, very weak or failed states, even after signing or ratifying agreements, are often technically or politically unable to exercise their sovereignty obligations within their 12- or 200-mile EEZ limits, let alone adhere to commitments to protect the high seas under UNCLOS.

International Piracy

UNCLOS provides an important framework for the prohibition of high seas piracy and global, armed robbery under international law. Under this convention, all states are required to cooperate in combatting piracy at all levels, whether global, regional, subregional, bilateral, or national. Piracy has been a historical problem on the open seas since the age of discovery and international navigation in the 16th century. At that time, slow-moving, undefended sailing ships were sitting ducks for seasoned pirates.

In the 19th century, concerted efforts of affected seafaring nations made piracy a universal crime, and all states had the joint responsibility to suppress or combat this crime. Eventually, antipiracy concerns found their way into formal laws such as the 1958 Geneva Convention on the High Seas, which later culminated in the 1982 UNCLOS conditions. The latter outlined specific obligations of states to repress piracy. The convention was signed and ratified by 157 states.

Under UNCLOS, high-seas piracy is an illegal act. All states have the right and the responsibility to seize and prosecute those responsible for acts of piracy, whether it occurs on the high seas or in the 200-mile EEZ, which borders the high seas. The law on piracy does not explicitly include incidents within a coastal state's immediate coastal waters. It assumes that states will utilize their own maritime protection, legal and criminal justice systems, and coastguard resources to patrol their territorial waters for pirates. It also assumes the capacity of states to enforce international law through prosecution of pirates in their national courts.

But what happens when weak or failed states cannot police their own 12–nautical mile territorial waters, or their 200-mile EEZ? What happens when they are unwilling or unable to assist in combating piracy on the high seas? In particular, what happens when states may be unable or unwilling to stop pirates using their coast and waters as staging points to attack other ships in their EEZ and on the high seas? This is the case in Somalia.

The Example of Somalia

After 2008, there was an upsurge in pirate attacks off the East African coast in the vicinity of Somalia. In 2008, pirates seized more than 60 ships off the coast of Somalia and exacted millions of dollars as ransom money for release of the ships and crews. These pirates appeared to be part of organized Somalia-based syndicates, benefitting from general lawlessness and lack of oversight in the country. Piracy groups with local and international connections were able to thrive due to poor national governance within Somalia, and weak capacity and political will to ensure maritime security and safety in its near and wider coastal environments.

Accordingly, Somalia became a source of concern for the United Nations Security Council, which adopted Resolution 1816 to address the threat of Somali-based piracy, recognizing the country's lack of capacity to prevent or interdict pirates operating in its waters. This resolution, under agreement with the Transitional Federal Government of Somalia, authorized the international community to undertake national-type policing within Somalia's waters.

Crewmembers from the cargo ship *Motor Vessel Polaris* board the U.S. guided-missile cruiser *Vella Gulf* in the Gulf of Aden to identify suspected pirates apprehended by the *Vella Gulf*, on February 11, 2009. The American cruiser was the flagship of Combined Task Force 151, a multinational force conducting counterpiracy operations in and around the Gulf of Aden, Persian Gulf, Indian Ocean, and Red Sea. (U.S. Navy)

In response, the European Union (EU) launched Operation Atalanta in 2008, under which European naval fleets started policing the East Africa coast to stave off pirate attacks launched from the coastline of Somalia. An additional Resolution 1851 also authorized EU states to operate on the land area of Somalia to prevent the planning of attacks staged from the country.

The United States also created a Contact Group on Somali Piracy to establish coordination mechanisms for intelligence sharing and collaboration with other nations, and with the shipping and insurance industries. In the history of maritime law, these two United Nations resolutions on Somalia have granted the most extensive security powers ever conferred upon states to deal with piracy through concerted action. Since the passing of these resolutions, piracy attacks from Somalia have been greatly reduced.

International Organizations

There are already many international organizations that address matters related to ocean governance. The main ones include the International Maritime Organization; the International Maritime Bureau; the United Nations Convention for the Suppression of Unlawful Acts Against The Safety of Maritime Navigation; the 1989 Basel Convention on the Control of Transboundary Movements of Hazardous Wastes and their Disposal; the International Maritime Organization's 1993 Code for the Safe Carriage of Irradiated Nuclear Fuel, Plutonium, and High-Level Radioactive Wastes in Flasks on Board Ships; and the International Tribunal for the Law of the Sea (ITLOS). The latter is responsible for international arbitration among states in the case of violation of international agreements. However, what matters most is not the quantity of organizations but their effectiveness in coordinating activities and achieving concrete results in monitoring and enforcement.

Interpreting the Law of the Sea

The following global examples demonstrate the difficulties of interpreting or enforcing UNCLOS regulations. The first one involves the seizing of an Argentinian navy ship by the West African country of Ghana in response to a conflict between a financial hedge fund in New York and Argentina. The second one involves a conflict in Chile's waters regarding the transportation of hazardous materials by a ship of another nation.

Seizing of an Argentinean Navy Ship

The background to the seizing of the Argentinian navy ship involves a long-standing feud between a U.S.-based hedge fund and the state of Argentina. The hedge fund was trying to collect Argentinian debt by seizing the country's assets, which in this case was the navy ship, the *Libertad*, that was docked in Ghana. The events unfolded as follows: the U.S. hedge fund asked Ghana to impound the ship as part of its quest to recover the debt owed by Argentina 10 years earlier. The local court in Ghana ruled in favor of the hedge fund and ordered the ship's seizure, upon which Argentina submitted the case to ITLOS. ITLOS ruled that Ghana had no jurisdiction to hold the ship because it was a military vessel, and because such an action could set a precedent for potential armed conflict and hamper peaceful relations among all states.

Precautionary Principles versus the Right to Safe Passage

The second example illustrates the interpretation by Chile of UNCLOS regulations regarding the transport of hazardous materials on ocean-going vessels. The context is as follows: Article 22 of UNCLOS sets forth that coastal states require ships carrying nuclear and hazardous materials to use special sea-lanes and traffic separation mechanisms. It does not, however, allow coastal states to suspend innocent passage of these ships, as long as they take special precautionary measures to avoid accidents and maintain strict safety standards.

When ships are navigating in the EEZ of other countries, UNCLOS prescribes that once they comply with the laws and regulations of that particular coastal state, they should receive safe passage. At the same time, another convention, called the International Convention Relating to Intervention on the High Seas in Cases of Marine Pollution by Substances other than Oil (the Convention), authorizes coastal states to protect marine environments by taking any precautionary steps needed to prevent or mitigate dangers to their coastlines from pollution or threat of pollution by substances other than oil.

On the basis of this Convention, the Chilean Maritime Authority in 1995 cited precautionary steps or principles to deny a British ship passage through its EEZ. The British ship, the Pacific Pintail, was carrying high-level nuclear waste from France to Japan, passing through Chile's EEZ on its route around Cape Horn. The Chilean Navy intercepted the ship and forced it to return to high seas by threatening it with military action. The Chilean government regarded its action as legitimate under the Convention and denied safe passage to the ship, a decision which appeared to contradict UNCLOS's rule of ships' rights to innocent passage. This case indicates that international laws may have inconsistencies or may be inconsistently interpreted by states.

A similar incident occurred in 1992. A ship called the *Akatsuki Maru*, travelling from France to Japan and carrying almost 2 tons of plutonium, faced fierce resistance on its route from over five coastal countries. They prohibited the ship to enter their EEZs, even though Japan maintained that these restrictions violated the right of safe passage according to UNCLOS. The environmental organization, Greenpeace (which protests against whaling in international waters), also organized demonstrations against the ship, leading to repeated clashes between Japanese patrol boats and Greenpeace vessels.

Camille Gaskin-Reyes

REFERENCES AND FURTHER READING

International Maritime Organization. "International Convention Relating to Intervention on the High Seas in Cases of Oil Pollution Casualties, 1969." http://www.imo.org/en/About/Conventions/ListofConventions/Pages/International-Convention-Relating-to-Intervention-on-the-High-Seas-in-Cases-of-Oil-Pollution-Casualties.aspx.

Payoyo, Peter Bautista, ed., *Ocean Governance: Sustainable Development of the Seas*. Tokyo: United Nations University, 1994.

Treves, Tulio. "Piracy, Law of the Sea, and Use of Force: Developments Off the Coast of Somalia." *European Journal of International Law* 20, no. 2 (2009): 399–414.

United Nations. "Convention on the Law of the Sea." http://www.un.org/depts/los/convention_agreements/texts/unclos/unclos_e.pdf.

United Nations Environment Programme (UNEP). "Regional Seas Program" www.unep.org/regionalseas/.

World Bank Environment Department. *Lessons for Managing Lake Basins for Sustainable Use*. Washington DC: World Bank Publications, 2005.

World Health Organization. "The Right to Water." http://www.who.int/water_sanitation_health/en/righttowater.pdf.

World Meteorological Organization (WMO). "The Dublin Statement 1992" http://www.wmo.int/pages/prog/hwrp/documents/english/icwedece.html.

CASE STUDIES

UNCLOS has provisions for landlocked nations to access passage to oceans through the waters of neighboring coastal countries, but history shows us that regional diplomacy and agreements between landlocked countries and their coastal neighbors are often difficult to manage. Before UNCLOS, there were no international legal and enforcement mechanisms to regulate passage of landlocked nations or their exit to oceans. Instead, states had two options: negotiate passage with neighbors or invade them to gain exits to seaports or oceans.

One of the case studies in this chapter shows how Bolivia, a landlocked state, after it lost its exit to the Pacific Ocean through a war with Chile, provoked clashes with Paraguay in the 1930s to gain access to the Atlantic Ocean through Paraguay's river ports. Since there was no international governance framework at the time, Bolivia, desperate for a way to the ocean, saw the bordering Chaco region as an exit to the Atlantic. This armed conflict was called the Chaco War.

Another case study discusses the use of the oceans and seas for illegal trafficking of arms, drugs, and people in violation of UNCLOS regulations; this example highlights the dilemma of monitoring and enforcing the Law on the high seas. A third case study illustrates the conflict between Japan and Australia over Japan's whaling practices. This case was presented by Australia to the International Court of Justice (ICJ) for resolution.

Case Study 1: Water Diversions, Rain Shortages Dry Up a Major Iranian Lake

Lake Oroumeih in northwestern Iran is rapidly disappearing. The image on the left, taken in August 1985, shows the lake near full capacity, with minimal presence of shallow water and salt deposits represented by lighter tones. In contrast, the image on the right, taken in August 2010, depicts a great reduction in the lake's water levels, with a significant increase in shallow salt deposits. There are increased agricultural areas around the lake, showing that water from the lake and rivers is being diverted for use in agricultural irrigation. Both images were taken from Landsat satellites.

Lake Oroumeih, also referred to as Urmia, is 87 miles from north to south and 52 miles east to west, making it the largest lake in the Middle East and the third-largest in the world. The lake and its surrounding wetlands are home to a number of different animal species, including amphibians, reptiles, mammals, and a rare species of brine shrimp. Around 200 species of birds, many of them migratory, live near the lake, including ducks, egrets, flamingos, and pelicans.

In addition, the watershed produced by Lake Oroumeih provides vital water to surrounding farms and the 6.4 million people who live nearby. In 1971, the lake was designated a Wetland of International Importance by the Ramsar Convention. However, since 1995, when the surface area of Lake Oroumeih was at a 40-year high of 2,355 square miles, the lake's surface area has been steadily decreasing, reaching a record low of 914 square miles in August 2011 and a drop in water level of 23 feet. A United Nations study found the lake to be at 60 percent capacity in 2011 compared to what it was in 1980, with the entire lake expected to be dry by 2013.

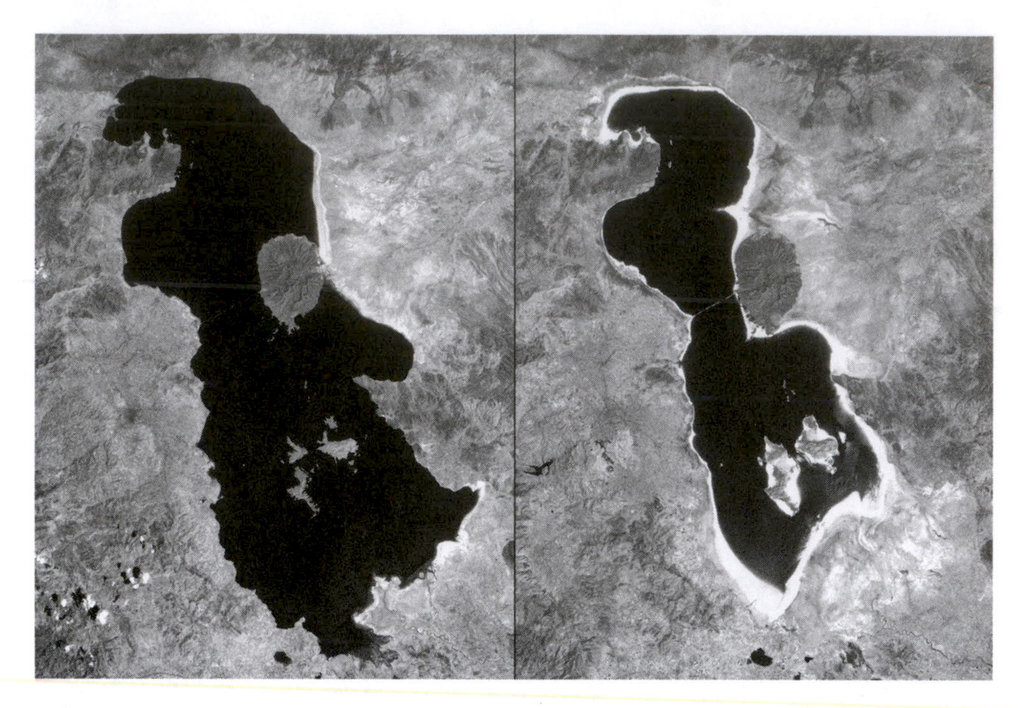

Left: August 1985. Right: August 2010. Iran's Lake Oroumeih (also spelled Urmia) is the largest lake in the Middle East and the third largest saltwater lake on Earth. But dams on feeder streams, expanded use of ground water, and a decades-long drought have reduced it to 60 percent of the size it was in the 1980s. Lighter tones in the 2010 image represent shallow water and salt deposits. Increased salinity has led to an absence of fish and habitat for migratory waterfowl.

Images taken by the Thematic Mapper sensor aboard Landsat 5. Source: USGS Landsat Missions Gallery, U.S. Department of the Interior / U.S. Geological Survey.

http://climate.nasa.gov/state_of_flux#Lakeshrinkage_Iran.jpg

The lake's significant decrease in water level is believed to be the result of many different factors. Since Lake Oroumeih has no significant outflow, the water supply will only decrease through evaporation or if less water is put into the lake. Thus, the main causes of the lake's depletion are believed to be decreased rainfall due to climate change, increased diversion of water upstream for use in agricultural irrigation, groundwater extraction, and the construction of dams along various rivers that feed the lake.

The severe water decrease in Lake Oroumeih has had major consequences. Salt levels in the lake have increased significantly, making it difficult for the lake's brine shrimp population to survive. Since birds primarily come to Lake Oroumeih to eat the shrimp, their demise has had a devastating impact on the area's ecosystem, as the shrimp also eat algae in the lake, keeping it under control. Tourism in the area has declined significantly since the diverse bird populations are one of the biggest draws. The lake's disappearance would have a devastating impact on the area's vegetation and local agriculture as the lake would become a salt bed. This would in turn inhibit plant growth and make crop irrigation more difficult as well as raise the possibility of "salt-storms," which can have severe health effects on humans and animals exposed to the salt.

In order to save Lake Oroumeih, farms in the area will have to divert less water away from the lake for agricultural irrigation and water will need to be imported from outside the basin in order to significantly raise water levels and decrease salinity. However, both solutions pose difficulties. Farmers and residents rely heavily on the watershed from the lake in order to sustain themselves, and with persistent droughts, water in the basin region is very difficult to come by. If water were to be imported from other sources, it would be a complicated and costly measure that would likely require the cooperation of other countries and take a significant amount of time to put into place. A measure that balances the needs of local residents with the logistical limitations of importing water would be ideal, but as of now the survival of Lake Oroumeih is still very much up in the air.

Renee Dubie

REFERENCES AND FURTHER READING

Karbassi, A., G. Bidhendi, A. Pejman, M. Bidhendi. "Environmental Impacts of Desalination on the Ecology of Lake Urmia." *Journal of Great Lakes Research* 36, no. 3 (2010): 419–424.

Pengra, Bruce. "The Drying of Iran's Lake Urima and Its Environmental Consequences." United Nations Environment Program. http://na.unep.net/geas/getUNEPPageWithArticleIDScript.php?article_id=79.

Zarghami, M. "Effective Watershed Management; Case Study of Urmia Lake, Iran." *Lake and Reservoir Management* 27, no. 1 (2011): 87–94.

Case Study 2: Illicit Trafficking of People and Products on the High Seas

Globalization and increased interconnectivity of markets and transportation modalities have led to the rise of vast global smuggling and trading networks across the seas and oceans of the world. International waters are a magnet for lawlessness and illegal and unregulated activities and crimes. Beyond the 200–nautical mile limit of national jurisdiction, waters beyond national limits are in an international zone, where crimes can occur with impunity, and enforcement of rules and regulations is weak.

Modern, Illicit Ocean Transport Networks

In the past, oceans were the international theater for exploration, trade, the transport of slaves and looted goods, old style piracy, and battles for new territories. In today's world, traffickers and syndicates of all sorts have turned to the high seas to ply their illegal trade in drugs, toxic substances, endangered species, smuggled natural resources, arms, and trafficked persons. In addition, intentional dumping of pollutants, illegal fishing, and other maritime crimes occur through the actions of deviant brokers and criminal rings.

These illicit actions are facilitated by the fact that more than 40 percent of the planet is covered by international waters that belong to everyone but no one in particular, and there is no central body to oversee events that occur on the high seas.

UNCLOS contains provisions to manage international waters, protect the ocean environment, and prohibit overfishing and the trafficking of illicit substances. Almost every country has ratified UNCLOS, and it is up to individual states and entities to demonstrate commitment to UNCLOS regulations.

Tragedy of the Commons

Human stewardship of the oceans is poor, and compliance of many nations or groups with UNCLOS, weak. Oceans exemplify the "tragedy of the commons," which is the depletion of common property by individual users who follow short-term self-interests to the detriment of long-term interests of all users. The high seas bear the cumulative impacts of deviant behaviors and multiple decisions of actors carrying out countless activities with impunity.

There are specialized bodies of the United Nations that monitor some regulations of UNCLOS, mainly when there is a conflict among states that require legal adjudication. The problem is complicated by the fact that there are dozens of overlapping authorities that deal with international water governance, but few coordination mechanisms and no overarching body to consolidate a patchwork quilt of organizations. It is no surprise that high seas trafficking, pollution, the illicit transport of goods and endangered species, and predatory fishing practices continue unabated.

Drug Trafficking

Drug trafficking on the seas involves many different smuggling networks, some organized in global criminal rings, others run by national competing cartels that crisscross the oceans to access international markets, mostly in North America and Europe. There are also different axes of drug movements like North to South, East to West and West to East, and more complex triangulation patterns such as West-East-North, to evade national authorities and enforcement agencies.

In the Americas, the main drug-producing and transshipment countries are located in Central and South America (mostly cocaine and marijuana), and in some parts of the Caribbean (marijuana). The main markets for illegal drugs are the United States, Canada, and Europe. The most direct way is not always the most effective way for traffickers to move product. Some routes run from South America across the Atlantic to West Africa for transshipment, and from there, further North to Europe or via Asia around the Pacific Ocean to North America.

In the Pacific and Atlantic Oceans around Central and South America, and throughout the Caribbean Sea, global drug cartels and national, regional, and local brokers and smugglers have built up powerful networks and complex transportation arrangements, including the use of seaplanes for island-hopping or triangulated transshipment routes to consumer centers using a variety of transportation modalities.

Many cartels have developed symbiotic relationships with political, state, and local authorities in shipment and transshipment countries to ease the passage of narcotics and illegal goods through the system. Corruption greases the wheel that moves illegal products. Narcotics are transported in a variety of vessels, using several means: sea planes, individual couriers, speedboats, fishing boats, luxury yachts, and magnets attached to hulls of tankers and container ships drugs can also be hidden in legitimate products in cargo holds of ships. Submarines have been discovered attempting to carry drugs under ocean waters.

In the last decade, West Africa has become a main intermediary or transshipment region in the cocaine trade between Latin America and Europe. There are estimates that about 25 percent of Europe's annual cocaine consumption, about 140 tons of cocaine (valued at about $1.8 billion), gets there via transshipment in West Africa. In addition to cocaine, West Africa is a transit point for heroin from Asia to North America, as well as a producer, transshipper, and exporter of cannabis and amphetamines (Ellis 2011).

This trans-African drug network (ironically, these routes are almost the reverse of the triangular slave trade centuries ago) is the result of a number of factors: the opening up of new European markets by cocaine producers in South America, more effective drug interdiction of Caribbean smuggling routes and the closure of some transshipment points, West Africa's lax institutional structures and regulations, and corruption and collaboration of vested interests with smuggling rings.

In West Africa, smugglers started transshipping heroin to the United States in the 1950s; in the 1970s Nigeria- and Ghana-based traders began sending African-grown marijuana, heroin, and cocaine to Europe and the United States. Since then, West Africa has become a known transshipment point. While drugs are carried to Europe by personal couriers or stashed on flights, there are also reports of intercepted ships carrying cocaine, heroin, and marijuana from West Africa to Hamburg and other European ports. These in-transit trafficking nodes have created smaller local markets of narcotic users in West Africa.

To try to confuse drug enforcement officials in North America and Europe, drug traders occasionally switch smuggling routes, e.g., sending narcotics by sea to Taiwan or Hong Kong and from there across the Pacific to North America or around the world back to Europe. In the last decade, sizeable sea shipments of drugs from points in the Caribbean,and South or Central America have been intercepted in or close to the waters of Nigeria, Cape Verde, Senegal, Mauritania, Guinea-Bissau, Liberia, Sierra Leone, Ghana, and Benin.

West African–based traders often combine narcotics smuggling with illegal diamond or arms trafficking.

Arms Trafficking

In addition to narcotics trafficking, there are arms trafficking networks around the world, which exploit or even fuel conflicts among nations. Somalia, a haven for international piracy, is also a transshipment point for arms. Arms smuggling has been reported in the Gulf of Aden, with Somalia at the heart of this network. This illicit trade involves government officials and private traders in East Africa and the Arabian Peninsula. Routes and supply lines in this area rely on a gamut of agents, corrupt members of governments, entrepreneurs, arms dealers, corporations, brokers, bankers, clansmen, militants, and warlords. The easy accessibility of war materiel supports the wider arena of armed conflict and violence in the Middle East and the Horn of Africa (Black 2011).

Transport routes for arms entering and leaving the area around Somalia are the regular routes for commercial ships across the Gulf of Aden. Weapons are offloaded or onloaded at Somali and other ports, transported in smaller vessels called dhows and loaded onto larger ships offshore in the Gulf. While there has been some effort by governments around the Gulf of Aden to control the arms flow to and from Somalia, and to enforce the 1992 UN Arms Embargo and the 2004 Nairobi Protocol, these measures are largely uncoordinated and ineffective in reducing illegal flows and transshipment of arms.

People Trafficking

The illegal smuggling, transportation, and trafficking of migrants on boats from places of poverty or hotspots of warfare to countries with economic opportunities or more peaceful environments, is not a new phenomenon. Globalized transport networks, poverty, depravation and the rise of conflict zones (including water conflict zones) have created lucrative business opportunities for traffickers and the steady increase in the transnational smuggling of migrants, refugees and job seekers. Traffickers often lure and exploit migrants with false promises of a better life, thus helping to create markets for their lucrative business (Kristof 2015).

It is important to distinguish between people smuggling and human trafficking: smuggling involves the transport of migrants across borders, oceans, and countries based on some kind of understanding (often informal) between the smuggler and the smuggled on the price and conditions of pay. These conditions may or may not be honored by smugglers during the course of the transaction, since the smuggler is in the position of power in the transaction. Often, the smuggled are held hostage and confined against their will until all payments are made; smugglers are known to abandon smuggled people in perilous conditions at the first sign of interdiction.

While smuggling is illegal and a serious crime, the active trafficking of people is a more egregious criminal activity. It involves the transportation of individuals from one country to another completely against their will or under false pretenses. Trafficked individuals usually end up in forced labor situations, kidnapped for ransom, or forced into sex work under horrendous conditions. They are often locked into slavery-type situations from which they are unable to free themselves. Many NGOs around the world have dedicated themselves to freeing children and teenagers from the bondage of traffickers and brothels.

In China, in the province of Fujian, sophisticated smuggling networks have developed over the last three decades to transport migrants from China to Chinatowns in New York and other cities. Well-connected brokers known in China as snakeheads set up lucrative enterprises to smuggle migrants across oceans and land routes for the hefty price of about $70,000 or more per migrant. These routes often involve ocean transport of passengers across the Pacific Ocean to Latin American ports, where they are disembarked and smuggled across land routes to the United States.

In order to evade detection, smuggler and trafficking networks take necessary precautions such as issuing fake passports or fake visas for migrants to access transshipment countries on long, often-circuitous routes. These routes can be so dangerous that not all migrants survive the passage; many perish in storms or are caught and detained by government authorities along the route. Those who reach their destinations in major U.S. cities have to pay the rest of their fees to their smugglers, usually through help from family and friends, or by working off their debts in forced labor (Keefe 2011).

The Growing Problem of Refugees and Migrants

In Africa and the Middle East, the exodus of migrants and refugees fleeing drought, armed conflicts, poverty, and ethnic violence has snowballed in the last five years, and led to further development of overland and ocean smuggling routes. In the last five years, huge flows of people have been leaving Middle Eastern countries such as Syria and Iraq to risk dangerous voyages from North Africa and Turkey across the Mediterranean Sea to Southern Europe, and from there further north to more affluent countries like Austria, Germany, and Denmark.

In the last five years, lucrative rings of smugglers have emerged, particularly in Turkey, to transport migrants in precarious boats and dinghies with few or no safety procedures along perilous water routes. Many migrants have lost their lives in storms or through the capsizing of overcrowded boats, and their plight has been the subject of international discussions and programs in the media. The flood of migrants has so overwhelmed European officials that many have been forced to close their land borders or facilitate a quick pass-through within Europe to prevent migrants from staying in their countries.

Camille Gaskin-Reyes

REFERENCES AND FURTHER READING

Black, Andrew. "Weapons for Warlords: Arms Trafficking in the Gulf of Aden." In *Deviant Globalization: Black Market Economy in the 21st Century*, edited by Nils Gilman, Jesse Goldhammer, and Steven Weber. New York: Continuum International Publishing Group, 2011.

Ellis, Stephen. "West Africa's International Drug Trade." In *Deviant Globalization: Black Market Economy in the 21st Century*, edited by Nils Gilman, Jesse Goldhammer, and Steven Weber. New York: Continuum International Publishing Group, 2011.

Keefe, Patrick Radden. "Snakeheads and Smuggling: The Dynamics of Illegal Chinese Immigration." In *Deviant Globalization: Black Market Economy in the 21st Century*, edited by Nils Gilman, Jesse Goldhammer, and Steven Weber. New York: Continuum International Publishing Group, 2011.

Kristof, Nicholas. "Meet a 21st-Century Slave." *New York Times*, October 25, 2015. http://www.nytimes.com/2015/10/25/opinion/sunday/meet-a-21st-century-slave.html.

Case Study 3: Whaling Conflicts between Japan and Australia at the International Court of Justice

Many decades ago, the international tide of world opinion began to turn against the slaughter of whales on the high seas. Notwithstanding the growing unpopularity of whaling and growing restrictions on the practice, Japanese whalers continued to ply the high seas to harvest and process whales in high-tech factory ships. This practice sparked international controversies and protests from other nations and environmental groups.

Greenpeace is one of the international nongovernmental organizations that led a battle for world opinion and physical confrontations on the seas to stop Japanese whaling ships from hunting whales in Antarctica. Japan argued that under the International Convention for the Regulation of Whaling (ICRW), it had the right to carry out research on whales and was not violating international whaling regulations.

Japan is party to the ICRW. In 1982, the ICRW set zero whaling catches for all countries, with two exceptions: it granted permits for aboriginal whaling and it allows scientific research under Article VIII. In 1994, the ICRW continued to press for restrictions on whaling by creating the Southern Ocean Whale Sanctuary. All nations were prohibited to carry out whaling in the sanctuary area. However, Japan continued to uphold the notion that since it had a cultural right to whaling and a scientific interest in research activities, it was exempt from the ban. Consequently, it continued to carry out two whaling programs, one in the Southern Ocean and one in the North Pacific Ocean.

On the other hand, Australia and many other nations declared their opposition to Japan's whaling activities, upholding the view that Japan was in effect using the research argument as a shield for its commercial whaling activities in the Southern Ocean. Furthermore, Australia claimed that Japan was selling whale meat on a commercial scale, acting in bad faith, and ignoring the ICRW's zero-catch limit for commercial whaling and the ban on hunting humpback and fin whales in the Southern Ocean Sanctuary.

The Case Goes to the International Court of Justice

The conflicts between Japan and other countries, Australia in particular, escalated to the point of international arbitration. In 2010, Australia brought a case against Japan to the United Nations International Court of Justice (ICJ) for allegedly violating international whaling rules. While Australia accepted the premise that under specific circumstances it would be lawful to kill whales for scientific research, it was emphatic in its claim that Japan had not focused its whaling activities on research, and had breached its international obligations to preserve marine mammals and the marine environment.

The ICJ convened a number of hearings and reviews of independent expert opinions on the science of whaling and whaling research. It took many years to deliberate on the different points of view and legal opinions of the lawyers representing both nations. During the hearings, Australia called two additional technical experts, and Japan summoned one, to bolster both countries' claims and counterclaims.

Japan Found in Violation

In 2014, four years after the case was presented, the ICJ pronounced a decision. The ruling was against Japan, which was ordered by the ICJ to stop whaling in the Southern Ocean. The ICJ pronounced that Japan had violated three regulations of the ICRW: the moratorium on commercial whaling, prohibitions on factory ships to render whales, and the outright whaling ban in the Southern Ocean Sanctuary. The ICJ ordered Japan to revoke any permits or licenses to kill whales.

In order to arrive at its opinion, the court based its judgment on a detailed review of ICRW regulations, including the conditions under which nations were given permission to research whales. It ruled that Japan could not adequately defend its position that its primary interest in harvesting whales was scientific research. The Court indicated that, after killing thousands of whales, Japan had not released or published the results of peer-reviewed scientific research for the benefit of the world's advancement on knowledge about whales. It concluded that Japan had not adequately furnished justification to support its position on the benefits of harvesting whales for research purposes.

As part of its ruling, the ICJ raised the issue that Japan had not adequately attempted to use nonlethal options to facilitate whale research, which violated ICRW regulations that nations should use state-of-the-art technologies for nonlethal whale research. The court also raised a number of concerns about the research methodologies used by Japan, including questions about the duration of the studies to produce scientific data, sample sizes and statistical significance of the results, flaws in the research methodology related to the fin whale, weaknesses in benchmarking and framing the concept of the studies, failure to adjust study protocols, and lack of communication about research results.

The ruling of the court was that Japan had violated its global responsibilities to the principles of the ICRW by using factory ships to catch, kill, and process fin whales. The court based its decision on the legal foundation governing whaling as well as the scientific analyses bolstering Australia's claims. The court reconfirmed that the objectives of the ICRW emphasized conservation of whale stocks and sustainable management of the whaling industry. It concluded that Japan's program in the Southern Ocean did not meet the parameters of a robust research or conservation program. However, the Court did not pronounce on the legality of Japan's whaling program in the northern Pacific.

The ICJ's ruling also did not touch upon the conflict between those parties or countries opposed to hunting of whales and those who accepted whale hunting as part of their culture or wildlife management programs. Therefore, its decision did not affect Norway or Iceland, two other countries engaged in whaling. Even though these countries were also signatories to the ICRW, they objected to the moratorium on whaling activities.

The court's ruling provoked reactions from Japan, from Greenpeace (its adversary on the land and the high seas), and Australia, the nation that had successfully brought the case against Japan. While Japan expressed disappointment about the court's ruling against it, by a vote of 12–4, it pledged to implement the ICJ's decision. Japan also indicated it would study its future options and alternatives. Greenpeace welcomed the ICJ's decision and stated that its long struggle against Japan's whaling boats had been vindicated. Australia was pleased with the outcome of its long legal battle and the effective ban on Japan's whaling in the Southern Ocean, and invited Japan to collaborate on nonlethal research on whales.

Camille Gaskin-Reyes

REFERENCES AND FURTHER READING

Ganderton, Tom. "Japan's 'Research Whaling' Ruled Illegal by International Court of Justice." Greenpeace. March 31, 2014. http://www.greenpeace.org/international/en/news/Blogs/makingwaves/japan-research-whaling-ruled-illegal-International-Court-of-Justice/blog/48725/.

International Court of Justice. http://www.icj-cij.org/homepage/.

International Whaling Commission. "Key Documents." https://iwc.int/convention.

Lee, Jane J. "Japan Halts Whaling Program in Response to International Court Ruling." *National Geographic*, April 2, 2014. http://news.nationalgeographic.com/news/2014/03/140331-whaling-japan-international-court-ocean-animal-conservation/.

ANNOTATED DOCUMENT

Berlin Rules: Principles of International Law Governing the Management of All Waters

Background

Beginning in 1996, the International Law Association (ILA) started reformulating the Helsinki Rules to incorporate international environmental law and international human rights law into water management. The project concluded in August 2004 with the formulation of the Berlin Rules on Water Resources. The Berlin Rules represented a new framework of international water laws that placed more emphasis on ecological integrity, sustainability, public participation, and minimization of environmental harm. In spite of its adoption by the ILA, there is still no mechanism in place to enforce the Berlin Rules. The following section includes a summary and excerpts of selected articles of the Berlin Rules.

Summary

The Berlin Rules confirm the right of persons affected by decisions on water use to have a say in those decisions. The duty of states is to take the appropriate steps to assure civic participation, given the international recognition that all people have right of access to water. This duty includes the obligation of states to provide information to the public on nationally and jointly managed water resources.

The rules pay attention to the rights and duties of nations and states engaged in joint management of water resources. States are encouraged to avoid wastage and duplication of efforts, and to cooperate in the collective management of water resources, whether they are internationally shared or are national sources such as surface or ground water.

The rules recommend that states integrate the management of water of international drainage basins into the use of nonwater resources such as the mining and industry sectors in order to minimize pollution and foster compliance of those industries with environmental laws or regulations.

The Berlin Rules emphasize the management of water under sustainable principles. They encourage countries to safeguard and improve the quality of water supplies in harmony with established goals of peace and economic and social welfare. They recognize the need to balance the goals of economic growth with key social and environmental values, and highlight the obligation to avoid wastage, destruction, or depletion of water resources to the detriment of future generations.

According to the rules, nations have an obligation to minimize environmental harm—to their citizens and to other states and communities sharing transboundary waters. This obligation derives from general principles of international environmental law, including the principle of equitable utilization, a gold standard for states to manage shared waters. The rules reiterate the rights of states to an equitable and reasonable share of waters when it is a shared public good, as well as duties and obligations to other nations to show fairness in the use of joint waters.

Some articles of the rules call upon nations to avoid significant harm to the water reserves of other states as a result of economic and other activities, especially when such activities occur in the waters of an international drainage basin.

According to the rules, the legitimate use of water resources by a state depends on the consent of the governed, which means that people should have a voice in how water is used. States are encouraged to seek informed consent of their citizens through direct participation, representation, or consensus-seeking methods.

Given the importance of water in people's lives, the rules reaffirm that the internationally recognized right of people to participate in decisions affecting their lives and their welfare should also apply to decisions concerning water sources. This means that the population should be informed about their rights and duties as citizens and stakeholders in the management and conservation of water. According to the rules, the information provided by the state should be easily understandable to all people involved.

Some articles of the rules draw upon international human rights law and recognize that the rights and interests of indigenous communities living in water basins should be considered in the development of plans, projects, and other activities in these areas. The inclusion of this point mirrors international law, which recognizes and respects the rights of indigenous peoples to self-manage their water resources as set forth in the ILO Convention Concerning Indigenous and Tribal Peoples in Independent Countries (#169).

The articles also affirm the right of indigenous groups under international law to prompt adequate, just, and effective compensation as well as any other appropriate measures to address damage from activities carried out in international drainage basins. The rules affirm that affected or dispossessed people have the right to fair and just compensation if they are displaced from their homes or lose their property and cultural assets. If the use of indigenous areas is proven to be in the interest of the greater good of the entire population or the protection of the basin as a whole, the rules indicate that the principle of fairness should apply in compensation measures.

The rules express the need for stricter standards to preserve the ecological integrity of entire aquatic ecosystems and the ecological health of watercourses, taking into account seasonal variations and natural patterns of flow. There is an obligation to preserve minimum stream flows to ensure that states and communities in downstream areas are not deprived of water.

Some articles speak to the disruption of the biological integrity of an ecosystem through the introduction of alien species. While not all species might be highly detrimental to all ecosystems, the rules mention the need to avoid invasive species that negatively affect local ecosystems. The legal basis of this requirement is the Biological Diversity Convention and other environmental agreements signed by nations. The introduction of invasive species in watercourses is prohibited if they are expected to have detrimental effects on ecosystems. The articles do not mention genetically modified organisms.

International environmental law generally prohibits the introduction of hazardous substances into the environment. Where it becomes absolutely necessary to use such substances, according to the rules, these substances should be treated to reduce their toxicity before release into water ecosystems. States are required to take appropriate measures to assure that private actors do not dump hazardous substances into water.

Given the imperative of environmental protection, some articles call upon nations of the world to step up measures to prevent or correct the pollution of internationally shared waters. This obligation applies to the management of the waters within a state's control as well as outside national jurisdiction. It requires countries sharing international water basins and resources to treat and dispose of wastes, pollutants and hazardous materials with the best available techniques to minimize environmental harm from pollution sources. While the "polluter must pay" principle is not mentioned as a goal, the concept of avoidance of harm to others and the obligation to clean up is included in the rules.

Some articles address the issue of water quality and guidance regarding the obligation of each nation to set water quality goals within its territory or jurisdiction. The assessment of environmental impacts of proposed water projects is also important in both domestic and shared waters. The World Bank—in sync with the Berlin Rules—requires prior environmental impact assessments before considering a loan or grant, regardless of whether the project would have transboundary effects.

Camille Gaskin-Reyes

Excerpts of Selected Articles

Article 4: Participation by Persons
States shall take steps to assure that persons likely to be affected are able to participate in the processes whereby decisions are made concerning the management of waters.

Article 5: Conjunctive Management
States shall use their best efforts to manage surface waters, groundwater, and other pertinent waters in a unified and comprehensive manner.

Article 6: Integrated Management
States shall use their best efforts to integrate appropriately the management of waters with the management of other resources.

Article 7: Sustainability
States shall take all appropriate measures to manage waters sustainably.

Article 8: Minimization of Environmental Harm
States shall take all appropriate measures to prevent or minimize environmental harm.

Article 12: Equitable Utilization
1. Basin States shall in their respective territories manage the waters of an international drainage basin in an equitable and reasonable manner having due regard for the obligation not to cause significant harm to other basin States.

2. In particular, basin States shall develop and use the waters of the basin in order to attain the optimal and sustainable use thereof and benefits therefrom, taking into account the interests of other basin States, consistent with adequate protection of the waters.

Article 16: Avoidance of Transboundary Harm
Basin States, in managing the waters of an international drainage basin, are expected to avoid and prevent acts or omissions within their territory that cause significant harm to another Basin State, given the rights of neighboring countries to make equitable and reasonable use of those waters.

Article 18: Public Participation and Access to Information
1. In the management of waters, States shall assure that persons subject to the State's jurisdiction and likely to be affected by water management decisions are able to participate, directly or indirectly, in processes by which those decisions are made and have a reasonable opportunity to express their views on programs, plans, projects, or activities relating to waters.

2. In order to enable such participation, States shall provide access to information relevant to the management of waters without unreasonable difficulty or unreasonable charges.

3. The information subject to access under this Article includes, without being limited to, impact assessments relating to the management of waters.

4. In providing information consistent with this Article, States need not provide access to information that would compromise:

a. Intellectual property rights, including commercial or industrial secrets;
b. Rights of individual privacy;
c. Criminal investigations or trials;
d. National security; and

e. Information that could endanger ecosystems, historic sites, and other naturally or culturally important objects or locations.

Article 19: Education
States shall undertake education at all levels to promote and encourage understanding of the issues that arise under these Rules.

Article 20: Protection of Particular Communities
States shall take all appropriate steps to protect the rights, interests, and special needs of communities and of indigenous peoples or other particularly vulnerable groups likely to be affected by the management of waters, even while developing the waters for the benefit of the entire State or group of States.

Article 21: Duty to Compensate Persons or Communities Displaced by Water Projects or Programs
States shall compensate persons or communities displaced by a water program, plan, project, or activity and shall assure that adequate provisions are made for the preservation of the livelihoods and culture of displaced persons or communities.

Article 24: Ecological Flows
States shall take all appropriate measures to ensure flows adequate to protect the ecological integrity of the waters of a drainage basin, including estuarine waters.

Article 25: Alien Species
States shall take all appropriate measures to prevent the introduction, whether intentionally or otherwise, of alien species into the aquatic environment if the alien species might have a significant adverse effect on an ecosystem dependent on the particular waters.

Article 26: Hazardous Substances
States shall take all appropriate measures to prevent the introduction of hazardous substances into the waters subject to its jurisdiction or control.

Article 27: Pollution
1. States shall prevent, eliminate, reduce, or control pollution in order to minimize environmental harm.

2. When there is a relevant water quality standard established pursuant to Article 28, States shall take all appropriate measures to assure compliance with that standard.

3. States shall ensure that wastes, pollutants, and hazardous substances are handled, treated, and disposed of using the best available techniques or the best environmental practices, as appropriate to protect the aquatic environment.

Article 28: Establishing Water Quality Standards

1. States shall establish water quality standards sufficient to protect public health and the aquatic environment and to provide water to satisfy needs, in particular for:

a. Providing drinking water of sufficiently good quality for human health;
b. Preserving ecosystems;
c. Providing water for agriculture, including irrigation and animal husbandry; and
d. Providing for recreational needs with due regard for sanitary and aesthetic requirements.

2. Standards established under this Article shall include, among others:

a. Specific quality objectives for all waters within a State's jurisdiction or control, taking into account the uses of the particular waters; and
b. Specific quality objectives applicable to a particular basin or part of a basin.

Article 29: The Obligation to Assess Environmental Impacts

1. States shall undertake prior and continuing assessment of the impact of programs, projects, or activities that may have a significant effect on the aquatic environment or the sustainable development of waters.

2. Impacts to be assessed include, among others:

a. Effects on human health and safety;
b. Effects on the environment;
c. Effects on existing or prospective economic activity;
d. Effects on cultural or socio-economic conditions; and
e. Effects on the sustainability of the use of waters.

Source: International Law Association. Berlin Conference. Water Resources Law. 2004. Available online at http://www.internationalwaterlaw.org/documents/intldocs/ILA_Berlin_Rules-2004.pdf. Used by permission.

PERSPECTIVES

Could Conflicts over Water Actually Promote Peace and Stability?

Overview

There is a simple truth: water resources are finite, yet vital for all aspects of human society and ecosystem functioning. A debate has emerged over whether water resources are an instigator of conflict or whether the stresses over water promote negotiation and compromise. Compounding the problem is climate change and its impact on the spatial and temporal distribution of water, making it more difficult to store and transport.

Further, as more of the planet moves to the free market model, there is the debate over whether water is a universal human right or just another tradable resource commodity.

Agriculture accounts for the highest percentage of water resources use worldwide; estimates range between 70 percent and 90 percent (Siebert et al., 2010). Geographically, there are over 280 internationally shared surface water resources worldwide (TFDD 2015). A recent infographic designed by UNESCO indicates that there are upwards of 3 billion people who lack access to safe drinking water and another 2.5 billion without access to sanitation (UNESCO 2015). A growing human population, with a sizable number already at risk for safe water access and a need for increased agricultural production in an era of unpredictably changing climate, results in conflicting needs and uses of available water resources (Dabelko 2008).

There are three schools of thought as to where the mounting complexity surrounding water resources may lead: 1) Countries will go to war over water, 2) water tensions will result in a global hegemony of control and distribution, and 3) water as vital to human survival will promote negotiation, compromise, and global stability. The last scenario clearly suggests that water resources can offer a diplomatic platform for peace and stability. This essay will explore these ideas and their bases, and give a few case study examples.

Water and War

In the early 1990s the Cold War officially ended with the dismantling of the Soviet Union and its communist-inspired economic and social ideology. As a result, securities scholars began to shift their focus to other threat sources (Buzan et al. 1998). Natural resources, such as oil and water, were identified as potential destabilizers and conflict starters. Typically, when experts refer to water wars, they point to a few prime examples, such as the Brahmaputra River (China and India), the Jordan River (Israel/Jordan/Palestine), Africa's Nile River basin (shared by 11 countries), and the Tigris and Euphrates Rivers (Turkey/Iraq) (Gleick 1993; Homer-Dixon 1991; Brown 2013; and Kaplan 1994). Each of these cases features existing geopolitical tensions or existing conflict.

The Pacific Institute, a not-for-profit environmental research organization, developed a Water Conflict Chronology that seeks to document all water-related conflict (Gleick 2015). Water, the institute argues, may not be the sole cause of a conflict, but it is often a contributing factor. The civil war in Sudan is cited as an example of a war partly caused by water and drought. Conflict between Israel and its neighbors is cited as another struggle over securing water resources. In another case a conflict arose in Bolivia called the Cochabamba Water War, where local farmers protested government proposals to commodify the water resources. Ultimately, after tense protests involving rocks and Molotov cocktails, the farmers were able to overturn these efforts (Shultz 2005).

Hydro-hegemony: Coerced Control of Water Resources

Some scholars argue that water tensions invariably promote instability, yet instead of direct violence result in domination and coercion. This perspective is termed "hydro-hegemony." Hegemony is a geopolitical term used to describe dominance of one group/country/ideology over another. The root of this perspective is that there will not be direct, violent conflict over water, but rather a diplomatic domination with implied threat. Water resources do not lead to violence or fair cooperation, but rather regional asymmetry related to power structures and information sharing between the stakeholders in question (Zeitoun 2006; Wegerich 2008).

Examples of existing hydro-hegemony include water resources shared between the United States and Mexico, or the Nile River, which is dominated by Egypt. Other examples include Israel-Palestinian relations, China's water domination in east Asia, or the hydropolitics in the Aral Sea basin, which is still dominated by Russia. In these cases, it is recognized that one country has more power than another and uses that power to exploit water resources, sometimes with full consent, however derived, from the other stakeholders.

In the case of the United States, the country has used treaties and water development schemes to dominate water resources with its neighbor, Mexico. The United States has more power—economically, diplomatically, and/or militarily—to force the country of Mexico to comply with its decisions regarding a shared water resource, such as the Colorado River. The result is that, while the United States develops and uses the resource, Mexico must make due with what is left over.

Water for Peace and Stability

Finally, the idea that water conflict can result in increased global peace and stability stems from the idea that water resources bring stakeholders to the table, that water issues actually necessitate negotiations, diplomacy, and compromise rather than perpetual conflict (Wolf 1998; 2004). Contrary to the two "conflict" schools of analyses, this research emphasizes that there are many more shared water resources where coordination and stability are the norm. For instance, Oregon State University's Transboundary Freshwater Dispute Database has catalogued over 280 shared, stable and peaceful basins (TFDD 2015). As Wolf argues, despite water scarcity, climate change impacts, or any other physical threat to water supply, countries resort to peaceful negotiations when there is institutional capacity to do so (Wolf 2003). Institutional capacity includes existing precedent including international agreements, treaties, organizations that manage water resources, and/or international bodies that manage water resources. This capacity provides a regular platform to allow for open dialogue that avoids conflict and promotes peace and stability (Wolf 1998; Kramer 2008; Veilleux 2013).

Recent negotiations on the Nile River between Egypt, Sudan, and Ethiopia are an example of peaceful diplomacy (Davison and Feteha 2015). Ethiopia started construction of a large-scale dam on its portion of the Nile River, the Blue Nile, unilaterally. The existing Nile treaty signed during the colonial era states that only Egypt and Sudan are legally guaranteed 100 percent water and there should be no upstream development on the water resources. Ethiopia was never colonized and did not sign this treaty. Ethiopia's move to develop a dam is legal, but would disrupt the flow of the Nile for a period of time to fill the reservoir behind the dam. Because of this, diplomatic talks between the Egyptian, Ethiopian, and Sudanese governments began in 2013, and in 2015 resulted in preliminary conclusions that allow Ethiopia to continue with the dam, while guaranteeing Egypt a portion of the generated electricity as well as technical say in the filling of the reservoir.

The future of shared water resources is a hot topic of debate. While some find that shared water resources will result in violence, instability, or asymmetrical domination within and between countries, others argue that water resource tensions offer the ability and opportunity for stakeholders to communicate and negotiate. Whether a country is calling for war over the resource, forcing the signing hand of another sharing the resource, or finding comprehensive solutions together, the importance of this water in politics, economics, and environment cannot be overstated. With increasing demand and dwindling supplies, water resource conflicts will continue; how we choose to handle the issue as a global community remains to be seen.

Jennifer C. Veilleux and Mark Troy Burnett

Perspective 1: Transboundary Water Management—Some Success Stories

When a river, sea, lake, drainage basin, or any other body of water crosses international borders, it is commonly assumed that conflict will result. This is especially the case when water resources overlap with already tense borders. In arid Central Asia, upstream Kyrgyzstan and Tajikistan are planning to build dams that could limit downstream Uzbekistan's water resources. Uzbek president Islam Karimov seethed that the upstream countries' presidents "forget that the Amu-Darya and Syr-Darya are trans-border rivers" (Lillis 2012). And in Africa, the now-ousted president of Egypt, Mohammed Morsi, threatened action over a $4.2 billion hydroelectric dam Ethiopia has proposed to build upstream on the Nile. "If it loses one drop, our blood is the alternative" (Al Jazeera 2013), he declared, referring to the river that makes agriculture—and indeed, human civilization—possible in Egypt.

Despite these headline stories, even as drought and increasing population have strained global water supplies, cooperation rather than conflict still persists in water management—including in the Middle East and North Africa (De Stefano et al. 2010). As such, we may be able to paint a rosier picture of hydropolitics than the media commonly portrays. Stories of cooperation do not sell as many newspapers as stories of bloodshed over watersheds, but that does not mean they do not exist. Indeed, highlighting stories of cooperation is essential to show that successful international management is possible—and necessary.

As Irina Bokova, director-general of UNESCO, urged, "With transboundary river basins and aquifer systems representing almost half the earth's surface, water cooperation is vital for peace" (UNDESA 2014). In total, there are 276 trans-boundary river basins in the world, with 256 of those being shared by two to four countries (UNESCO 2013). Indeed, with freshwater resources being so vital for human life, transboundary resources can even bring countries to the table to work towards a mutually beneficial solution—perhaps showing, as the United Nations has suggested, that "water more often unites than divides people and societies" (UNDESA 2014).

Successful Examples

In Europe, transborder water management has been fairly successful for a number of reasons. Compared to Africa, Asia, and South America, Europe enjoys relatively low tensions, while the existence of the European Union (EU) has brought countries closer together. In fact, the EU Water Framework Directive (WFD) has established a legal framework to enhance water quality status in member states. They are each individually required to obtain good status in all of their surface water by 2015 and all of their groundwater by 2027 (ICPDR 2009). In order to meet these requirements, EU countries have had to work together, such as along the Danube River Basin.

Located in Central Europe, the Danube River Basin crosses more countries than any other basin in the world—19 in total, including both EU and non-EU members. The Danube River Basin Management Plan, adopted in 2000, emerged directly out of the EU WFD, which requires member states to "endeavour to produce a single river basin management plan" (ICPDR 2009). Although the non-EU countries of Bosnia and Herzegovina, Montenegro, Serbia, Slovakia, and Ukraine are not legally obligated to join or to improve their water quality, they have still provided data in the same way on everything from wastewater collecting systems to the number of protected areas. This sort of cooperative work helps harmonize standards of water quality and access across entire river basins, a method which can go a long way towards managing what is ultimately an international environmental resource. The International Commission for the Protection of the Danube River (ICPDR) has turned into "one of the largest and most active international bodies of river basin management expertise in the world," and it may be able to export its knowledge to other countries seeking advice. As nature does not respect political borders, successful management must inevitably be transborder, too.

Challenges

Transferring management advice to other contexts, however, is fraught with challenges. Referring to issues in India, an International Water Management Institute (IWMI) briefing warned: "The problems that river basin institutions in the developed world successfully address—such as pollution, sediment buildup in rivers and the degradation of wetlands—are not the top priorities for Indian policy makers and people" (Vidyanagar 2002). Indeed, in Asia, no organization exactly like the ICPDR has yet emerged. Other more loosely organized groups like the Mekong River Commission in Southeast Asia may be signs of better management to come, but then again, this commission only includes the downstream countries of Thailand, Laos, Cambodia, and Vietnam. The upstream countries of China and Burma are merely observers. Without both upstream and downstream countries coming to the table as around the Danube, fully integrated management will not be possible.

Thus, some caution should be exercised in thinking that tussles over water management can ultimately promote peace and stability. In the Middle East, while geographer John Allen determined that conflicts over water have actually lessened since the 1970s, this is largely because countries in the region have been able to import "virtual water" from other places around the planet. By importing grain grown abroad, Middle Eastern countries are essentially purchasing crops grown with water elsewhere, lessening the burden on their own resources (Allan 2002). Seen in another light, however, the international trade of water has served as a temporary Band-Aid that has allowed countries to avoid solving water management conflicts in their own neighborhood. Uzbek president Karimov, the leader who is so concerned about the dams his neighbors on higher ground are building, warned, "Everything can be so aggravated that this can spark not simply serious confrontation but even wars" (Lillis 2012). When tensions become so high that outright war rather than cooperative management of a water basin seems more appealing, even virtual water will not be an apt substitute. Yet, hopefully, statements such as Karimov's will remain heated rhetoric and the global trend of cooperation rather than conflict will prevail.

Mia Bennett

Perspective 2: Countries Do Not Go to War over Water

Water resources are finite and necessary, and as such represent a potential for conflict. Scholars and practitioners have outlined a hierarchy of needs related to water resources: ecosystem functioning, domestic household use, small-scale agriculture, commercial agriculture, industrial use, municipal use, and national projects (such as a dam). Users in this suite of needs include flora and fauna, individuals, communities, cities, parks, governments, and corporations. Such a range of uses and users for finite water resources can lead to conflicting uses that change access, quality, and quantity of the resources for all involved. In many cases, a fine balance between stakeholders exists that may be disrupted when a change event occurs such as a drought, a population increase, an economic policy shift, a governmental change, or a dam development project. Change can lead to potential conflict scenarios over shared water resources. However, given communication, diplomacy, and existing frameworks for management, there is an opportunity for water conflict to turn into opportunities for peace and stability.

Historically, there have been no documented international violent conflicts exclusively over water resources. Indeed, countries have frequent disputes over shared water, but governments will typically negotiate and compromise. When a water resource crosses international boundaries, interest for diplomacy in the form of a treaty or other equivalent agreement is high because international relations have much more visibility on the world stage. However, there is no international water law, so when countries meet to discuss water sharing ideas the decisions are highly contextual. Countries that share water resources have different economic, social, cultural, and political goals for and uses of the water resources. Some level of sharing water resources is almost always agreed upon in some fashion, at some point, though the outcomes are not always fair.

Discussions Are Important

Some scholars observe that relationships on shared water resources—such as China being upstream from Laos, or the Navajo Nation within the territory governed by the U.S. federal government—are imbalanced, or hegemonic, relationships. However imbalanced, water sharing allows for the opportunity of asymmetric powers to engage, something that could lead to expanding the diplomatic conversations to other topics. Recently, in the Nile River basin, diplomatic meetings that originated about shared water resources gave the opportunity for Egypt and Ethiopia to discuss common terrorist threats and how to manage these together. Water resources discussions between Israel and the Palestinian state have continued despite surrounding violent conflict in the region. Agreements made official through written documents are considered peaceful and stable agreements by nature, and in the case of water, by necessity.

Treaties and other agreements about water resources often include elements about using the shared water resources that are peripheral. The treaty between the United States and Canada regarding the Columbia River watershed includes text pertaining to electricity generation and flood control, and amendments scheduled to go into effect in 2024 will include consideration for riparian ecosystems. The highest consumption of water resources worldwide is for agriculture, but one of the highest uses is for power generation—directly in generating hydropower and indirectly to create steam to drive turbines in coal, oil, and nuclear power plants, as well as cooling in these plants. Water is used in industrial processes from creating plastic goods to jeans to metal works. One scholar posits that everything we consume and use, from the moment we wake up in the morning to going to sleep at night, has water imbedded in the product. This is called virtual water. For this reason, water agreements may include trade agreements for goods or energy, transportation routes, and a myriad of diplomatic decisions that may not make it into the water agreement but are fostered by the conversation.

Conflict Resolution

Water conflicts have been found to occur at local and regional levels, and subsequently are often resolved through local or regional mechanisms. When there are institutional mechanisms and diplomatic norms to manage resources (such as treaties, agreements, traditions, or organizations), the likelihood for negotiation and conflict resolution is much higher. This is because of an established method of communications and understood power dynamics sets a precedent for how to communicate across stakeholders' interests. Organizations that manage water usually have dispute resolution methods by which to handle conflicts when they arise. These methods can include fines for users, specialized lawyers and experts for court proceedings, and meetings for ongoing communications between potentially conflictive parties.

An example is the acequia system in northern New Mexico. The acequia canal system is a water management system that includes an institutional structure of negotiation and dispute resolution. Participants in the water compact meet frequently and the system even includes a mayordomo—a yearly, rotating leadership/overseer position. The mayordomo oversees water flow through a series of engineered water systems, is in charge of inspiring users to keep their sections clear of obstruction, and fines users who take more than their designated share.

When conflicts of use occur, salient issues of economic, cultural, social, and environmental reasons are elevated. Sometimes these issues may lead to discoveries that cause some serious policy changes and water-use culture changes. The recent drought in California has included new dialogue about people's personal water bills, the filling of swimming pools, the watering of golf courses, and the types of crops grown. Farmers have considered changing from water-intensive crops, such as cotton, to less intensive crops. Towns are exploring new methods for how water consumption is metered. Homeowners may decide to zero-scape their previously green lawns and industries may decide to implement water efficient technologies. In 2015, to allay fear and mitigate conflict, the state of California enacted a comprehensive policy to encourage reduction and efficiency in usage.

Conflicts, while potentially explosive, often serve to promote awareness, responsibility, and conservation. Water disputes can lead to independent studies and important questions, such as: "Is this water use sustainable?" "Is this water use necessary?" and "How can we maintain current use with less so that other stakeholders may also benefit from our shared resources?" Often what starts out as a resource dispute with a winner-take-all attitude results in a win-win scenario, due to the forced dialogue between potentially competing parties.

Tensions over water resources will continue as population grows and water and land development increase. Doubtless, the disputes will be exacerbated by the unknown hydrologic impacts of climate change. If past precedent is any indication, though, there is hope that society will, in general, use water resource disputes as opportunities for negotiation, compromise, innovation, peace, and stability.

Jennifer C. Veilleux

REFERENCES AND FURTHER READING

Al Jazeera. "Egypt Warns Ethiopia over Nile Dam." June 11, 2013. http://www.aljazeera.com/news/africa/2013/06/201361144413214749.html.

Allan, J. A. "Hydro-Peace in the Middle East: Why No Water Wars? A Case Study of the Jordan River Basin." *SAIS Review* 22, no. 2 (2002): 255–272.

Brown, Lester. "The Real Threat to Our Future is Peak Water." *The Guardian*, July 10, 2013. http://www.theguardian.com/global-development/2013/jul/06/water-supplies-shrinking-threat-to-food.

Buzan, Barry, Ole Wæver, and Jaap De Wilde. *Security: A New Framework for Analysis*. Boulder, CO: Lynne Rienner Publishers, 1998.

Dabelko, Geoffrey D. "An Uncommon Peace: Environment, Development, and the Global Security Agenda." *Environment: Science and Policy for Sustainable Development* 50 (2008): 32–45.

Davison, William, and Ahmed Feteha. "Egypt, Sudan Edge Toward Cooperation on Ethiopia's Nile Dam." Bloomberg. http://www.bloomberg.com/news/articles/2015-03-24/egypt-and-sudan-edge-toward-cooperation-on-ethiopia-s-nile-dam.

De Stefano, Lucia, Paris Edwards, Lynette de Silva, and Aaron T. Wolf. "Tracking Cooperation and Conflict in International Basins: Historic and Recent Trends." *Water Policy* 12, no. 6 (2010): 871–884.

Gleick, Peter. Pacific Institute Water Conflict Chronology Database. 2015. http://worldwater.org/water-conflict/.

Gleick, Peter. "Water and Conflict: Fresh Water Resources and International Security." *International Security* 18 (1993): 79–112.

Homer-Dixon, Thomas F. “On the Threshold: Environmental Changes as Causes of Acute Conflict.” *International Security* 16, no. 2 (Fall 1991): 76–116.

International Commission for the Protection of the Danube River (ICPDR). Danube River Basin District Management Plan. Document Number IC/151. Vienna: ICPDR Secretariat, 2009.

Kaplan, R. D. “The Coming Anarchy.” *Atlantic Monthly* 273 (1994): 44–76.

Kramer, A. *Regional Water Cooperation and Peacebuilding in the Middle East*. Initiative for Peacebuilding, 2008.

Lillis, J. “Uzbekistan Leader Warns of Water Wars in Central Asia.” EurasiaNet.org. September 7, 2012. Available at http://www.eurasianet.org/node/65877.

Pachauri, R. K., and A. Reisinger. “IPCC Fourth Assessment Report.” Geneva: Intergovernmental Panel on Climate Change (IPCC), 2007.

Shultz, Jim. “The Politics of Water in Bolivia.” *The Nation*, January 28, 2005. http://www.thenation.com/article/politics-water-bolivia.

Siebert, S., J. Burke, J. M. Faures, K. Frenken, J. Hoogeveen, P. Döll, and F. T. Portmann. “Groundwater Use for Irrigation—A Global Inventory.” *Hydrology and Earth System Sciences* 14, no. 10 (2010): 1863–1880.

Transboundary Freshwater Dispute Database (TFDD). Oregon State University. 2015. http://www.transboundarywaters.orst.edu/.

UNESCO, World Water Day Infographic. 2015. http://www.unesco.org/new/en/world-water-day.

UNESCO. “UN Water World Water Day 2013: International Year of Water Cooperation.” 2013. Available at http://www.unwater.org/water-cooperation-2013/water-cooperation/facts-and-figures/it/.

United Nations Department of Economic and Social Affairs (UNDESA). “International Decade for Action ‘WATER FOR LIFE’ 2005–2015.” October 24, 2014. Available at http://www.un.org/waterforlifedecade/water_cooperation.shtml.

Veilleux, J. C. “The Human Security Dimensions of Dam Development: The Grand Ethiopian Renaissance Dam.” *Global Dialogue* 15, no. 2 (2013): 1–15.

Vidyanagar, Vallabh. “The Challenges of Integrated River Basin Management in India: Issues in Transferring Successful River Basin Management Models to the Developing World.” International Water Management Institute IWMI-TATA Water Policy Program. Gujarat, India: International Water Management Institute, 2002, 1–6.

Wegerich, K. “Hydro-hegemony in the Amu Darya Basin.” *Water Policy* 10, no. 2 (2008): 71–88.

Wolf, A. T. “Conflict and Cooperation along International Waterways.” *Water Policy* 1, no. 2 (1998): 251–265.

Wolf, A. T. *Regional Water Cooperation as Confidence Building: Water Management as a Strategy for Peace*. Berlin: Adelphi Research, 2004.

Zeitoun, M., and J. Warner. “Hydro-hegemony: A Framework for Analysis of Transboundary Water Conflicts.” *Water Policy* 8, no. 5 (2006): 435–460.

6 WATER AS THE BASIS FOR SOCIAL AND ECONOMIC LIFE

OVERVIEW

Water, air, food, and shelter are basic needs for human survival. Water is part of the origin and the scaffolding of life. It supports a myriad of social, economic, and cultural pursuits: from daily tasks to agriculture, industry, tourism, health, religious rituals, and other products and services essential for human welfare. The expansion of agriculture and fisheries—due to world demand—and the upsurge of manufacturing and world trade have impacted the quantity and quality of fresh and marine water resources in most countries.

This chapter delves into the range of water-related activities and the benefits and challenges of human use of water resources in a globalized world. The case studies explore the controversies of water privatization, the key role of the Panama Canal as an interoceanic passage, and the perils of waterborne diseases for human health and wellbeing. The perspectives section discusses the challenges and obligations of providing clean water for the world's population.

Emerging Water Stress

For the most part, people have always considered water as a renewable resource that flows and reflows as part of the natural cycle of evaporation and precipitation. After all, in most places on the earth, it falls from the sky in different forms of precipitation, is an integral part of social, cultural, and economic life, and, most of all, it is the liquid without which we cannot survive.

Since demand for water has increased globally and in many places outpaced supply, nations and citizens across the globe have started to grasp that there are limits to what once appeared to be endless streams of water in the atmosphere, above the surface, or below the ground.

People in both developed and developing countries are experiencing water stress related mainly to population growth, overexploitation of aquifers, ground- and surface-water pollution, climate change, glacial melt, and increasingly erratic precipitation patterns. There are serious concerns about the adequate and equitable supply of clean water for future generations, especially in drought-ridden countries and the poorest communities.

The reduced availability of clean drinking water has given rise to the growth and high economic returns of the global bottled water industry. The heightened interest of investors in the acquisition of public water utilities, including the purchase of private land for the value of its aquifers and the rising price of water as a commodity, all reflect the economic value of water, particularly in countries without a safe water supply.

Water Consumption

To grasp the importance of water for everyday life and livelihoods, we can divide water consumption according to economic activities. About 70 percent of global freshwater use goes to agriculture, 20 percent to industry, and about 10 percent to household and public use (UNEP). It is hard to think of any human activity that does not involve water. A list that immediately comes to mind includes: planting, drinking, cooking, cleaning, irrigating, fishing, boating, cooling, mining, manufacturing, recreating, building, worshipping, designing, painting, transporting, and healing, but this list is not exhaustive.

The greatest use of water, however, is for agriculture. Humans need water to grow and irrigate crops for food and fodder for livestock. Agriculture started around 12,000 years ago, when early societies made a crucial transition from nomadic to sedentary life. The single most important feature in this shift was the introduction of water management techniques for agriculture, in particular the use of irrigation in the domestication of crops.

This development made all the difference for civilization. Plant domestication and irrigation in arid areas reduced the uncertainty of life and starvation for hunters and gatherers. Instead of wandering in clans to hunt and gather wild nuts or berries, people could remain in one place, grow crops, store surplus food, domesticate animals, and build their societies.

The use of irrigation by the Sumerians in arid Mesopotamia was an important starting point for sedentary life. It was here that irrigation techniques developed, resulting in increased crop yields, surplus food production, and grain storage in this vulnerable, mostly dry area. The presence of the Tigris and Euphrates Rivers was an indispensable factor for the Sumerians to harness and channel water to fields.

In addition to food security, irrigation enabled trade and exchange of agricultural and nonagricultural goods. Surplus food also freed up time for the growth of more stratification in Sumerian society, which included farmers, artisans, scribes, traders, priests, rulers, and city dwellers. This advancement in turn led to the development of written language and literacy, the hallmark of emerging civilization.

Ironically, the harnessing of irrigation that developed Sumerian society also contributed to its demise. The society eventually collapsed due to the negative impacts of overirrigation: rising water tables, soil waterlogging, salinization and crop destruction, and the inability of the land to support further population growth in the area.

As agriculture was developing in Sumeria 12,000 years ago, 4 million people lived on the entire planet. In 1000 BCE—due to agricultural expansion—the world's population grew to about 50 million, and a thousand years later, to 230 million. In 1950 it was 2.5 billion, and in 2012 it had surpassed 7 billion (Sachs 2008, 60). The challenge remains—now and in the future—to balance population growth with food security, ecological equilibrium, and more equitable allocation of food in the poorest nations and communities worldwide.

Increase in Cultivated Land Area Means Greater Demand for Water

In line with population growth, there has been a remarkable increase in the land area under crop cultivation (especially grains) and the extension of pasturelands around the world. The expansion of irrigation was accompanied by innovations in seeds and crop varieties, new water and farming technologies, the extension of large, more mechanized farms, and more intensive application of fertilizers and pesticides. In 1800 there were approximately 8 million hectares under irrigation in the world; today there are over 300 million hectares. About 20 percent of the world's cultivated land is currently irrigated (Chellaney 2013, 73).

The highest growth in irrigation was in the 1960s, parallel to the use of petroleum by-products for fertilizers (natural gas) and pesticides (petroleum), widespread innovations in biotechnology, and the construction of dams, central to large-scale irrigation schemes (but also methane producers). These technological advances led to improvements in global food supply, but the flip side of the doubling of irrigation in the last four decades was increased depletion of groundwater and higher fossil fuel use for the agricultural sector.

While irrigation and higher crop yields put more food on the table globally, they did not eradicate hunger in all parts of the world due to distribution constraints and the inability of the poorest people and the landless to buy food. Nonetheless, since the 1960s, large irrigation projects have played a critical role in poverty alleviation and economic growth in Asia. Irrigation has made it possible for many water-deficit Asian countries to reduce the uncertainties of drought and erratic crop production, and to liberate themselves from dependency on seasonal rains, thereby leaving famine on the doorstep.

The Water Haves and Have-Nots

The world is divided into water-rich and water-poor countries. As a general principle, the natural endowment of countries with plentiful rainfall is a big factor in the success of agriculture and economic development. North America and most countries in Europe have the good fortune to have essentially won the water lottery. These continents are located in temperate climate zones with relatively long rainy periods, enabling a greater reliance on natural precipitation.

Hence, there is generally more rain-fed agriculture in industrialized countries than in developing nations, but not in all parts of developed countries. For example, the United States has a water-deficient belt in Texas, Arizona, and southwest California, which is experiencing major drought-related impacts on farming, and fierce competition for surface and groundwater resources.

Across the world, China and India are disadvantaged nations in terms of water endowment. Together, these two countries make up 37 percent of global population, but have only about 10 percent of the world's water and 19 percent of global arable land. Accordingly, we find almost 72 percent of the world's irrigated area in Asia, compared to 15 percent in the Americas. Overall, the countries with the most irrigated areas in the world are India, China, and the United States, all of which are experiencing water stress in some form or another, and are also subject to the impacts of climate change and further water scarcity.

Irrigation Issues

Notwithstanding irrigation's benefits for global food production and reducing famines, its negative effects have become more and more evident. Modern farmers using irrigation usually have to rely on the purchase of costly chemical fertilizers and pesticides (made from petroleum and natural gas) not available to poor farmers. Irrigation users also depend on the purchase or leasing of huge plots of land, farm machinery, wells, water pumps, and different types of agricultural processing and packaging equipment.

While agriculture is the sector of the economy that feeds the world, it is also the sector that uses the most water. Large-scale agriculture is associated with high quantities of greenhouse gas emissions through fertilizer use, livestock production, and reliance on mechanical and technological inputs. Some countries have become extremely dependent on dams and reservoirs to store and provide water for irrigation. There are two main problems with reservoirs: 1) they produce methane, if forests are flooded to fill them, and 2) severe drought causes higher evaporation of reservoirs and reduces water for irrigation.

The expansion of agriculture and irrigation systems has strained groundwater reserves in countries as culturally different as the United States and India, or as geographically apart as China and Chad. Huge irrigation schemes require high volumes of surface water or the pumping of groundwater. Excessive use of wells provokes withdrawals of water at a faster place than can be replaced by rainwater infiltration into the soil. But locked aquifers—water reserves sandwiched between impervious underground layers of rocks—are finite; once they become depleted, they cannot be replenished by rain or water runoff.

Inefficiencies in irrigation also cause water loss through excessive evaporation, water seepage, and deficient water pumping, transmission, and application methods. Millions of hectares of cropland are ruined in the world yearly as a result of obsolete irrigation techniques or leaked water. Drip-feed irrigation, which directs water droplets directly to roots of plants, wastes less water and is more effective than sprinkle or flood irrigation. However, this technology is costly and only used in 4 percent of global irrigated acreage.

As the experiences of the Sumerians have instructed the world (but have not always been heeded), irrigation can have negative side effects. In modern times, the same process has played out many times: inadequate drainage, evaporation of irrigated fields, waterlogging, and salt deposits. The problem is compounded by soil contamination with pesticides and fertilizers. These substances eventually find their way to the surface, pollute ground and surface water, and severely degrade land and marine ecosystems. For example, over the last 50 years, about 24,000 villages in northern and western China have been partially or completely abandoned due to soil pollution, salinization, and increased desertification (Barlow 2013, 17).

Drawing Reserves from Humanity's Water Bank

The process of unsustainable water depletion for agriculture (and other activities) can be compared to drawing cash reserves from a bank account and continuing to spend without depositing enough money to cover ongoing purchases and future expenses. We are drawing heavily on humanity's water bank.

Increasing water depletion is creating competition, which is likely to provoke a race for the remaining resources and a downward spiral even further. The net effect could be the equivalent of a run on a bank: heavy water users would rush to drill wells, impound surface water, or buy up enough water reserves to secure a share, before it dries up. Other members of society who could technically not store water or pay for water would be severely affected. Competition of this nature poses huge risks for the security of all water users in the long term and requires better water management to benefit all communities.

From the United States to Cameroon, water users, whether they are large farmers, small communities, individuals, agro-industrial concerns, industries, oil concerns, or tourism operators, continue to drill more wells and tap ground water for irrigation, household, and commercial purposes. This reduces water supply for local and neighboring wells and cross-border communities, where aquifers extend across national boundaries. Aquifers that have provided reliable streams of water for decades are in danger of going dry, especially locked aquifers, which cannot be replaced, not even by rainwater.

The Ogallala Aquifer

The drawdown of groundwater from such a locked aquifer is occurring in the United States. This is the case of the Ogallala Aquifer, a huge underground lake that extends from the eastern slope of the Rockies to the Texas Panhandle. For decades, it has provided water for agriculture and meatpacking industries on the plains. Currently, there are over 200,000 wells from this source, irrigating over 8.2 million acres. Its water levels are dropping fast, three feet per year since 1991 (Barlow and Clarke 2002, 16). But there are serious concerns that the aquifer could run dry in the 21st century, affecting billions of dollars in the food and livestock industry in states above the aquifer.

The 2011 sale of 16 trillion liters of Ogallala groundwater to private water suppliers for $103 million has caused additional worries about water hoarding (Barlow 2013, 80). Other fears were signaled by environmental and community groups regarding a proposal to build a Keystone XL pipeline to carry bitumen, an oil product from

Alberta, Canada, to refineries in Texas over some areas of the aquifer. This project was stopped, however, after a thorough environmental review by the U.S. government in 2015.

Global Water Shortages and Drought

Globally, water demand and reduced supply are creating mini-hotbeds of conflict among countries, municipalities, and users. In Africa, water shortages and droughts are causing famine, out-migration of affected people, and increasing conflicts among communities, as in Darfur, Sudan. In California, large-scale mechanized production and irrigation of fields, droughts, and market demand for water-thirsty and lucrative tree crops, i.e., almonds, have exacerbated water shortages and caused a review of state water policy, greater emphasis on conservation, and consumer action, and has made farmers rethink production modes and crop choices.

Agricultural Exports Are Water Exports

Even arid or semiarid countries with water problems are tapping their limited ground and surface water to grow crops for exports to other arid countries or to water-endowed countries. In a sense, water-deficient countries are exporting water stored within crop exports—also called virtual water—taking advantage of lucrative niche or organic markets in other countries. Seen from this angle, global trade in food is also about global trade in water, since it takes a lot of water to grow food for production and exports. The fact that a huge share of water goes to the agricultural sector raises concerns about the sustainability of continuous water withdrawals, barring further technological and efficiency improvements.

There is a social impact perspective as well. Water embedded in export crops in arid, developing countries uses up internal water resources of the producing country and may affect domestic food supply and food prices, since small farmers may lack water to grow food for themselves and the domestic markets. This increases the price of local produce for poor nonagricultural and urban consumers as well.

In a world of threatened water supplies, wealthy countries are in a stronger financial position to import water-intensive agricultural products from others, and to conserve their water resources. For example, Saudi Arabia and Japan import most of their food crops (and water footprint) from other countries. Due to widespread globalization and the importance of trade, more developing countries are relying on agricultural exports, using up huge quantities of water for their processing.

Some examples of virtual water exports are asparagus from Peru, food and beef from Ethiopia, and nonfood crops such as cotton from India. Peru relies on irrigation of its virtually rainless coastal desert to produce asparagus for export to North America and Europe. Ethiopia, a water-deficit, poor country—with its own unresolved food security problems—is building dams to support agricultural exports, and at the same time raising concerns in Egypt over potential reduction of downstream Nile waters.

India, too, is irrigating land at high capital costs, drawing on groundwater to produce edible and nonedible cash crops for the world, while it faces major challenges of

A farmer walks near an irrigation canal in Kadamtali about 41 miles east of Gauhati, India, August 11, 2009. (Anupam Nath/AP Photo)

ensuring adequate nutrition for its poorest and marginalized groups. To put things in perspective, it takes 37 gallons of water use before a consumer brews a cup of coffee, 634 gallons to eventually produce a hamburger, and 634 gallons to make a cotton shirt (Barlow 2013, 166). This indicates the potential water supply bottlenecks related to crop production, meat production, and virtual water exports.

There are social equity considerations regarding the growth of export agriculture in developing countries. While exports may boost overall economic growth, they benefit groups differently. Large-scale pumping and irrigation projects require costly investments, i.e., wells, pipes, pumps, tractors, seeds, fertilizers, pesticides, and complex credit and marketing arrangements, which small farmers cannot afford. Large projects also displace or marginalize small farmers, who cannot pay for the rising costs of land and water. The rising costs of inputs and land also drive up the price of local food.

Even in developed countries, agricultural production and exports have an impact on water consumption. For example, in California it takes 16 gallons of water to produce a handful of almonds, 42 to produce three mandarins, and 2 gallons to produce six pistachios. This state produces more than one-third of the vegetables and two-thirds of all the fruits and nuts grown in the United States, with farmers using about 80 percent of California's water to grow these products for both internal consumption and exports (Buchanan et al. 2015).

Genetically Modified Crops

The introduction of genetically modified (GM) crops is a recent but growing part of agricultural production. Every year more farmers around the world are cultivating GM crops. Most GM crops are in the United States, Brazil, Argentina, Canada, and India. In 2012 and 2013, developing countries planted more hectares of GM crops than industrialized countries. The most common ones are rice, corn, soybean, and cotton. Marketing campaigns of large GM corporations have been successful in targeting farmers in developing economies.

One feature of GM crops is the ability of some strains to resist drought, salinity, heat, and pests. These characteristics are attractive to countries facing low rainfall, drought, floods, or intrusion of saltwater into the soil. As the planet warms, weather events are likely to become more erratic, and pests are likely to multiply. GM crops offer possibilities to address these problems and expand food supply. Pest resistant crops may also reduce pesticide use and runoff of effluents into water systems and oceans. However, questions remain about the higher costs of GM crops, which could be prohibitive for small farmers, and their longer-term impacts on native species, i.e., cross-pollination of GM genes to non-GM plants and possible mutations of natural species.

Water in Agro-industry and Cattle Ranching

Water is essential for the meatpacking, animal feed, and chicken industries within the food supply chain. However, these activities generate tons of liquid and solid waste as by-products (ammonia, methane, phosphates, nitrates, and antibiotics), which can enter groundwater sources and contaminate surface water, entire watersheds, and coastal areas.

An average beef cow produces 11 times more phosphorus in its waste than a human; a hog generates 10 times more. Water runoff containing high phosphorus levels from livestock leads to nutrient overloading of lakes, eutrophication (oxygen loss), and rapid growth of algae. This process smothers surface water, kills fish and water organisms, and affects fishing, boating, swimming, and tourism. Examples of nutrient overload are found in the Gulf of Mexico and Lake Winnipeg in Canada, the world's 10th largest lake.

While agro-industries in the United States have attempted to reduce the disposal of effluents into water sources in compliance with environmental law, it is difficult for regulatory agencies to fully monitor and oversee the discharge of pollutants due to the huge quantities of farms and processing plants and their dispersion over widespread, rural areas. Millions of gallons of water are used on a daily basis to feed, slaughter, process, and clean animals, activities which produce huge amounts of sewage, wastewater, and sludge, and run the risk of polluting lagoons, bays, watercourses, and oceans.

Water and Bio-fuel

Large quantities of water are also needed for biofuel production. It is estimated that it takes about 2,500 liters of water to grow enough corn to produce one liter of corn ethanol for biofuel (Chellaney 2013, 67). The growing demand for ethanol in the United States has been a factor in the spread of irrigation of cornfields in areas with insufficient rainfall. Policy decisions to replace gasoline with more corn ethanol in order to reduce greenhouse gases (GHG) should assess all environmental trade-offs: while ethanol may substitute fossil fuels, higher corn production for ethanol is likely to extend irrigation and use more ground or surface water.

Water and Industry

In addition to agriculture and agro-industries, the manufacturing of consumer products and the provision of a number of consumer services require water as an essential input. Industry uses about 20 percent of the world's freshwater supplies (Barlow and Clarke 2002, 7). Demands on water are growing, as larger developing countries such as China, Brazil, India, and Mexico catch up on the production and export of industrial products.

Use of water for industrial production and the manufacturing of consumer products are expected to double in the next decade, if current growth trends persist (Barlow and Clarke 2002, 7). The purchase of industrial products and luxury consumer goods is directly related to people's incomes and purchasing power. It has been observed that the increasing affluence of Chinese consumers is a factor stimulating the consumption of luxury items.

Increased globalization and the move to more open trade regimes (including free trade agreements and zones), are key factors driving the rise of manufacturing and exports. Some industries and services have been established in the border zone between Mexico and the United States following the North American Free Trade Agreement (NAFTA), but the downside of this development has been increased emissions, air and water pollution in the border area, and health consequences for residents on both sides of the frontier.

Other industries have been outsourced to China, Vietnam, and other Asian countries, creating many water bottlenecks in those countries. China is particularly affected. It is a water-stressed country, and its production of low to higher-end industrial and consumer products for world markets has helped generate an increasing demand for water. China's export model, in addition to its population growth and domestic demand for water, has been a major factor in the country's water supply bottlenecks and pollution problems.

Chinese manufacturers use huge quantities of water for coal production, which is mainly used to fuel the manufacturing process. Its industrial development also uses large quantities of deionized freshwater in computer manufacturing for local and export markets. China's need for agricultural products to feed its population has sparked the increase in agricultural exports from other countries to China, as well as the export of luxury items such as cars, designer apparel, and brand name consumer products from industrialized countries to China. These trends place further demands for water on countries producing exports for the huge China market.

Water and Industrial Effluents

Up until the early 1970s, many U.S. industries such as textiles, coal mining, iron and steel, and pulp and paper plants regularly dumped effluents in waterways and contaminated groundwater sources as well. The Cuyahoga River in Cleveland, Ohio, caught fire many times up until the 1960s due to the enormous amounts of industrial effluents and chemical pollutants in the river. In 1972, the U.S. Clean Water Act introduced regulations to control pollution, clean up waterways, and punish polluters.

In many developing countries, there are inadequate regulatory and enforcement structures that fail to effectively address water pollution and prevent severe waterborne health and sanitation problems. Due to lax environmental protection in developing countries, a number of manufacturing plants often leave developed countries with robust environmental regulations to set up shop where there are fewer strict laws and less enforcement, and where countries are vying for investment and less likely to enforce penalties.

The release of toxic substances into the air, water, and ground has become regulated in the United States through the Environmental Protection Agency (EPA). Industries such as mining, pulp and paper, oil and gas, and silicone production can no longer contaminate rivers and other sources of water without consequences. However, the industrial sector still continues to use large volumes of water, although more and more businesses are seeking ways to recycle water or use it more efficiently in production. For example, the pulp and paper industry uses up to 190,000 gallons of water per ton of paper or rayon; silicone wafer production requires up to 4.5 million gallons of water per week (Black 2004, 9).

The harbor of Cleveland, Ohio, with the mouth of the Cuyahoga River in the foreground, plays an important role in the urban economy. (U.S. Army Corps of Engineers)

Water and Energy

The energy sector is a big user of water, and also has the capacity to pollute water. Not only is there a problem of oil spills in coastal and marine environments, but spills or pipeline leaks on land can also have major environmental impacts. Within the oil industry, a major area of water use is hydraulic fracturing, or fracking. Hydraulic fracturing is the practice whereby a mixture of water, sand, and chemicals is injected underground and horizontally into rock formations to release trapped natural shale gas.

In the shale-gas operations of Pennsylvania, for example, 4 million gallons of water are needed for each fracked well, compared to about 100,000 gallons for a conventional gas well. Fracking also releases methane gas and other contaminants, which are pumped out from the rock into the surrounding areas. There have been increasing reports of groundwater contamination in the oil and shale gas boom state of North Dakota.

The nuclear industry, a subsector of the energy industry, also uses large quantities of water for the cooling of reactors. For this reason, nuclear plants are often built besides lakes or oceans. However, these locations can make them vulnerable to tsunamis and intensive flooding, as observed with the 2011 Fukushima accident in Japan.

Water and Tourism

Many tourism, sports, and recreation industries and services rely on water. These activities are related to ocean and river cruising, water sports, ocean resort development, recreational swimming, diving, waterskiing, and snorkeling industries. Sand, surf, and sun attractions in tropical countries are based on the image and the marketing of tropical, blue-green, sparkling waters, combined with palm trees, white sands, and pictures of colorful coral reefs (even if the water may be polluted or the coral reefs may be stressed).

Tourism and recreation-oriented services provide employment and are big contributors to the economies of many countries, particularly developing countries in tropical areas that are competing for the global tourist dollar. However, the sustainability of the tourism, travel, and recreation-based industries requires the maintenance of clean, uncontaminated ocean water, unpolluted beaches and coastal areas, and the protection of pristine coral reefs for snorkeling and diving. Without clean water, there is no attractive beach. The maintenance of clean waters and pristine coastal ecosystems goes beyond the scope of the tourist industry itself it is dependent on other factors causing source pollution and inadequate regulation and enforcement of measures to curb contamination.

In colder climates and highland areas, water in frozen form is an important part of winter sports and also a big economic contributor to local economies. These activities include: downhill and cross-country skiing; ice fishing, glacier exploration, and ice walking; sledding and snowboarding; and ice-skating, curling, and ice hockey sports. Winter sports attract millions of participants, as well as television, online, or on-site viewers in Canada, the United States, Switzerland, Austria, Russia, and other countries with a tradition of these industries. They require a consistent supply of snow and ice. However, due to the effects of climate change—warmer temperatures and snowmelt—many ski resorts have been forced to make snow artificially, raising the costs of the service. The future of many ski and winter sports–oriented resorts may be in jeopardy with further increases in snowmelt.

Water as an Industry

The growth of industries that revolve around water as a product in its own right merits attention. Due to water shortages, increasing pollution of ground water, and the marketing of water and sports drinks in sports, there has been a rising trend in bottled water and beverage industries. In addition, drought and the reduction of clean potable water have also seen the growth of desalinization plants in arid or rainfall-deficient areas. These plants produce freshwater from ocean water through a process of reverse osmosis, which extracts the salt from the liquid. They require huge amounts of energy at high costs, and are usually out of the financial or technical reach of cash-strapped, impoverished, and water-poor developing countries.

However, with increasing water scarcity, rising water costs and the growing emphasis on water conservation and recycling, the conversion of ocean water into useable water for industrial, household, and agricultural purposes is expected to become a viable option in the future. As technology improves, desalinization costs are likely to decrease. In addition, alternative sources of energy (renewables such as solar and wind) can be harnessed to fuel the desalinization process, particularly in hot or windy places.

Other industries directly related to water as a consumer product are becoming more important. They include the manufacturing of water tankers, water barges, and huge floating water bags or skins to transport water by sea, as well as the industries to produce pumps, pipes, water containers, and bottled water and water-based beverages. In Greece, private tugboats are already deployed to pull massive floating polyurethane bags throughout the Aegean Sea. These bags are part of the water-manufacturing complex, and are indispensable to supply parched islands with freshwater from the Greek mainland.

The prospect of lucrative water transportation and sales ventures has also sparked a number of ideas for the development of water transfer enterprises around the world. These include proposed schemes to transport meltwater from Greenland to water-thirsty countries, or to ship water by bags from Turkey to Turkish-occupied, parched North Cyprus. Even if some of these ideas still have not come to fruition, the fact is that private corporations around the world are increasingly involved in the delivery of private potable water services to consumers, as well as in the private management or outright privatization of former public water utilities (see case study 1).

The growing role of the private sector in the provision of water to urban and rural consumers, and the costs involved, has ramped up the debate about whether water is a commodity or a human right. The bottled water and beverage industry (beer, soda, and sports and energy drinks) is a high user of water. Bottled water and drinks have become extremely popular globally, mainly due to their convenience for sports and other activities. Inadequate access of consumers to clean or piped water as well as consumers' fears in developing countries about the safety of water have driven global sales. People in industrialized countries consume about 10 times more water daily than those in developing countries (UNEP).

International corporations are involved in the lucrative bottled water sector; their stocks are actively traded on global stock markets. A few companies control 70 percent of the world's bottled water market. They are also involved in the production and sale of water bottles and water containers. Estimates of annual revenues from bottled water amount to about 40 percent of the oil industry's revenues. In the 1970s, the annual volume of bottled water traded around the world was 300 million gallons. By 1980, this had risen to 650 million, and in 2000, to 22 billion gallons. Global sales of bottled water ranged from between $26 to $36 billion in 2000 (Barlow 2013, 143).

There are concerns about groundwater being extracted by water corporations for their processing, bottling, and sales operations as well as the environmental impacts of inappropriately disposed of plastic containers. Not all countries have laws to regulate the recycling of plastic bottles and containers, which are not biodegradable and cause harm to marine life. Throughout the world, travelers can find plastics in landfills and on beaches, strewn around cities and villages, or dumped into watercourses, wetlands, and oceans.

Camille Gaskin-Reyes

REFERENCES AND FURTHER READING

Barlow, Maude. *Blue Future: Protecting Water for People and The Planet Forever*. New York: The New Press, 2013.

Barlow, Maude, and Tony Clarke. *Blue Gold: The Fight to Stop the Corporate Theft of the World's Water*. New York: New Press, 2002.

Black, Maggie. *The No Nonsense Guide to Water*. Oxford, UK: New Internationalist Publications, 2004.

Buchanan, Larry, Josh Keller, and Haeyoun Park. "Your Contribution to the California Drought." *New York Times*, May 21, 2015. http://www.nytimes.com/interactive/2015/05/21/us/your-contribution-to-the-california-drought.html?_r=0.

Chellaney, Brahma. *Water, Peace and War: Confronting the Global Water Crisis*. Lanham, MD: Rowman & Littlefield, 2013.

Environmental Protection Agency (EPA). "Summary of the Clean Water Act." http://www.epa.gov/laws-regulations/summary-clean-water-act.

Ohio History Central. "Cuyahoga River Fire." http://www.ohiohistorycentral.org/w/Cuyahoga_River_Fire?rec=1642.

Rothfeder, Jeffrey. *Every Drop for Sale: Our Desperate Battle Over Water in a World About to Run Out*. New York: Putnam, 2001.

Sachs, Jeffrey D. *Common Wealth: Economics for a Crowded Planet*. New York: Penguin Press, 2008.

Schwartz, Nelson D. "Investors Are Mining for Water; The Next Hot Commodity." *New York Times*, September 24, 2015. http://www.nytimes.com/2015/09/25/business/energy-environment/private-water-projects-lure-investors-preferably-patient-ones.html.

United Nations Environment Programme (UNEP). "Vital Water Graphics." 2008. http://www.unep.org/dewa/vitalwater/article48.html.

CASE STUDIES

These case studies illustrate the dilemma of trying to transfer public ownership and management of water to private hands in Bolivia; the role of waterways in the development of international trade and commerce; and the issues related to waterborne diseases and their implications for people's health.

Case Study 1: Cochabamba Water for Sale?

The Cochabamba water privatization experience illustrates how attempts to privatize municipal water resources in 2000 resulted in a water war with significant political and social consequences. The story begins in Cochabamba, Bolivia's fourth-largest city, located in a high Andean valley. Prior to 2000, most people outside Bolivia had heard very little about Cochabamba, but in 2000 events unfolded that would soon grab the world's attention.

Water Tariff Hikes and Impacts

In 2000, a 17-year-old Bolivian student was shot dead by military police while protesting an increase in water rates due to the privatization of Cochabamba's municipal water system. This fatal event occurred during a period of widespread rioting in the city. Protesters had seized the city center, blocked roads, and launched a labor strike that brought the city's economy to a halt. As a result of the 35 percent water-rate hike imposed by the private company newly in charge of water—this was more than low-income families paid monthly for food—the poor could no longer afford to pay water bills.

The Cochabamba experience spoke to the core of a heated debate over water as a human right or a commodity. With privatization, water in Cochabamba became unaffordable for the poor. The process started with a World Bank project. In 1999, the Bolivian government, guided by the advice of financial institutions such as the World Bank and the International Monetary Fund (IMF), contracted with a private consortium, the Bechtel Corporation (known as Aguas del Tunari) to manage Cochabamba's public water and sewage system.

As part of its project, the World Bank recommended bringing in a private company to administer Cochabamba's decrepit, underfunded, and mismanaged public water management system. The World Bank argued that this step would improve water efficiency and enhance customer service. However, the World Bank's loan to Bolivia did not include provisions for cross-subsidies of tariffs or any form of financial assistance to poor households to pay for the increased water costs. As part of the deal, Bechtel negotiated conditions with the Bolivian government to raise water rates each year to fund its investments and receive a guaranteed profit on its investments.

Privatization Woes and Citizen Protests

Reforms in the water sector were necessary to improve the system in Cochabamba, but at what cost? In the late 1990s, the system had already fallen into disrepair.

Between 50 and 60 percent of water supply was lost through leakage from rusted pipes before water even reached consumers. Much of the water that did pass through the pipes went to the wealthiest consumers who could pay for their household water connections. To prepare the ground for the transition from public to private management, Bolivia passed a national law allowing the licensing of water services to private companies for distribution to consumers.

The 40-year, $2.5 billion private concession was intended to provide water and sanitation services to the city's residents and generate electricity and irrigation for agriculture. As part of the contract, the company agreed to absorb $30 million in the debt of the public utilities company from the Bolivian government, improve infrastructure, and build a new water reservoir. To fund these construction projects, the agreement with the government allowed the consortium to raise water rates of residents immediately after the contract came in force.

Steep increases in tariffs—without immediate improvements in the quality of the system—provoked a huge public outcry. People who had already built their own rooftop water tanks and pipelines were charged for their installations. Citizen protests quickly turned into violent street conflicts, culminating in the fatal shooting of the student mentioned above.

In the eyes of Cochabamba's residents, particularly the poor, who could not afford sudden and steep rate hikes, the government's policy to turn over responsibility for water to a private concern appeared to be a violation of their human right to water and a breach of public trust. Their protests became rallying cries between proponents and opponents of privatization in Bolivia. Different points of view were played out in the conflict. Was water a human right or a tradable commodity? What was the priority of human needs versus a corporate bottom line?

Retreat of the Government

How did this conflict eventually play out? Following $20 million of property damage from the riots, countless injuries, one death, and mass unrest in Bolivia, the government was forced to terminate the Bechtel concession and take back public control of the water system. After the Cochabamba project was scrapped, Bechtel took out a $25 million lawsuit against the government for breach of contract, but legal action was later withdrawn.

The government made the decision to scrap the project to avoid further escalation of the conflict, but there were political ramifications of the water crisis. A massive, broad-based wave of resistance developed in Cochabamba, led by a coalition of workers, peasants, nongovernmental organizations, and concerned citizens. It was successful in forcing the government to rescind the privatization plan, and it led to the departure of Bechtel and eventually the demise of the pro-privatization government. After Bechtel's pullout, there was no institution left to run the water company, and the Bolivian government then handed over management of water to the workers of the former public utility and the communities.

Upon accepting their new responsibility, workers and communities reelected a new board of directors for the water utility and established new principles for water allocation such as greater efficiency, eradication of corruption, fairness to workers, and social justice for poor water consumers. One of the first actions of the new management was to build a huge water tank to provide free water for the poorest residents in the city. The new utility management also worked with 400 small communities throughout Cochabamba to lay their own pipes to access water for the first time at the household level.

Key Lessons

Cochabamba was not an isolated event. It also provided important lessons. In the last decade or so, the privatization and trading of water has become more prevalent throughout the world, but not without popular protests. Privatization has increased due to many factors: reduced water supply due to depletion or pollution of freshwater; weak water governance and corruption of public water utilities; growth of private water companies; conflicts among competing water users; unmaintained infrastructure; inequitable water allocation between elites and the poor; and lack of involvement of poor communities in water planning and decision-making.

As the Cochabamba case study showed, the objective of private water companies to maximize returns on investments clashed with the demands of underserved poor communities for water and their inability to pay for water connections and increased tariffs. Sound institutions, cooperative public/private partnerships, and functioning regulatory entities would help resolve these differences and foster access of all citizens to clean affordable water.

In many developing countries, poor communities are forced to buy expensive but dirty water from tankers and intermediaries. Often, the poor pay disproportionately more for water than rich residents in the same city, since poor communities have no bargaining power to negotiate prices set by monopoly water sellers from the outside.

Who owns water, how much clean water residents should receive, and at what price, are matters of survival that societies will continue to address as water becomes more scarce. This was true for Bolivia and will also be the case for other countries, where water is scarce and becoming even scarcer due to climate change and melting glaciers.

Currently each Bolivian receives on average 41 liters of water daily, 20 percent below the minimum for subsistence. The poorest Bolivians, including those in Cochabamba, obtain even less water than the average resident. The lack of clean water causes hundreds of Bolivian deaths and thousands of illnesses every year through waterborne diseases. In 1992 a cholera epidemic caused the death of 500 people in Cochabamba (Rothfeder 2001, 100). It is not enough to provide water to communities; it has to be clean and affordable if it is to make a difference in people's lives and improve their health status.

Camille Gaskin-Reyes

REFERENCES AND FURTHER READING

Barlow, Maude, and Tony Clarke. *Blue Gold: The Fight to Stop the Corporate Theft of the World's Water*. New York: New Press, 2002.

Olivera, Oscar. *Cochabamba! Water War in Bolivia.* In collaboration with Peter Lewis. Cambridge, MA: South End Press, 2004.
Richter, Brian. *Chasing Water: A Guide for Moving from Scarcity to Sustainability.* Washington, DC: Island Press, 2014.
Rothfeder, Jeffrey. *Every Drop for Sale: Our Desperate Battle Over Water in a World About to Run Out.* New York: Putnam, 2001.
Salzman, James. *Drinking Water: A History.* New York: Duckworth Overlook, 2013.

Case Study 2: Cutting through the Oceans: The Panama Canal and the Proposed Nicaragua Canal

Panama's development since Spanish conquest is related to its geography. The role of the Isthmus of Panama as a passageway between the Atlantic and the Pacific Oceans—and later home to the U.S.-built Panama Canal—is etched into the country's history. Spanish explorers first came to Panama in 1501; the Spaniard, Vasco Núñez de Balboa, first approached the area from the Atlantic and in Panama first caught sight of the Pacific Ocean in 1513.

Once the Spanish colonists saw both oceans, they grasped the importance of the Isthmus. They saw it as a way to transship gold and silver from Spanish colonies (today Bolivia and Peru) on the Pacific Coast across Panama overland (on foot, in canoes, and by mule) to the Atlantic Coast, and from there reload the cargo for ship transport to the motherland. This land crossing between the oceans was crucial because it saved Spanish ships time, and they avoided the perils of sailing around the tip of South America and facing the treacherous weather of the Strait of Magellan.

In 1581, the Spaniards considered an alternative passage of goods through Nicaragua, also located at a narrow crossing point across Central America. However, Panama remained the preferred route. It continued to play a vital role as a passage between the oceans and a well-trodden conduit for colonial goods and metals from Spanish colonies to Spain for many centuries. After the turn of the 18th century, as Latin America became free from Spanish rule, Panama became a province of Colombia but maintained its transit role.

Getting to the Gold Rush in California

Up until the mid-1800s, Central America and the Isthmus of Panama had remained a relatively undeveloped area. However, in the 1840s, something dramatic occurred in California: the gold rush. This proved to be an important boon to spur Panama's development. Gold seekers in the United States, lured by the promise of gold, looked for alternative and faster routes from the Atlantic coast to California that did not involve taking the precarious land trek across the country or sailing around the tip of South America.

People soon discovered that the gain in time and distance via Panama was impressive. The voyage from New York to San Francisco around Cape Horn, in southern Chile, took one month and was 13,000 miles long—with considerable peril. Via Panama it was a more efficient trip of 5,000 miles, even with its dangers. The Panama passage involved the risk of contracting cholera and other diseases, the discomfort of slogging through mud and malaria-laden dense jungles, and the perils of enduring excruciating heat, landslides, and thunderstorms.

Travelers were so anxious to get to the gold in California that they were willing to undergo considerable hardships to cross Panama, whether by mule, by canoe, or on foot. As a result of the growing number of Panama crossers, the U.S. government hatched an idea to build a railroad across the isthmus, starting in 1850. An agreement was signed between the United States and the Colombian authorities which granted the United States exclusive rail transit rights across the province of Panama.

The railroad was completed in five years at a cost of 8 million dollars and at a huge human cost. It is estimated that about 6,000 railroad workers—many recruited from Caribbean countries—died during its construction. At its completion, it was the world's first transcontinental one-track railroad, spanning almost 48 miles. Railroad travel across Panama became linked to the increase in passenger steamship travel, which departed from both coasts of the United States. The route also provided mail service; the cost of a one-way ticket to California through Panama or vice versa was $25 in gold.

French Interest in a Panama Canal

But more was yet to come. After the railway was in operation, the French started to show interest in the construction of a water-based canal across Panama to carry passengers and cargo. In 1878, the Colombian government authorized the French, under the leadership of Ferdinand de Lesseps, who had built the Suez Canal earlier, to dig a canal through Panama. Even though the United States was also interested in a Panama canal at the time, it began exploring the option of constructing a canal through Nicaragua in Central America. Both ventures began feverishly, the French digging in Panama, the United States in Nicaragua. However, a financial crisis hit in 1893, and both groups were forced to suspend digging.

Nonetheless, de Lesseps never lost interest in the Panama Canal project, and in 1880 announced a plan to resume building a ground-level canal (like the Suez Canal) without locks. After nine years of digging, millions of dollars of investments, and a loss of about 20,000 lives of Caribbean workers (due to malaria, yellow fever, and other tropical diseases), the French initiative went bankrupt. Lesseps abandoned construction; investors lost their money (McCullough 1977, 20). It was time for U.S. action.

U.S. Action on the Panama Canal

Throughout the period of incipient French construction, the interest of the United States in a canal across Panama had remained dormant. In 1902, however, the United States made a bold move to take over construction. The U.S. Congress voted in favor of the canal and signed a treaty with Colombia's foreign minister to start the works. However, this treaty did not meet approval from Colombia's congress, which rejected the U.S. offer.

In response to this refusal, President Theodore Roosevelt countered by dispatching U.S. warships and troops to Panama to support a breakaway of Panamanian provincial rebels from Colombia. In 1903, the Panamanian separatists, supported by military involvement of the United States, declared independence from Colombia, thus paving the way for the United States to become the canal builders of Panama.

A country was born. A canal was being built. The newly independent Panama named a French engineer (previously involved in the Lesseps project) to negotiate and sign a treaty with the United States in 1903. This treaty gave the United States control of a 10-mile area called the Panama Canal Zone to build and operate the canal in perpetuity. The treaty also provided a one-time 10-million-dollar payment to Panama for the land and an annual payment of $250,000 (U.S. Department of State). The treaty was considered by many Panamanians to be an insult to Panamanian sovereignty, since a French citizen had signed the treaty with the United States in the name of Panama.

Formidable Feat

Once the U.S. took control of the Panama Canal Zone, construction began in earnest. There were technical problems to solve, i.e., digging the passage through the isthmus, building locks, and addressing medical issues such as control of malaria, which had plagued the Lesseps canal from the start. The United States also had to resolve financial issues and labor shortages. The new operating entity of the canal raised funds and imported thousands of additional workers from the Caribbean, while at the same time solving the malaria and yellow fever problem through spraying and public sanitation measures.

The canal took 10 years to build. It was a formidable feat. The project dug and moved an estimated 100 million cubic yards of dirt, the equivalent of constructing a new Great Wall of China from San Francisco to New York, or building 63 pyramids the size of the Great Pyramid of Cheops in Egypt. It finally opened in 1914, about 401 years after the Spaniards had first glimpsed Panama. Like the railway and the French attempts to build the canal, U.S. construction took a huge toll on the health and lives of canal workers, although the death rate was lower compared to the previous projects.

The American construction of the canal moved mountains and rechanneled rivers. The Chagres River was dammed to create an artificial lake 85 feet above sea level. To facilitate the movement of ships from one coast to the other and to resolve differences in height, the project widened rivers and built two sets of locks. In 1914, the steamer *Ancon* made its way through the locks and passed through the entire canal. Its passage marked the official opening transit of the waterway attended by dignitaries from the United States and other countries. Thereafter, it became an important international route under exclusive sovereign management and jurisdiction of the United States.

Over the years, U.S. sovereignty over a swath of Panama called the Canal Zone would become a bone of contention between Panama and the United States. Political and social tensions built up numerous times, often leading to minor skirmishes in the Canal Zone. In 1977, a turn of events reversed the in-perpetuity clause of the canal agreement. President Carter of the United States and President Torrijos of Panama signed an agreement to revert the canal and the Canal Zone to Panamanian sovereignty and ownership in 1999.

Return of the Panama Canal to Panama

Since 1999, Panama Canal governance has successfully remained in Panamanian hands. A new project to extend the canal to accommodate larger size tankers, mega-container ships, and cruise ships (called post-Panamax) has been initiated. The extended canal will include a third set of wider and deeper locks at an estimated cost of 7 billion dollars. The project was started in 2007 and is scheduled for completion in 2016.

Nicaragua's Rival Canal Proposal

In spite of the billion-dollar canal expansion in Panama, Nicaragua (Panama's earlier rival for canal building) announced plans in 2014 to go ahead with its own canal construction, to be financed by Chinese investors. The proposed canal is to be located a few hundred miles north of the Panama Canal. The concessions granted to the investors in Nicaragua include construction of two ports on the two coasts, a railroad, an oil pipeline, roads, and a trade-free zone. The proposed Nicaragua canal is expected to be three times longer than the Panama Canal, taking advantage of several river courses, which will be widened (Watts 2015).

Similar to Panama, a series of locks will enable ships to cross from one coast to the other in Nicaragua, at a price of at least 40 billion dollars. Construction is proceeding in spite of environmental concerns about the proposed dredging of Lake Nicaragua, Central America's largest source of freshwater. Other issues are related to potential impacts of sedimentation, disposal of excavated earth, and risks of saline intrusion of the water table.

The Nicaragua Canal, if built, would have the capacity to accommodate the largest container and cargo ships in the world, whose size has been steadily rising due to increased Asian–North American trade and global commerce in general. It is expected to revive Nicaragua's old dream of building a canal and also provide alternatives to or competition with the expanded Panama Canal.

Other Project Ideas

The expansion of the Panama Canal as well as the Nicaragua canal proposal have stimulated other project ideas in Central America to construct deep-water ports and rail links connecting the Atlantic and Pacific coasts: there are calls for so-called dry (rail and road) canals in Guatemala, Costa Rica, and El Salvador, but no construction has started. The melting of the Arctic Ocean has brought to the forefront the possibility that the Northwest Passage across the Arctic may become viable for future commercial shipping and shave off some miles from the current Panama Canal Asia-to-Europe route. This alternative and the implementation of any other planned Central American passageways might also increase competition for the Panama Canal.

Camille Gaskin-Reyes

REFERENCES AND FURTHER READING

Anderson, Jon Lee. "The Comandante's Canal." *New Yorker*, March 10, 2014.

McCullough, David. *The Path between the Seas: The Creation of the Panama Canal 1870–1914*. New York: Simon & Schuster, 1977.

U.S. Department of State. Office of the Historian. "Building the Panama Canal, 1903–1914."

Watts, Jonathan. "Land of Opportunity—and Fear—along Route of Nicaragua's Giant New Canal." *Guardian*, January 20, 2015. http://www.theguardian.com/world/2015/jan/20/-sp-nicaragua-canal-land-opportunity-fear-route.

Case Study 3: Waterborne Diseases and Human Development

Clean drinking water is essential to sustain human life. In industrialized societies, most people take it for granted that water that comes out of a tap is safe for drinking and is free of harmful pathogens. This is not the case in many developing countries of the world, where waterborne illnesses are serious problems and can be fatal. In fact, millions of people avoid tap water—even if it is available—due to concerns about its quality. This case study explores the consequences of unclean water for the status of public health.

The problem of unclean potable water has led to an explosion of the bottled water industry throughout the world, particularly in low-income countries. It is customary to see water vendors in African cities selling small plastic bags or pouches with water or water tankers supplying low-income marginalized settlements with water. However, there is no guarantee that these water alternative sources are safe, given the lack of clean water and the inadequacy of potable water standards, regulations, and monitoring procedures.

Waterborne Diseases

Waterborne diseases are a serious problem globally, but are more pronounced in the developing world, particularly in African, Asian, and Latin American countries. They are caused by a wide range of microorganisms, biotoxins, and other contaminants, which lead to crippling and often fatal outbreaks of illnesses. The most common ones are cholera, schistosomiasis (carried by snails), and gastrointestinal diseases.

However, there is some good news. The number of children dying from intestinal diseases around the world—which are closely linked to contaminated water, poor hygiene, and inadequate sanitation—has fallen in the last 15 years from 1.5 million deaths in 1990 to 600,000 in 2015 (World Health Organization).

Now for the bad news: globally 2.5 billion people still lack access to improved sanitation, 1 billion people practice open defecation and around 750 million do not enjoy any source of clean drinking water. At least 1.8 billion people have to use drinking water polluted by human waste (World Health Organization). Compounding this problem is the fact that millions of people in developing countries either cannot afford soap or do not have clean water to wash their hands, a practice which would help block the spread of diseases.

In developed countries, there are strict standards for safe drinking water. However, water can still become contaminated, if natural disasters and floods occur and cause sewage pollution of potable water, or if pollutants seep into ground and surface water sources. Water can carry arsenic, lead, and heavy metals from industrial effluents; once these toxins get into the water supply, they enter the human body and cause severe damage. Humans also suffer enormous health impacts by eating fish contaminated by mercury and other poisonous substances, including microplastics ingested by fish in polluted waters.

In agricultural areas, animal wastes are a potential threat to the cleanliness of ponds, lakes, and reservoirs. There are also risks to human health through the transmission of avian influenza through water and sewage. In coastal environments, unsafe management of shellfish and fish stocks in aquaculture ponds as well as water contamination through sewage can lead to cholera outbreaks.

Water Standards

In the United States, the Safe Drinking Water Act (SDWA) is the principal federal law that ensures the quality of drinking water. The SDWA was first passed by Congress in 1974 and subsequently amended in 1986 and 1996. Under the SDWA, the Environmental Protection Agency (EPA) is charged with setting the standards for water quality and overseeing compliance with these standards by states, localities, and water suppliers. Public water systems are obliged by law to demonstrate that drinking water meets the standards through periodic monitoring and testing. Only the EPA and state-certified laboratories are allowed to conduct water quality analyses.

The EPA standards list all potential water contaminants and the maximum contaminant levels allowed in primary drinking water. These contaminants may come from different sources: microorganisms (viruses and bacteria), disinfectants (cancer-causing substances), inorganic chemicals (arsenic and metals such as lead, arsenic, copper, and mercury), organic chemicals (benzene), and radioactive materials (radium).

The Arizona Department of Health Services lists 13 common waterborne diseases, their symptoms, and inception periods. The Centers for Disease Control and Prevention (CDC) in Atlantia, Georgia, and the National Institute of Environmental Health Sciences (NIEHS) also provide public information on water, sanitation, and health matters, as well as outbreaks of waterborne diseases. Such clear standards, enforcement mechanisms, and public information updates are not readily available in many developing countries.

Climate Change: Impacts on Human Health

Climate change is likely to increase the incidence of diarrheal disease worldwide, lead to more intense weather conditions such as increase in temperatures, floods, and storms, and increase waterborne diseases. For example, droughts can cause higher concentrations of pathogens, which can contaminate surface water and overwhelm the effectiveness of water treatment plants. In many countries, very old treatment plants already pose a risk: they are either ineffective or working at limited capacity to treat wastewater.

In addition, climate change provokes alterations in ocean and coastal ecosystems, including the rise in acidic levels. The runoff of chemical and other contaminants from the land degrades freshwater sources, especially in areas where the population uses untreated surface water for drinking, bathing, and washing. Increased frequency of natural disasters is likely to cause flooding of water and sewage facilities and lead to cross contamination of water and sewage pipes. In lakes, the spread of harmful algal blooms is already a noticeable factor in the potential spread of toxins and pathogens.

Water as a Development Goal

Given the importance of clean water to human well-being and survival, it is no surprise that the Millennium Development Goals (MDGs) of the United Nations placed emphasis on clean drinking water and sanitation as a goal. However, this was one area where achievement was lagging due to inadequate funding and political and technical bottlenecks, which prevented the improvement of drinking water and sanitation services in developing countries.

Water and sanitation improvements are included in the Sustainable Development Goals (SDGs) of the United Nations, approved in 2015. They fall under Goal 6, which ensures clean water and sanitation. The SDGs have a time period of 15 years, from 2015–2030, and are intended to replace the past MDGs with the goals of ending poverty, protecting the planet, and ensuring prosperity and well-being for all. The SDGs emphasize the connection between poverty and people's lack of access to water, basic social services, affordable energy, protected ecosystems, and income generation.

People who contract waterborne diseases lose productivity. They cannot work at full capacity and often lose income while they are ill. In countries with limited social safety nets, poor public health facilities, or no health insurance programs, many poor communities or families carry a heavy disease burden. They have few cash reserves or backup systems, and can sink into poverty, destitution, and homelessness. It is not uncommon for entire families to be affected by cholera and unable to harvest crops, fish, or go to work.

Most times, the very poor cannot afford a doctor or the cost of medicine, or even transportation to medical clinics, if they are available. For these reasons, they languish longer in their ill state, often succumbing faster to diseases than affluent groups in the same country. This is particularly true in rural areas, where poor families typically do not have reliable sources of drinking water and lack access to medical facilities and community health workers. They also have inadequate toilet facilities, draw water from contaminated wells and rivers, and are often forced to buy polluted water at high prices.

Breaking the Vicious Cycle

Breaking the vicious cycle of waterborne diseases and poverty in developing countries requires investments in public health, safe drinking water, physical infrastructure to pump water from wells, water treatment plants, pipes, and household connections to transport clean water to households. In urban centers, where population growth is rising, it is a huge technical and financial challenge to supply water to marginalized and widely dispersed squatter households.

Breaking the vicious cycle also requires a well-functioning public health system, effective organizations, trained and equipped health and sanitation professionals, robust laws, and regulatory agencies to safeguard water quality standards, as well as enforcement mechanisms to make polluters pay for contaminating water sources.

Due to the scale and intensity of the problem worldwide, a number of international development agencies (World Bank and regional development banks), nongovernmental organizations and foundations such as UNICEF, the Gates Foundation, Save the Children, Doctors Without Borders, Oxfam, Partners in Health, the World Health Organization (WHO), and the Pan American Health Organization (PAHO) are working to help provide improved water and sanitation services for people around the world. However, these efforts are not enough to address the magnitude of the problem.

Camille Gaskin-Reyes

REFERENCES AND FURTHER READING

Arizona Department of Health Services. www.azdhs.gov.
Bill and Melinda Gates Foundation. www.gatesfoundation.org/.
Centers for Disease Control and Prevention. www.cdc.gov.
Doctors Without Borders. www.doctorswithoutborders.org/.
Environmental Protection Agency (EPA). "Safe Drinking Water Act (SDWA)." https://www.epa.gov/sdwa.
National Institute of Environmental Health (NIEHS). http://www.niehs.nih.gov/.
Oxfam. www.oxfamamerica.org.
Pan American Health Organization (PAHO). http://www.paho.org/hq/.
Partners in Health. www.pih.org
Save the Children. www.savethechildren.org.
United Nations Children's Emergency Fund (UNICEF). www.unicef.org.
World Health Organization (WHO). www.who.int/en.

ANNOTATED DOCUMENT

The Dublin Statement on Water and Sustainable Development (1992)

Background

In 1992, the International Conference on Water and the Environment (ICWE) was held in Dublin, Ireland. Hundreds of participants, including government experts from 100 countries and representatives of 80 international, intergovernmental, and nongovernmental organizations attended the conference.

The findings and results of that conference are called the Dublin Rules, which are codified in the Dublin Statement on Water and Sustainable Development, adopted on January 31, 1992, at the close of the conference.

The conference began with a Water Ceremony, which involved children from all parts of the world. The children made a moving plea to participants of the conference to play their part in preserving precious water resources for future generations.

In transmitting the Dublin Statement to a global audience, the conference participants urged all those involved in the development and management of water resources around the world to allow the message of the children as representatives of the future generations to guide their actions.

Summary

The Dublin Conference and the report summarizing the Dublin Rules was another conference in a long line of conferences that provided important recommendations for world action on water conservation.

Experience has taught us that verbal commitments alone rarely work without the support of substantial and immediate investments in water quality, public awareness campaigns, legislative and institutional changes, technology development, and capacity building of those in charge of water management around the world.

The main message of the Dublin Rules is that scarcity and misuse of freshwater pose a serious and growing risk to sustainable development and protection of the environment and ecosystems. The rules emphasize that human health and welfare, food security, industrial development, and marine and terrestrial ecosystems are at risk. They reiterate the urgent need to manage water and land resources more effectively than in the past.

The documents presented at the conference regarded the mismanagement of global water resources as a critical problem, with the capacity to affect the survival of many millions of people, unless nations of the world take immediate and effective action.

The rules call for fundamental new approaches to the assessment, development, and management of freshwater resources, and the need for greater political commitment and involvement from the highest levels of government to the smallest communities.

The rules recognize water as a global public good, its interdependent nature, and its importance for survival of all peoples in the world. They urge all governments to implement the specific actions recommended in the conference report through urgent programs to support water and sustainable development.

The conference report stressed that concerted action is needed to reverse the present trends of overconsumption, pollution, and the rising threats from drought and floods. These plans, called an Action Agenda, are included in the following excerpts of the Dublin Rules.

Camille Gaskin-Reyes

Selected Excerpts of the Dublin Rules and The Action Agenda

The following passages of the Dublin Rules set forth rules or codes for action at local, national and international levels, based on the following guiding principles:

Principle No. 1:
Fresh water is a finite and vulnerable resource, essential to sustain life, development and the environment.

Since water sustains life, effective management of water resources demands a holistic approach, linking social and economic development with protection of natural ecosystems. Effective management links land and water uses across the whole of a catchment area or groundwater aquifer.

Principle No. 2:
Water development and management should be based on a participatory approach, involving users, planners and policy-makers at all levels.

The participatory approach involves raising awareness of the importance of water among policy-makers and the general public. It means that decisions are made at the lowest appropriate level, with full public consultation and involvement of users in the planning and implementation of water projects.

Principle No. 3:
Women play a central part in the provision, management and safeguarding of water.

This pivotal role of women as providers and users of water and guardians of the living environment has seldom been reflected in institutional arrangements for the development and management of water resources.

Acceptance and implementation of this principle requires positive policies to address women's specific needs and to equip and empower women to participate at all levels in water resources programs, including decision-making and implementation, in ways defined by them.

Principle No. 4:
Water has an economic value in all its competing uses and should be recognized as an economic good.

Within this principle, it is vital to recognize first the basic right of all human beings to have access to clean water and sanitation at an affordable price.

Past failure to recognize the economic value of water has led to wasteful and environmentally damaging uses of the resource.

Managing water as an economic good is an important way of achieving efficient and equitable use, and of encouraging conservation and protection of water resources.

The Action Agenda

Based on these four guiding principles, the conference participants developed 10 recommendations for an Action Agenda to enable countries to address water resources problems comprehensively. The major areas for action and follow-up from the Dublin Rules are listed as follows:

Alleviation of poverty and disease

At the start of the 1990s, more than a quarter of the world's population still lacked the basic human needs of enough food to eat, a clean water supply and hygienic means of sanitation. The Conference recommends that priority be given in water resources development and management to the provision of food, water and sanitation to the un-served millions.

Protection against natural disasters

Lack of preparedness, often aggravated by lack of data, means that droughts and floods take a huge toll in deaths, misery and economic loss. Economic losses from natural disasters, including floods and droughts, increased three-fold between the 1960s and the 1980s.

Development is being set back for years in some developing countries, because investments have not been made in data collection and disaster preparedness. Projected climate change and rising sea levels intensify the risk for some countries and threaten the security of existing water resources. Damage and loss of life from floods and droughts can be reduced by disaster preparedness actions.

Water conservation and reuse

Current patterns of water use involve excessive waste. There is great scope for water savings in agriculture, industry, and domestic water supplies. Irrigated agriculture accounts for about 80% of water withdrawals in the world. In many irrigation schemes, up to 60% of this water is lost on its way from the source to the plant. More efficient irrigation practices will lead to substantial freshwater savings.

Recycling could reduce the consumption of industrial consumers by 50% or more, with the additional benefit of reduced pollution. Application of the 'polluter pays' principle and realistic water pricing will encourage conservation and reuse.

On average, 36% of the water produced by urban water utilities in developing countries is 'unaccounted for'. Better management could reduce these costly losses.

Combined savings in agriculture, industry, and domestic water supplies could significantly defer investment in costly new water-resource development and improve the sustainability of future supplies. More savings will come from multiple use of water. Compliance with effective discharge standards, based on new water protection objectives, will enable successive downstream consumers to reuse water which presently is too contaminated after the first use.

Sustainable urban development

The sustainability of urban growth is threatened by water depletion and degradation. After a generation or more of excessive water use and reckless discharge of municipal and industrial wastes, the situation in the majority of the world's major cities is appalling and getting worse.

As water scarcity and pollution force development of ever more distant sources, marginal costs of meeting fresh demands are growing rapidly. Future guaranteed supplies must be based on appropriate water charges and discharge controls. Residual contamination of land and water can no longer be seen as a reasonable trade-off for the jobs and prosperity brought by industrial growth.

Agricultural production and rural water supply

Achieving food security is a high priority in many countries. Agriculture must not only provide food for rising populations, but also save water for other uses. The challenge is to develop and apply water-saving technology and management methods, and, through capacity building, develop better institutions and incentives for the rural population to adopt new approaches for both rain-fed and irrigated agriculture. The rural population must have better access to potable water and sanitation.

Protecting aquatic ecosystems

Water is a vital part of the environment and a home for many forms of life on which humans depend. Disruption of flows has reduced the productivity of many such ecosystems, devastated fisheries, agriculture and grazing, and marginalized the rural communities.

Various kinds of pollution, including trans-boundary pollution, exacerbate these problems, degrade water supplies, require more expensive water treatment, destroy aquatic fauna, and deny recreation opportunities.

Integrated management of river basins provides the opportunity to safeguard aquatic ecosystems and make their benefits available to society on a sustainable basis.

Resolving water conflicts

The most appropriate geographical entity for the planning and management of water resources is the river basin, including surface and ground water within the basin. The effective integrated planning and development of transboundary river or lake basins has similar institutional requirements to a basin entirely within one country.

The essential function of existing international basin organizations is one of reconciling and harmonizing the interests of riparian countries, monitoring water quantity and quality, development of concerted action programs, exchange of information, and enforcing agreements.

In the coming decades, management of international watersheds will increase in importance. A high priority should be given to the preparation and implementation of integrated management plans, endorsed by all affected governments and backed by international agreements.

The enabling environment

Implementation of action programs for water and sustainable development will require substantial investments, not only in capital projects, but, crucially, in building capacity of people and institutions to plan and implement those projects.

The knowledge base

Measurement of components of the water cycle, in quantity and quality, and of other characteristics of the environment affecting water is an essential basis for undertaking effective water management. Research and analysis techniques, applied on an interdisciplinary basis, permit the understanding of these data and their application to many uses.

With the threat of global warming due to increasing greenhouse gas concentrations in the atmosphere, the need for measurements and data exchange on the hydrological cycle on a global scale is evident.

Data are required to understand the world's climate system and the potential impacts on water resources of climate change and sea level rise. All countries must participate and be assisted to take part in global monitoring, the study of the effects, and the development of appropriate response strategies.

Capacity building

All actions identified in the Dublin Conference Report require well-trained and qualified personnel. Countries should identify, as part of national development plans, training needs for water-resources assessment and management, and take steps internally and, if necessary, with technical cooperation agencies to provide the training, and working conditions to retain trained personnel.

Governments must assess their capacity to equip their water and other specialists to implement the full range of activities for integrated water-resources management. This requires provision of an enabling environment in terms of institutional and legal arrangements, including those for effective water-demand management.

Awareness raising is a vital part of a participatory approach to water resources management. Information, education and communication support programs must be an integral part of the development process.

Follow-up

Experience has shown that progress towards implementing the actions and achieving the goals of water programs requires follow-up mechanisms for periodic assessments at national and international levels.

All Governments should initiate periodic assessments of progress. At the international level, United Nations institutions concerned with water should be strengthened to undertake the assessment and follow-up process.

In addition, to involve private institutions, regional, and non-governmental organizations along with all interested governments in the assessment and follow-up, the Conference proposes, for consideration by UNCED, a world water forum or council to which all such groups could adhere:

It is proposed that the first full assessment on implementation of the recommended program should be undertaken by 2000.

UNCED is urged to consider the financial requirements for water-related programs, in accordance with the above principles. Such considerations must include realistic targets for the time frame for implementation of the programs, the internal and external resources needed, and the means of mobilizing such funds.

Source: International Conference on Water and the Environment. The Dublin Statement on Water and Sustainable Development. 1992. Available online at http://www.wmo.int/pages/prog/hwrp/documents/english/icwedece.html. Used by permission of the World Meteorological Organization.

PERSPECTIVES

What Steps Should Be Taken to Help Provide Access to Clean Water for More People in the World?

Overview

"Whiskey is for drinking; water is for fighting over."

—Mark Twain

Over the past several years, the lack of access to freshwater in many parts of the world has come to be regarded as the most serious and immediate environmental crisis facing the world. Other crises may draw more headlines—for example, pollution, overfishing of oceans, or global warming—but water scarcity is already a deadly crisis, responsible for thousands of deaths per year as people in developing countries are forced to drink from contaminated water supplies.

Competition Over Water

Competition over stressed water sources has played a large role in many world conflicts and is expected to be a trigger for many more tensions as the problem gets worse. Water scarcity is also one of the most prominent ways that global climate change is felt as droughts become more frequent and severe in many areas. For these reasons and others, such organizations as the United Nations (UN), the World Bank (WB), and the European Union (EU) are devoting increasing amounts of money and research time to the problem. Yet providing access to clean water to the millions who lack it is a formidable task with many components to account for and many obstacles to overcome.

The crisis disproportionately affects many parts of the developing world, places where governments and people are least equipped to cope with it. In some of those areas, such as arid parts of Africa, the problem is both a resource crisis, meaning that such water sources as rivers and aquifers are overburdened and drying up, and a service crisis, meaning that even the potential resources that exist are not being effectively delivered due to poor management, lack of funding, and governmental corruption.

Often the water that is delivered is contaminated due to pollution and a lack of proper sanitation and water-treatment facilities. In addition, climate change is exacerbating the resource crisis—increasing the frequency and severity of droughts in many semiarid parts of the developing world—while population growth is putting further strain on limited available resources and inadequate services.

The UN estimates that 90 percent of the 3 billion people projected to be added to the world population by 2050 will be in developing countries, most in areas that lack (or by then will lack) access to clean water. The increase in population will mean more demand for agricultural irrigation, which accounts for 70 percent of human water use. Beyond the figures provided by international water resource managers, the sight of thirsty faces and lands may best prompt world citizens to ask: what's the best way we can provide access to clean water for more people?

Water and Early Civilizations

Throughout history, human populations have had to take elaborate measures to compensate for nature's uneven distribution of the freshwater that people need in order to drink, bathe, and grow food. Elaborate distribution and irrigation systems were often among the most impressive engineering feats of the ancient world, devised by groups in the Middle East, China, India, Mesoamerica, and Europe. Many early civilizations developed in arid areas and could not count entirely—or at all—on rainfall to provide the water they needed.

Thus, early irrigation systems tapped into such fabled rivers as the Nile, Yellow, Indus, Euphrates, and Tigris, or springs such as Israel's spring of Jericho. Inevitably, conflicts over water sources arose. In the 19th century, water was the defining issue in the white settlement of the arid western United States, prompting a popular quote attributed to Mark Twain: "Whiskey is for drinking; water is for fighting over."

New Problems

Added to the age-old problems of uneven distribution, pressures on freshwater availability have been increasingly affected by population growth, increasing pollution, poverty, corrupt or ineffective governments, and climate change. All of these impact parts of the developing world where lack of infrastructure is already a problem.

While water scarcity has been a source of hardship and conflict through history, these factors are combining to pose what many authorities studying the problem consider the most severe crisis the world will face in the 21st century. For poor nations sharing overtapped rivers and aquifers, the threat of wars over water is real. In many cases, the role of water supplies in world conflicts in such areas as Darfur and the Middle East is underappreciated, according to many experts.

The distribution of water through evaporation and precipitation is more or less a constant amount—about 24,000 cubic miles of rainfall per year—but the distribution has always been inconsistent. In some countries, up to 90 percent of the year's rain may fall in a few major storms during a short monsoon season. Rainforests receive a huge amount of the world's rainfall; the Amazon River watershed accounts for one-fifth of the world's freshwater flow, but the area is largely made up of inaccessible forest (and the more that trees are cut down to access the area, the less rain falls due to the loss of evapotranspiration).

Increasing the Supply of Clean Water

For areas that need more water, there are a limited number of ways of getting it. Dams can be built on rivers and streams, and water can be stored in reservoirs. Wells can reach down and tap water stored in underground aquifers. Water can be brought in from somewhere else. In coastal areas, seawater can be desalinated. Another measure sometimes counted as a water source is conservation—people learning to cut back on unnecessary water use and adopting such methods as drip irrigation to cut down on waste. The water saved can then be counted as a sort of "new" supply.

Some of those methods can have severe drawbacks in certain areas. Building dams has become increasingly controversial, especially larger projects that are seen as damaging to the environment and that often displace communities of people who live in valleys that must be flooded to become reservoirs. Desalinating seawater is an expensive option and is in very limited use. Drawing on groundwater poses risks because in some areas it is essentially a nonrenewable source. Currently about 20 percent of the water people use comes from groundwater, and that proportion is rising. With continued extraction, groundwater tables in some areas become lower, requiring deeper wells and higher costs.

Export of Water?

There are proposals in water-rich Canada to export bulk quantities of freshwater. Canada, with 0.25 percent of the world's population, stores an estimated 20 percent of the world's total freshwater supply and circulates about 9 percent of the world's "renewable" freshwater—the water that flows in rivers and falls as precipitation. There are business interests that would like the country to allow freshwater to be transported to the U.S. Southwest, and possibly shipped overseas in tankers or huge bags that would be towed by ships to parched countries on other continents.

Proponents say that for Canada to market water is no different from Saudi Arabia marketing oil. These proposals have critics, however, who feel that water should not be treated as a for-profit commodity, and that selling it to developing nations could cause conflicts over who gets the water, possibly widening divisions between rich and poor in countries receiving the water. Many critics also oppose such proposals on environmental grounds. The Canadian government hasn't approved anything yet.

Poor Water Management

Another element of the looming crisis is poor management of water resources. The problem was dramatized in July 2009 when inhabitants in five villages in Nigeria's Kano state organized a protest, saying they had waited 15 years for the state government to provide wells. The villagers must carry water containers many miles and often take water from contaminated ponds, which led to six deaths from cholera in one month alone. Nigeria, Africa's most populous nation with 170 million people, is one of four West African nations in which fewer than half of the residents have access to safe drinking water, according to the United Nations (UN).

Corrupt and inefficient national and local governments and a lack of resources and technological expertise have hampered efforts in such places as rural Nigeria. The UN, World Bank, EU, and other international agencies have put forth models for alleviating such obstacles, including a concept called Integrated Water Resources Management, which involves coordinating different uses of water (agriculture, industry, home use, the environment) as well as all the different entities involved in construction, development, and delivery of any services that involve water. With better management, available resources can be used more efficiently and safely.

However, management issues are very difficult to solve, and it is even harder to fruitfully address the problems posed by population growth and climate change. It is therefore perhaps not surprising that a multinational public opinion poll conducted in August 2009 by an organization called Circle of Blue, which surveyed 1,000 people each in 15 countries, identified freshwater availability as the world's top environmental concern.

Dennis Moran and Mark Troy Burnett

Perspective 1: Infrastructure Is Fine, but We Need to Focus on Regulation Management and International Cooperation

Overexploited and polluted by human activities, water has become a fragile resource, in quantity and in quality, everywhere in the world. Rampant population growth, increasing urbanization, economic growth, and changing consumption put more and more pressure on water resources. During the last 50 years, world water consumption has tripled as the population has doubled. With a world population projected to be

8 billion by 2030 and 9 billion by 2050, the need will increase considerably; for example, it will be necessary to increase irrigated agricultural lands to feed this population.

It is necessary to take into account the increasing need related to the development of India and China, and in several areas of the world geopolitical problems can lead to tensions over the use of transboundary aquifers and rivers. Lastly, the effects of climate change on water resources are already being felt and will likely worsen such situations. This is why it is all the more necessary to make efforts to provide clean water to more people in the world.

Problems of Untreated Water

Developing and financing infrastructures for access to drinking water and sanitation for all is a prime objective. In developing countries, 85 percent of domestic wastewater and 70 percent of industrial waste are directly discharged untreated into the aquatic environment, where they pollute the water supply, leading to dire health consequences. Public authorities, either national or local, must develop infrastructural elements such as drinking-water production plants, wastewater treatment plants, and delivery networks, and make sure they are well operated and maintained. Wastewater treatment can be either collective (for urban areas) or onsite (for rural areas).

Local Water Management

As water must be locally managed, it is important that national authorities give to local authorities the necessary powers and skills to organize water and sanitation services. To optimize the development of infrastructures, objectives, deadlines, and actions should be defined in management plans. This requires political will and financing. For developing countries, access to clean water should be the priority for bilateral aid and international funding institutions.

Even in developing countries, it is absolutely necessary for decision makers to develop the economic autonomy of water services to guarantee the financing of investments and functioning in the long term. The viability of water utilities requires a detailed, balanced budget, whether the management system is public or delegated to a private operator. Income must come from water bills paid by the users by setting a price for the service, setting cooperative mechanisms to provide for the poorest, if necessary. Even if the water price is low, it is important to give an economic value to water to motivate people to conserve.

Need for Regulation

Revenue from water bills should be used to invest in improvements so that "water pays for water." Providing water for human consumption should be the top priority, and allocation for other uses (irrigation, industry, etc.) should be defined through participative processes to take account of local needs.

Water resources should be better regulated, in quality and quantity, following a cross-sectoral approach called integrated water resources management at the river-basin level. In most cases, the legal framework should be strengthened by setting standards for water use and water pollution, and by enforcing compliance. Then it is necessary to adopt a comprehensive approach toward all pressures on water resources, thus allowing everyone to have access to safe water and sanitation, ensuring agricultural production and industry, preserving water resources and aquatic environments, and managing floods and droughts. These interests often compete and can no longer be individually solved in a sectoral way.

Need for Coordination

As stated by the International Network of Basin Organizations, we need coordination among the different water uses, between upstream and downstream, between quantity and quality, and between surface water and groundwater. Since water has no national and administrative boundaries, river basins are the most logical territories for organizing water management. Integrated water management at the river-basin level means defining specific administrative arrangements, river basin management plans, programs for action, networks for monitoring water status, financing mechanisms, and other elements. Users' participation should be organized within specific bodies, for example, through river basin committees, gathering representatives of state and local authorities and of the different categories of users.

Need for Capacity Building

Capacity building is also a key factor. Developing infrastructure alone is not sufficient. Providing clean water to more people in the world is also a question of governance and institutional capacities. It implies improving countries' entire organization toward water management, through policy reforms and clarification of responsibilities. It also implies improving the skills of technicians, engineers, and decision makers. The technical nature of water professions requires very precise qualifications, and personnel costs represent the highest item of expenditure in water supply and sanitation utilities. Basic and continuing vocational training for all personnel involved is central for the successful management of water utilities.

Need for International Collaboration

Finally, international cooperation should be reinforced. Cooperation should be strengthened between countries sharing the same river or aquifer. There are about 263 transboundary rivers and several hundred transboundary aquifers. As co-riparian countries share responsibility, the signing of international agreements and the creation of international commissions should be supported where they do not exist.

In Europe, the Water Framework Directive represents a very important step for the joint management of water resources: it constitutes an operational framework for multilateral coordination on a river-basin scale with common objectives and principles that can be exported to other parts of the world. The 1992 United Nations Economic Commission for Europe (UNECE) Convention on the protection and use of transboundary watercourses and international lakes, also known as the Helsinki Convention, developed key principles for cooperation across a broader geographical area than just the European Union (EU)—in Central Asia, for example. The 1997 United Nations Watercourses Convention, which entered into force in 2014 but has only been ratified by 36 nations, also provides a global framework for transboundary water cooperation.

International cooperation should also be reinforced through assistance programs to Africa, Asia, and Latin America. The financial resources devoted by international and bilateral donors to cooperation programs and capacity building should significantly increase and prioritize water issues. To conclude, common cause is necessary to meet water challenges, and both developed and developing countries have a stake in avoiding a crisis. The role of the European Union and the United States is therefore particularly essential.

Coralie Noël

Perspective 2: Infrastructure First, Then Better Management

Although water is abundant on Earth, less than 3 percent of all water on the planet is freshwater, and the majority of that freshwater is trapped as ice in glaciers and ice sheets. According to the World Bank and UNICEF, nearly 750 million people around the world do not have access to safe drinking water. Additionally, the world's population is increasing ever more rapidly. The world's population has grown from 1 billion in 1804 to 6 billion in 1999 to over 7 billion in 2016, thanks to better medicine and better agriculture.

Issues in Developing Countries

The fastest population growth is happening in developing nations. These same countries are in the greatest need of improved access to clean water. Further, water consumption is increasing worldwide. Currently, an average of 5–8 gallons per person per day of clean water is sufficient to meet basic human needs in the developing world, as opposed to such developed nations as the United States and Germany, which have rates of 100 and 34 gallons per person per day, respectively.

As developing nations industrialize, standards of living rise and water consumption increases. As consumption increases, the amount available to other animals and plants decreases. This in turn limits ecosystems' ability to naturally filter pollutants from water, ultimately posing health risks to humans. Considering all of these factors, the problem of providing access to clean water to more people around the world is a complex and multifaceted one. A variety of strategies should therefore be employed to address the issues of freshwater access and quality.

Local Involvement

The first step toward increasing access to clean water begins at the local level. People living in communities without basic water and sanitation infrastructure need to be educated about the links between water quality and human health in order to promote a change in behavior. For example, in many parts of the developing world, especially in crowded slums, water sources are often polluted by human waste, which leads to deaths from preventable diseases. Nongovernmental organizations (NGOs) such as the World Bank would be well suited to train local leaders on these issues.

Local officials, such as council members and teachers, could then provide training to the general public through meetings and schools. All citizens need to be educated on water-quality issues, as it may take years to change behavior at the community level. Utilizing established local leaders would increase local buy-in to the program.

Partnerships Needed

Governments, NGOs, and private companies also need to work together to extend basic water-supply infrastructure to the people currently without it. When building this infrastructure, it is essential that it be cost efficient, as these communities are generally very poor and will need to be able to afford the necessary operation and maintenance costs. Increasing access to clean water will help to improve the economies of the developing world by increasing productivity by decreasing illness.

An example of this type of partnership is the World Bank's project to increase rainwater harvesting in drought-stricken municipalities in India. For this project, the World Bank partnered with a university and various government officials to develop and construct a rainwater harvesting system that connected tanks on each house to larger community-wide storage tanks. These tanks lead to a village tank, which is used to replenish the well. In regions that sit atop aquifers, low-cost tube wells are a quick way to increase domestic water supplies. This technology is common in Africa and Asia, and the World Bank has been involved with funding some of these projects. If a tube well program is implemented, however, appropriate groundwater extraction regulations need to be established to prevent depletion of the aquifer.

Local Government Efforts

Once basic infrastructure is constructed, local governments or councils need to be created to manage and protect the new water supply systems. A study in the former Soviet Union showed that community-based management of infrastructure is more effective than a centralized government management approach. Water users who are also responsible for maintenance are more proactive in routine maintenance, and community members are less likely to vandalize infrastructure when they have an important role in its upkeep.

Additionally, any fees associated with maintaining infrastructure could be more easily collected by community members. Along with the creation of local governing bodies, additional education is needed on water conservation in order to promote the long-term sustainability of the freshwater resource for the community.

While local government should take the lead in managing water resources, there are important steps that need to be taken at higher levels of government. National and regional water quality standards must be established, as water is a common pool resource, meaning that it is accessed by many users. Surface water travels from place to place, and pollution from users upstream impacts users downstream. Groundwater moves at a much slower rate than surface water, but here, too, pollution in one region of an aquifer can impact water quality in other regions.

Water Quality Standards

Many developed nations have long-standing water quality standards that could provide models for the developing world to follow. The major obstacle to these large-scale standards is funding. There is a lack of money in the developing world, and environmental quality monitoring and regulation is not a top priority for spending when budgets are limited. This is why education at the local level in all communities is critical. If everyone maintains clean freshwater supplies locally, then there is not as great a need for top-down water quality regulation.

Providing access to clean water to more people is important from many standpoints. From a humanitarian perspective, all humans should be able to access clean water, as it is essential for life. In this age of globalization, it is in everyone's best interest to help maintain high standards of hygiene and sanitation, as diseases can rapidly spread around the world. Further, expanding access to clean water can help to stabilize communities and improve global security. Finally, from an environmental perspective, maintaining good water quality will keep ecosystems healthy and allow them to provide the ecosystem services that humans often take for granted long into the future.

Nathan Eidem

REFERENCES AND FURTHER READING

Barlow, Maude. *The Blue Covenant: The Global Water Crisis and the Coming Battle for the Right to Water*. New York: The New Press, 2009.

de Villiers, Marq. *Water: The Fate of Our Most Precious Resource*. Boston: Mariner Books, 2001.

Falkenmark, Malin, and Johan Rockstrom. *Balancing Water for Humans and Nature: The New Approach in Ecohydrology*. London: Earthscan, 2004.

Global Environment Monitoring System (GEMS/Water). United Nations Environment Programme (UNEP). http://www.gemswater.org.

International Water Management Institute (IWMI). http://www.iwmi.cgiar.org.

Kelly, Elisha, and Gideon Fishelson. *Water and Peace: Water Resources and the Arab-Israeli Peace Process*. Westport, CT: Praeger Publishers, 1993.

Newton, David E. *Encyclopedia of Water.* Westport, CT: Greenwood Press, 2003.

Pearce, Fred. *When The Rivers Run Dry: Water—The Defining Crisis of the Twenty-First Century.* Boston: Beacon Press, 2006.

Shiva, Vandana. *Water Wars: Privatization, Pollution, and Profit*. Boston: South End Press, 2002.

Simon, Paul. *Tapped Out: The Coming World Crisis in Water and What We Can Do about It*. New York: Welcome Rain Publishers, 2002.

UNESCO. "World Water Assessment Programme (WWAP)." http://www.unesco.org/water/wwap.

World Water Council. http://www.worldwatercouncil.org.

7 WATER, HUMAN SETTLEMENTS, AND THE GROWTH OF CITIES

OVERVIEW

This chapter analyzes the importance of water for human settlement and survival, as well as ongoing water challenges in today's cities and urban neighborhoods. The nurturing of early civilization and the genesis of permanent settlements was linked to the availability of water sources such as rivers, estuaries, deltas, lakes, oases, fjords, straits, or coastlines.

These sources provided freshwater and exits to the sea to support human livelihoods, and develop agriculture, industry, services, trade, and transportation networks. Availability of water continues to be a driving force for agriculture and a key factor for urban expansion in today's world. The case studies examine new trends in water-efficient construction, water stress in the Mexico City basin, and water bottlenecks in African cities. The perspectives section looks at the issue of California's water scarcity and its impact.

Water and Cities

Over 145 capitals of the world are located on rivers. Every major city in Europe once developed around a river, for example: London on the Thames, Rome on the Tiber, Paris on the Seine, Berlin on the Spree, Budapest on the Danube, Moscow on the Moskva, Madrid on the Manzanares, and Amsterdam on the Amstel.

Outside of Europe, major river systems such as the River Plate in Latin America, the Mekong River in Asia, and the Nile, Zambezi, and Congo Rivers in Africa have played significant roles in the settlement and development of urban areas, and the transportation of people and goods throughout continental areas before and after colonial times.

In North America, the city of New York is located on the Hudson; Washington, D.C. is on the Potomac; and Toronto, Canada, is on Lake Ontario, a major lake in the Great Lakes system. The Great Lakes/Saint Lawrence Seaway enabled the growth of major urban and industrial settlements in Canada and the United States, and continues to function as a major water transportation artery extending 2,340 miles to the Atlantic Ocean.

Water and the Earliest Settlements

As far back as 5,000 years ago, long before the rise of the most important European cities, water sources and river valleys were the foundation of the earliest urban-centered civilizations across the world. These were located in certain parts of China, Mesoamerica, the Central Andes, Mesopotamia, the Indus Valley, and Egypt.

The ability of people in these societies to harness water, especially in arid areas, was instrumental for the construction of sophisticated aqueducts, wells, water storage systems, and irrigation canals, as well as the domestication of plants and animals, the production of surplus food, and the consolidation of power in cities or religious centers.

Surplus food enabled trade and the conquest of new areas; it also supported settlements for nonagriculturalists such as priests, rulers, warriors, potters, and scribes. This advance in turn provided the foundation for institutions and an entrenched ruler or priest caste, cementing the cultural and political structures for permanent communities and empire building.

In the Indus River Valley the first cradle of civilization arose in the period of 3000–1500 BCE with the development of ancient urban centers such as Harappa and Mohenjo Daro. The Harappan civilization was one of the earliest to build towns and establish trade networks with other areas. It developed the technical capacity to construct wells and irrigation works using lift and surface canals as far back as 2600 BCE (Ponting 2008, 343). These innovations produced the surplus food to support city centers and nonagricultural occupations.

Around 3000 BCE—independent of the Indus Valley development—the Sumerian civilization reached its highpoint in Mesopotamia (today's Iraq). The fulcrum of early Sumerian city building was located around the urban centers of Uruk, Ur, and Lagash. These city hubs emerged due to their location around the Tigris and Euphrates Rivers. Uruk is considered to be one of the oldest cities in the world. A 10-kilometer-long wall surrounded the city at its heyday. In 3000 BCE Uruk contained huge temple mounds and public buildings, and 40,000 people, twice the size of ancient Athens, Greece, at its peak (Ponting 2008, 61, 296).

The Tigris and Euphrates Rivers were the most crucial factor in the development of Sumerian towns and civilization. They provided fertile alluvial soil and water for early farming and production of grains to support higher population in the area. At first, agricultural production was confined to the banks of these two rivers, since rainfall was practically nonexistent in parts of Sumer. The spring thaw in the mountains produced floods in early summer, but rivers were at their lowest in fall, when crops needed water the most.

Fortress ruins on the Euphrates River, once the cradle of civilization. (Joel Carillet/iStockPhoto)

To resolve this problem, rulers in the town centers organized huge armies of manual labor, including slaves to build dams, reservoirs, and irrigation canals. This step led to the expansion of agriculture, more surplus food, trade, and the consolidation of urban-based power. As a result of irrigation, town rulers expanded farming and fishing activities to stretch further into river basin and delta marsh areas, enabling more trade.

Surplus food allowed elites to concentrate power in Sumerian cities, where the rulers lived. They exercised their authority over the working population, and conquered other groups through taxation, wars, tribute, and forced labor. Higher food production liberated more and more people from agricultural work. Writing developed around 3000 BCE. Records indicate that city administrators kept detailed accounts of agricultural production and trends. As salt accumulation became a problem in Sumerian fields, the data showed declining crop yields and eventual replacement of wheat by more salt-tolerant barley. The town-based civilization would eventually collapse when the agricultural base crumbled.

Egyptian Ceremonial Centers

In Egypt, the emergence of civilization and ceremonial centers was closely linked to the domination of the waters of the Nile River and its seasonal rhythms of flooding. As in Mesopotamia, this society was based on the rise of the religious and military rulers, the pharaohs. They lived in centralized ceremonial centers and exercised social and political control over the masses of working people. To address the uncertainty of seasonal Nile flooding and fluctuations in water flow and deposits of river sediments, rulers organized new water management techniques to improve harvests, which led to surplus production.

Surplus food and its storage in granaries sustained the masses to continue working and maintain the centers of power. The pharaohs were able to conscript and feed labor for the construction of huge pyramids to cement their prestige and dominance. These impressive works signified the power of the religious and urban elite over the masses of the people.

The rise and fall of each dynasty of pharaohs was linked to extreme fluctuations in Nile flood levels and ensuing crop failures or successes. Construction projects continuously imposed heavy burdens on workers and farmers, who had to support a growing number of priests, rulers, administrators, bureaucrats, military personnel, and nonagricultural workers. The toll on the working masses eventually resulted in revolts by the peasantry from 2250 BCE onward, culminating with the collapse of the Ramesside Dynasty.

Development of An-Yang Settlement in China

China was another place where urban settlements developed very early. A complex civilization known as the Shang evolved around the Wei and Yellow Rivers in 1800 BCE. The site of An-yang was the first capital of the Shang culture. Like the Sumerians and the Egyptians, the Shang achieved power through the appropriation of surplus agricultural production by urban elites, and the development of a large ceremonial, religious, and military center, dedicated to the worship of the deity Ti.

Civilizations in Mesoamerica and South America

Civilizations in Mesoamerica and South America evolved later than in Mesopotamia due to delayed human settlement of the Americas. Around 1200 CE, the Toltecs and the Aztecs (from today's Mexico) developed a complex civilization in Mesoamerica, which included the growth of densely populated settlements in the Valley of Mexico. As in the other examples of early civilizations, the growth of the Aztecs' capital, Teotihuacan, depended on the Aztecs' ability to organize large-scale city development, establish and maintain irrigation canals, and create agricultural surplus to sustain religious and military elites in urban ceremonial centers.

The city of Teotihuacan was a massive ceremonial center. It included temples, large-scale pyramids, and thousands of residential areas organized around the compounds of leaders and clans. Around 400 CE, Teotihuacan's population was 100,000, twice the size of Rome at its prime. Religious and military elites ruled the city and controlled the masses. They conscripted local labor and captured other groups to build structures in astronomical alignments, marking the rulers' interest in lunar cycles, rain frequencies, and harvest patterns. It is believed that Teotihuacan's demise around 750 CE was linked to the overuse of irrigation, soil salinity, and popular uprisings of the subjugated masses.

Further south in Mesoamerica, the Mayan civilization started its advance as early as 2500 BCE. Mayan centers first arose in the lowland tropical jungles of today's Guatemala and Belize, and were endowed with abundant water resources. Tikal and

Teotihuacán, located about 30 miles east of present-day Mexico City, was the largest and most important city in Mexico prior to the dominance of the Aztecs. With a population of about 150,000 at its peak (ca. 400-600 CE), the city comprised thousands of apartments, as well as temples, plazas, and palaces. Looming over the existing ruins of Teotihuacán is the Pyramid of the Sun, the largest stone pyramid in pre-Columbian America. Water systems and drainage features were a key part of the complex. (Corel)

Tenochtitlan emerged as the major urban centers of the Mayans. At its peak, Tikal had a permanent population of at least 50,000. It was highly structured, and contained steep pyramids of about 100 feet high with temples and palaces arranged around a vast ceremonial plaza.

Mayan structures were aligned to astronomical points and positions of the sun, moon, and planets. Religious and military elites lived in the urban center, while construction workers lived in the surrounding areas. Environmental problems provoked by deforestation, soil erosion, and silt accumulation in rivers and irrigation ditches eventually led to the collapse of the Mayan society and its great cities.

The Inca Civilization

Even further south in South America, the imperial Inca capital of Cusco began to emerge in 600 CE. The Inca Empire developed in the central Andes mountain range, independently from the Mesoamerican civilizations in the North. The Incas ruled all four corners of their empire from the capital of Cusco, conceived as a grid radiating outward to the cardinal points of the compass.

Territorial expansion and domination of the Incas were linked to militaristic and organizational skills and their ability to harness and dominate available water sources. The Incas used a range of tactics to control their subjects and bring new groups under their rule: strategic warfare and construction of military outposts; elaborate systems of courier communications; mobilization of forced labor; collection, storage, and accounting of tributes throughout their vast empire; and the control of water, irrigation, and terrace-based agricultural production.

Inca ruins on a terraced mountainside. The Inca used such terraces to increase the amount of arable land and manage water use in the steep and rugged terrain of the Andes mountains. (David Owens/iStockPhoto)

Their water control strategy was critical, given the lack of rainfall on the coasts in Peru. The Incas managed the flow of rivers down from the highlands to the arid coast, thereby dominating the food and water supply of the land and control of other groups. The demise of the Inca Empire would eventually occur with the arrival and domination of the Spanish in the 15th century, but shortly before European arrival, a power struggle within the Inca elite was already weakening the empire.

Water and Elites

As the above examples of past civilizations have indicated, the harnessing of water and the improvement of agriculture, surplus food production, and trade were key factors in the growth of militaristic, political, or religious elites. In all the cases discussed, these elites built urban-based empires and controlled the working population. Irrigation, water management, and agricultural production delivered excess food for nonagricultural members of society, and allowed the elites to requisition labor for construction of temples, tombs, and pyramids, and for military warfare. Over millennia, the rise of groups not directly involved in farming would become the basis for urban societies in the world.

The earliest cities were ceremonial centers dominated by priests, rulers, artisans, and administrators—built-in designs or grids mirroring religious, militaristic, or astronomical alignments; later urban centers reflected other functions. For example, the growth and layout of city centers such as Rome, Constantinople, Athens, Venice, and

Malacca reflected their strategic locations on trade routes or their rise as seaborne trading empires. No matter the orientation or function, the main basis of early cities and later permanent megacenters was the availability of water sources close by.

Water, Early Trading Routes, and Hubs

In Europe, around 1000 BCE, the growth of trade on the seas and the development of city and river ports such as Carthage, Athens, Constantinople, Marseilles, Naples, Venice, and Cologne led to the rise of seafaring groups such as the Greeks, Phoenicians, Romans, and Persians. Earlier trading hubs on the Mediterranean Sea eventually became cities, which flourished with the expansion of trading routes and networks.

Roman military success went hand in hand with the establishment of trading settlements at key locations such as London, Paris, and Cologne, located on major rivers. The prosperity of these cities rose and fell with changes in rulers, trade routes, and the ebb and flow of commercial fortunes. After the end of Roman rule, the drop in trade led to the decline of many of these cities, as people left or moved on to new areas of activity.

China was the first country in the world to begin the preindustrial process and to consolidate early urbanization, mostly due to its production of iron and its use of paper currency and currency exchange systems. Under the Sung Dynasty in the 11th and 12th centuries, cities such as K'ai-feng, Cha'ang, Loyang, and Hang-chou entered into a strong period of expansion. It is significant that these centers were located in close proximity to Southeast Asia trading networks. Around 1200 CE, Hang-chou's population was two million, 10 times larger than any other city on earth at the time. In the 13th century about 20 percent of Chinese was already living in cities (Ponting 2008, 299).

Until the 12th century, European cities such as Constantinople, Cordoba, Venice, Seville, and Palermo had benefitted from their proximity to or location on Mediterranean Sea trading routes (Novaresio 1996). However, after a slowing down of trade in the Mediterranean, growth in Europe shifted to northern European centers such as Paris, London, and Rotterdam, cities located on the banks of important European rivers (Seine, Thames, and Rhine).

The outbreak of the plague in Europe, or the Black Death, as it was called, slowed down city development between 1300 and 1550, but urban growth resumed after the mid-16th century. Factors that led to this upswing included the expansion of trading empires based on improvement of navigation and ship construction as well as the gateway location of European cities relative to ports and important ocean routes. During this period, European cities suffered from severe sanitation and water pollution problems, as raw sewage, solid, industrial, and animal waste, and other pollutants were dumped directly into city streets and rivers. Population growth and waves of villagers moving to cities also aggravated environmental and public health crises of cities in the 16th century onwards.

Waterborne Diseases and High Death Rates in Early Cities

Death rates due to dysentery, typhus, and other water- or vector-borne diseases were high in early English centers such as London and Sheffield. The contamination of

Plague victims blessed by a priest, from *Omne Bonum* by James le Palmer, England, 1360–1375. The Black Death killed up to a third of the population in Europe during its 15-year span. (The British Library)

rivers with human wastes and effluents meant that fewer sources of drinking water were available to urban inhabitants. A major problem that cities had to address then (and still do) was separating clean drinking water from sewage. In the early Indus Valley cities of the past, this problem had been resolved through the storing of freshwater in tanks away from water coming from lavatories and drains, but these lessons were not known or forgotten.

Most European cities did not master the art of keeping their water supply uncontaminated by pollutants until the 19th century. From the 14th century onward, rivers such as the Thames in London and the Seine in Paris became so polluted that city officials were forced to use artesian wells to tap groundwater or lay pipes from water sources outside of cities. These solutions proved to be inadequate; most houses had no lavatories, and most urban residents had no access to individual water or sewer connections.

European industrial development mostly occurred in cities located on or close to rivers. Rivers provided water for drinking and industrial activities such as smelting, tanneries, and textile mills. Rivers close to industries received the direct discharge of industrial by-products and became a lethal brew of polluted water; sewage; industrial, animal, and human wastes; dead animals; and garbage. These toxins caused periodic outbreaks of intestinal diseases and cholera in cities. In London, the stench of the River Thames and the city in hot weather permeated rich and poor areas, and even the House of Commons.

Even when technological developments in the early 19th century allowed the flushing of lavatories with water, human waste still ended up in rivers and open sewers due to the lack of wastewater and sewage treatment facilities. The first sewer in Moscow was not built until 1898; few residents were connected to a sewage system then. Paris continued to dispose of some sewage into the Seine as late as the 1960s, and until the 1980s some of Moscow's sewage continued to flow untreated into the Moskva River.

Water Issues in Cities of Developing Countries

While severe urban environmental, water supply, and sanitation problems are for the most part a phenomenon of the past in developed societies, they persist in most developing countries. Urbanization is rapidly developing throughout the entire world but more so in developing countries, outpacing the provision of water and sanitation services. Clean drinking water and the disposal of human and solid wastes are major challenges facing cities and agglomerations in Latin America, Asia, and Africa. Billions of people in the world still have no access to safe drinking water and sanitation; most live in urban centers.

In India, cities on the Ganges River, including the sacred city of Varanasi, continue to discharge untreated human waste, cremated and partially cremated bodies, animal carcasses, and industrial effluents into the river. The Ganges is severely polluted and carries many diseases. In Manila, the capital of the Philippines, most residences are not connected to a sewer system. Untreated sewage goes into the Pasig River, which also provides water to urban residents. In other major cities such as Rio de Janeiro in Brazil and Lagos in Nigeria, there are problems of water pollution and scarcity.

Rapid urban expansion has placed enormous pressures on the supply of clean water and construction of adequate waste treatment facilities. The more populated the city, the greater the demand for water resources, and the larger the area from which they must be drawn. Also, the more industrialized the city, the higher the potential for discharge of industrial waste and effluents into ground and surface water. In megacities such as Mexico City, the water table is very low due to excessive pumping of groundwater for urban supply (see case study 2 in this chapter), and water is sourced from other areas.

In megacities—cities with at least 10 million people—depletion and pollution of the closest surface or groundwater resources is a major problem. Increasingly, water has to be accessed from more expensive, distant sources. In addition, the construction of buildings, roads, parking spaces, industrial parks, sports arenas, bridges, shopping malls, highways, and airports in cities, means that asphalt and cement surfaces are

Each day, thousands of gallons of raw sewage flow into the Ganges River in Varanasi, India. In some locations along the river, it is estimated that pollution has reached 340,000 times the permissible level. (John McConnico/AP Photo)

increasingly built up, reducing the natural flow of surface water and options for aquifer replenishment.

The rapid growth of cities in developing countries has outpaced and overwhelmed the ability of city administrators and public health and sanitation experts to cope with the existing population, let alone address the demand of migrants leaving rural areas looking for better lives in the city. The financial, technical, and human resources needed to build, upgrade, and expand water and sewage systems to address population growth are prohibitive for low-income, cash-strapped countries undergoing urbanization pressures.

In times of drought, many cities have had to resort to rationing measures or outright cuts in water supply to urban dwellers, whether in Los Angeles, California, or São Paulo, Brazil. These measures are potentially unpopular and often risk the anger of politically organized urban citizens capable of orchestrating street closures and other protest measures, as occurred in the water scarcity crisis of São Paulo, Brazil, in 2014, which affected millions of people (Gerberg 2015).

Other cities are experiencing contamination of their water supply. An egregious example is seen in Flint, Michigan, in the United States, where city administrators in 2014 switched the water supply from a clean lake source to a polluted river in Flint, causing lead contamination of water pipes with potential impacts on the health of city residents, especially children.

In large cities of Africa, Asia, and Latin America, there are additional complications in the provision of water and sewage services to poor city residents. A large

Demonstrator shouts slogans against São Paulo's Governor Geraldo Alckmin during a protest against the rationing of water in São Paulo, Brazil, February 6, 2015. At that time Southeastern Brazil was suffering the worst drought in more than eight decades, and authorities were considering the introduction of water rationing in the city. (Andre Penner/AP Photo)

percent of the poor lives in marginalized, squatter neighborhoods; most are perched upon steep, deforested slopes. From a technical and economic perspective, it is difficult to provide water and sewer connections to poor people, either because they cannot afford to pay or because costs are too high to run pipes uphill. These zones are also prone to flooding, landslides, and natural disaster hazards, as well as cholera breakouts and other waterborne diseases.

A Polluted River Runs through It

The poorest urban dwellers in developing countries have little access to domestic, piped water, and often share communal standpipes with irregular, poor-quality flows of water. Some residents live without access to even standpipes and have to rely on polluted wells or streams, or the purchase of poor quality water from private water trucks. Often, the poor pay higher prices than richer residents for water in the same city because tankers and private vendors alone set the price. But even where poor households have access to piped water, the reality is that they receive it intermittently or its quality may be poor.

Poor neighborhoods have few options and financial resources to safely dispose of solid and liquid wastes and to address drainage and surface runoff problems. In contrast, high-income urban neighborhoods in the same city usually have access to and the ability to pay for residential water and sewage connections, regular garbage pickup service, and piped water for household use, landscaping, gardens, fountains, and swimming pools.

Urban health problems in developing countries stem from the precarious situation of water and environmental sanitation and the inadequate disposal of human waste, the same problems observed in London centuries ago. Waterborne diseases cause a high incidence of infant, child, and adult illnesses, as well as infant and child deaths in poor countries.

The lack of clean water in poor households makes hand washing and domestic cleaning tasks difficult, and contributes to unhygienic food preparation and the spread of diseases. Insufficient water for washing also causes skin diseases to proliferate; and stagnant pools of dirty water and poor drainage in slums encourage mosquito breeding and water-related vectors such as malaria and yellow fever. The situation in some cities is alarming.

All large cities in the world have experienced problems with water supply at one time or another. Even in developed countries, vulnerable segments of the urban population have difficulties coping with natural hazards and maintaining the supply of clean water in crises, as Hurricane Katrina in New Orleans demonstrated. In New Orleans, the poor, disabled, and elderly residents of the city, especially those in low-lying, flooded areas, suffered disproportionately from water and sanitation problems caused by the hurricane.

An increasing number of rivers or reservoirs close to major cities are experiencing water contamination problems. Water pollutants fall into three main categories: liquid organic wastes, liquid inorganic wastes, and waterborne pathogens. Liquid organic wastes include sewage, liquid wastes from industries, and pollutants from storm runoff. Liquid inorganic wastes usually contain heavy metals such as lead or mercury, and waterborne pathogens are disease-causing agents such as bacteria, viruses, and worms, which contaminate water sources and cause diseases such as cholera.

In Asia, rivers usually carry four times the world average of suspended solids and 20 times those of high-income countries (Hardoy et al. 2004, 107). They also contain three times the world's average for bacteria from human waste, more than 10 times the guidelines of the Organization of Economic Cooperation and Development (OECD) and more than 50 times the norms of the World Health Organization (WHO). Pulp and paper production in Asia produces more water pollution than the same industry in North America. These differences have to do with more lax environmental regulations and enforcement measures in Asia, and cleaner technologies and stricter environmental standards for industrial production in North America.

Cities and Climate Change

As the process of urbanization continues its relentless pace, cities all over the word are expected to face further challenges. Climate change, rising ocean levels, and more intensive storms are already threatening low-lying cities with millions of residents, and posing risks to urban water supplies. This is particularly the case of cities in Bangladesh and Nigeria, coastal states of the United States, and cities in Europe such as London, Venice, and Amsterdam. In recent years, innovations in building techniques and the development of environmentally sustainable construction standards have fostered water conservation and recycling, and greater energy efficiency practices in the built up environment.

Wastewater Recycling

Wastewater recycling (gray water) is increasingly used in buildings and in cities to reduce the outflow of effluents to sewage systems, and to clean used water and deploy it for manufacturing and other nonpotable water purposes such as landscaping and the flushing of toilets. The case studies that follow outline some examples of adaptive measures to conserve water in urban environments, achieve greater efficiency of water use, and deploy better water management methods to address urban water bottlenecks.

Camille Gaskin-Reyes

REFERENCES AND FURTHER READING

D'Altroy, Terence N. *The Incas.* Oxford: Blackwell Publishing, 2003.

Gerberg, Jon. "A Megacity without Water: Sao Paulo's Drought." *Time*, October 13, 2015. time.com/4054262/drought-brazil-video/.

Hardoy, Jorge E., Diana Mitlin, and David Satterthwaite. *Environmental Problems in an Urbanizing World.* London: Earthscan, 2004.

Hemming, John. *The Conquest of the Incas.* Orlando: Harcourt Brace, 1970

Novaresio, Paolo. *The Explorers: From the Ancient World to the Present.* New York: U.S. Media Holdings, 1996.

Overey, Richard, ed. *Hammond Atlas of World History.* Maplewood, NJ: Times Books, 1999.

Ponting, Clive. *A New Green History of the World.* New York: Penguin Group, 2008.

CASE STUDIES

The following case studies address water stresses and the need for conservation and better management of water in the urban environment. The first case study looks at the issue of growing water demand of cities and water supply bottlenecks, which prompted the development of new building standards that conserve water, optimize water use, and foster the recycling of gray water. The second case study discusses the severe water stress of Mexico City, a metropolis of 20 million people, which is experiencing water shortages, subsidence of the water table, problems of overexploitation of aquifers, and wastewater pollution problems. The third case study turns to the problem of selected African cities coping with shrinking water supply, increased urban growth, and water consumption, as well as increasing wastewater and sewage disposal problems.

Case Study 1: Cities and LEED Building Technologies

More people living in cities mean there are increases in built-up urban environments and growing demands on water and energy resources for city development. In metropolitan areas, the way we build and provide services, including clean, reliable supplies of drinking water, affects the quality of residents' lives. Therefore, one of the main challenges of urban planning and building in the city is how to optimize the use and reuse of water in a sustainable manner.

Metropolitan centers of the world are increasingly facing problems of water supply and higher costs to provide water to urban dwellers. The price of water in 2013 rose by nearly 7 percent in 30 large U.S. cities, which denoted a 25 percent increase since 2010 (Walton 2013). Many cities from Santa Fe to San Francisco have had to put in place drastic water rationing and conservation measures, including penalties for breaking the law. For this reason, a number of thirsty cities in the U.S. and around the world are now turning to the use of Leadership in Energy and Environmental Design (LEED) standards, which aim to conserve and improve efficiency of water use in city environments.

Drought and Vulnerability of Cities

In 2015, the worst drought since 1930 hit São Paulo, Brazil's biggest city, which has a population of over 20 million people in the greater metropolitan area. The drought arose due to lack of rainfall, but was worsened by a chronically leaking water system, deforested watersheds, polluted rivers, and population growth in the Sao Paulo area. These factors outstripped the capacity of the city to service the needs of residents. The drought lowered the capacity of city reservoirs and forced the water utility company to suspend daily water services for the urban population through a five-days-off-and-two-days-on system as part of a rationing plan (Romero 2015).

The situation was so critical that it was reported that officials at the city's water utility urged residents to go to other places until the problem was resolved. However, even before the taps ran dry, inhabitants of the city had already been experiencing sporadic water cutoffs in both affluent gated communities and sprawling squatter settlements of the city. The situation forced many residents to take the matter into their own hands by drilling their own wells around homes and buildings, or hoarding water at home in water-deprived neighborhoods.

Cities also experience water problems when they are located in vulnerable coastline places. After Hurricane Sandy ripped through New York and New Jersey in 2012, city planners woke up to the need for serious action to address urban vulnerabilities, given the projected rise of sea levels and the risk of more severe storms in the future. During the storm, Hurricane Sandy dumped huge volumes of water into the Montague Tunnel, a subway tube connecting Brooklyn and Lower Manhattan. Power companies such as Con Edison as well as the city's hospitals suffered over $1 billion in damage, and countless homes and buildings were damaged along coastal areas of New York and the Jersey Shore (Feuer 2014).

Since the storm a Hurricane Sandy Rebuilding Task Force has been formed within the Department of Housing and Urban Development (HUD) to address building standards, zoning laws, the construction of seawalls, bulkheads, and buffer zones, and the protection of power stations, subway tunnels, sewage treatment plants, hospitals, and public utilities. It is ironic that, in the 1960s, after the passage of Hurricanes Carol and Donna, New York had already considered proposals for building barriers, swinging gates, and walls in areas such as the Bronx, Jamaica Bay, and Coney Island. However, due to lack of funds these projects were abandoned. Sandy has revitalized interest in such projects.

Hurricane Sandy brought home some of the threats related to the growth of cities and built-up environments, and the need to make cities more resilient to the onslaught of large storms and natural hazards. The storm debunked the myth that urban expansion is possible anywhere and everywhere without major risks, as long as the demand for buildings and services can be satisfied.

Sustainable Design and Construction: LEED Standards

Enter the age of sustainable design and building, and the increasing importance of LEED (Leadership in Energy and Environmental Design) standards, which provide some innovative solutions to the problem (LEED). This relatively new set of standards has developed and spread not only in the United States, but also throughout the world. In response to climate change impacts, engineers, architects, city planners, developers, building owners, and private and public entities are increasingly championing concepts of water conservation, water harvesting and reuse, alternative energy sources, and higher energy efficiency in building technology.

Through the treatment and recycling of blackwater (raw sewage) to gray water, water can be reused in buildings' cooling towers, in toilet flushing, or in irrigation of rooftop and urban gardens. LEED standards have a rating system that ranks green building design and awards points according to key factors in construction and operation: water conservation, harvesting, and recycling; reduced energy consumption; reduced emissions and use of natural resources and materials; sustainable site location; and innovative green design.

Within this system, water recycling provides a number of important points to meet LEED certification. In the United States, the certification process is administered by the U.S. Green Building Council. While there is some debate as to whether all LEED-certified buildings are truly "green" in design, the practice of building according to LEED standards has caught on throughout the United States and overseas (U.S. Green Building Council).

Thousands of green builders in the United States are entitled to tax breaks and other incentives once certified by the Council. In 2015, the top ten cities of the United States ranked by total number of LEED projects were as follows: New York; Washington, D.C.; Chicago; Houston; San Francisco; Los Angeles; Seattle; Atlanta; and San Diego (Long 2015).

There are a number of LEED examples in the United States and around the world. A case in point is Levi's Stadium in Santa Clara, California, home to the San Francisco 49ers. This stadium, connected to the city's recycled water system, is the first stadium to use a drought-proof water source. This factor is particularly important, given the problems associated with water shortages and drought in the state of California. Recycled water accounts for 85 percent of all water used to operate the stadium, including irrigation of playing fields, a 27,000-square-foot green roof, toilet flushing, and water for cooling towers. Other features include energy-efficient systems and the use of solar power and recycled construction material to reduce the carbon footprint of the stadium. It is the first NFL stadium to obtain a LEED Gold rating from the U.S. Green Building Council (Levi's Stadium, 2014).

Another example in sustainable building is the Electrical Service Building located in the California city of Burbank at Burbank Water and Power's (BWP) EcoCampus (Burbank Water and Power). This building received a LEED Platinum certification, the highest level of awards. BWP is a community-owned utility serving the needs of Burbank since 1913 for electric and water services. It was the first utility in the United States to commit to achieving 33 percent of renewable energy by 2020. The construction project took a campus-wide integrated approach to building—from the solar-covered parking structure, to landscape design, to the incorporation of storm water capture for the recharge of ground water aquifers.

Commercial buildings are using LEED standards as well. Many hotel resorts have incorporated green laundry systems into their operations by cutting down on the use of hot water, and using ozone treatments and water reuse systems to recycle most of the water used for laundering into irrigation systems. Green laundry systems contain washers with two drains, which allow for dirty water and chemicals to be separated from cleaner water, and for that cleaner water to be recycled. The whole process is monitored through computers that measure water quality, volume, flow, pressure, and other factors.

In addition, many green residential buildings are using zero-discharge toilet systems, which completely recycle all blackwater with biological remediation techniques and ozone to sterilize water before reuse. This closed-loop blackwater recycling technology allows water to be used over and over in the same sewage system, leading to conservation of water and cost reduction for hotel and other building operators. In Manhattan, high-rise condominiums are constructing on-site water recycling systems to recycle water systems, flush toilets, cool towers, and irrigate building landscapes. Such systems are expected to substantively reduce drinking water consumption and wastewater discharge.

In other countries of the world, the use of LEED standards is spreading, mainly due to increasing water and energy bottlenecks in metropolitan areas. For example, in Australia, droughts have caused city officials to scramble for solutions in Sydney, one of Australia's main cities. Using LEED standards, developers of high rises have started installing blackwater recycling systems that draw and reuse sewage from the same building as well as city sewers. Technologies that use reverse osmosis membranes and ultraviolet light are also deployed to disinfect effluents and recycle them back to the building's cooling systems and to nonpotable water areas in the buildings.

In Songho, South Korea, LEED standards are being applied to the development of a "smart" city located on 1,500 acres of reclaimed land on the Yellow Sea. The project is expected to be the first South Korean design of a sustainable neighborhood. The features include LEED water conservation and wastewater management principles, centralized water heating and cooling systems in all buildings, and automatic solid waste collection and processing centers on-site (Cityquest 2014).

China, too, is going LEED. Given the increasing problem of urban air and water pollution, this technology is expected to help China focus on reducing carbon emissions and bring about a transformation of China's built environment. These changes should translate into improvements in the public health of China's urban population. China already has the world's largest urban population with 758 million people, and is expected to add an additional 292 million to cities by 2050. For this reason, China has started to take aggressive steps to address urban pollution and water scarcity. Beijing and Shanghai already rank among the top cities of the world using LEED standards (Gray 2014).

Camille Gaskin-Reyes

REFERENCES AND FURTHER READING

Burbank Water and Power. "Environmental Projects." https://www.burbankwaterandpower.com/.

Cityquest. "Songdo, South Korea: Conceptualized as an Ultimate Smart and Sustainable City." New Cities Foundation. December 28, 2014. www.newcitiesfoundation.org/cityquest-songdo-south-korea-conceptualized-ultimate-smart-sustainable-city/.

Feuer, Alan. "Building for the Next Big Storm: After Hurricane Sandy, New York Rebuilds for the Future." *New York Times*, October 25, 2014. http://www.nytimes.com/2014/10/26/nyregion/after-hurricane-sandy-new-york-rebuilds-for-the-future.html.

Gray, Christopher. "Can Green Building Make Chinese Cities More Competitive?" U.S. Green Building Council (USGBC). November 13, 2014. http://www.usgbc.org/articles/can-green-building-make-chinese-cities-more-competitive.

Leadership in Energy & Environmental Design (LEED). http://leed.usgbc.org/.

Levi's Stadium. "Stadium Becomes First U.S. Venue of Its Kind to Earn LEED Gold Certification." July 22, 2014. http://www.levisstadium.com/2014/07/stadium-becomes-first-us-venue-kind-earn-leed-gold-certification/.

Long, Marisa. "USGBC Releases the Top 10 States for LEED Green Building Per Capita in Nation." U.S. Green Building Council. February 4, 2015. http://www.usgbc.org/articles/usgbc-releases-top-10-states-leed-green-building-capita-nation.

Romero, Simon. "Taps Start to Run Dry in Brazil's Largest City." *New York Times*, February 16, 2015. http://www.nytimes.com/2015/02/17/world/americas/drought-pushes-sao-paulo-brazil-toward-water-crisis.html.

U.S. Green Building Council (USGBC). http://www.usgbc.org/.

Walton, Brett. "The Price of Water: A Comparison of Water Rates, Usage in 30 U.S. Cities." Circle of Blue. April 26, 2010.

Walton, Brett. "The Price of Water 2013: Up Nearly 7 Percent in Last Year in 30 Major U.S. Cities; 25 Percent Rise Since 2010." Circle of Blue. June 5, 2013. http://www.circleofblue.org/2013/world/the-price-of-water-2013-up-nearly-7-percent-in-last-year-in-30-major-u-s-cities-25-percent-rise-since-2010/.

Case Study 2: Mexico City: A Water-Stressed Metropolis

Mexico City extracts water from its aquifers more than twice as fast as it replenishes them (Malkin 2006). As a result, the area upon which the city is built is subsiding, causing it to deteriorate and sink. Centuries ago, Mexico City was the center of the Aztec empire. Today, it is a capital city with 20 million inhabitants—and its population is growing daily.

The capital's reaction to the water scarcity is to extract water elsewhere, from the highlands and from surrounding states. This is a very expensive solution. The Cutzamala system pumps 480 million cubic meters down 160 kilometers to Mexico City. The system carries a quarter of the city's water uphill (United Nations). Inequalities in Mexico City aggravate the problem. The system takes water away from the poorest inhabitants of the mountains, like the Mazahua communities. The capital of the country gets the water from certain states, like the State of Mexico, while local communities are denied access to clean water.

Heavily Populated

To understand the water problems in Mexico City, it is useful to look back in time. The basin has always been densely populated. During 300-750 CE, which was the height of the Teotihuacan culture, it had a population of 300,000 people. Later, when the Spanish conquest began, in 1519, the population was around 1.2 million people. After the conquest, the native population faced violence and sicknesses, which led to a dramatic decline.

Since the conquest, Mexico City has continued expanding as population continued to increase. During the 20th century, the country was under a regime that initiated the industrial revolution which benefitted the ruling elite. Under the dictatorship of Porfirio Diaz, from 1884 to 1911, factories and railroads were built, and the city went through a period of modernization. Those who largely benefited were the small but powerful upper classes, whose aim was to modernize the wealthiest areas of Mexico City.

The newly constructed railroads stimulated rural out-migration to the capital; droves of people from the rural areas flocked to the city looking for employment in the modern industries. Some of the smaller towns close to the capital were absorbed within the urban perimeter. In less than 100 years, the population of the urban conglomerate of Mexico City grew from 1 million to about 15 million. With the influx of more people, the city needed a drainage system to remove torrential surface runoff from the urban part of the basin, and most of the old lake beds dried up.

Overexploitation of Aquifers

To support the water needs of its growing population, Mexico City had to overdraw its aquifers. The overexploitation of the aquifer system lowered the center of Mexico City by about 30 feet between 1910 and 1990 (United Nations University 1992). Moreover, increased industrialized activity, combined with millions of vehicles in the city and the low wind speed in the basin, has continued to degrade the quality of the city's air to harmful levels.

Water availability in Mexico's capital has declined rapidly since 1950 due to even higher population increases, instability of rainfall regimes, and drop in aquifer levels (Geo-Mexico 2011). The municipal authorities of Mexico City are also concerned about severe problems in wastewater management. For example, for more than 10 years, a wastewater treatment project consisting of four treatment plants has been paralyzed due to disagreements among local, state, and federal governments (Malkin 2006).

Today Mexico City treats less than 10 percent of its wastewater, sending its sewage into rivers that irrigate farmland to the north (Notimex 2011).

Leaking Pipes and Creaking Systems

An additional problem is that about 40 percent of the water supply to the city is lost to leaks (Rios 2016). Because of the ongoing subsidence of the ground, the pipes are constantly breaking. Often, there is mixing of pipes with water and sewage, creating enormous health problems. Nonetheless, not everyone living in the city suffers from water shortages. People living in upper class neighborhoods of the metropolitan areas have access to running water every day of the week, while poorest parts of the capital lack services.

Water problems are not exclusive to Mexico City, but since it is the largest urban area, water scarcity is felt more acutely in the capital. In other parts of the country, other cities and villages face challenges such as the lack of financial resources for water-related projects, outdated water supplies and services, insufficient adaptation programs to climate change, overexploitation of groundwater, the outdated legal status of water institutions, and lack of awareness of the population on climate change impacts, on droughts and floods, and on water conservation methods.

The UN Water Agenda

According to UN Water, Mexico's government has signed on to achievement of the United Nations 2030 Water Agenda, which aims to achieve equitable supply and universal access to water services, among other initiatives. Within this context, the Mexican government's goal is to develop a long-term action plan for sustainable use of water resources. However, implementation of the 2030 Water Agenda requires annual investments of more than 4.16 billion USD (Conagua 2013), which are prohibitive.

An interim solution to Mexico City's water problem could be provided by the recent discovery of an aquifer that could deliver enough water for some of its 20 million residents, and reduce the city's need for water coming from distant regions (Fausset 2013). This step could buy some time for the government and other groups to find sustainable solutions.

As a long-term plan, Mexico's Payment for Hydrological Environmental Services Programme (PSAH for its acronym in Spanish) has been envisaged. Introduced in 2003, the project has evolved into one of the largest programs for the protection of forests. The Mexican National Forestry Commission (CONAFOR) offers payment to landowners that preserve urban forests for watershed protection and the supply of water for the city. From 2003 to 2006, more than 300 land cooperatives signed contracts to protect thousands of hectares of forest; since then, the program has been growing each year.

Participation of Civil Society

Civil society groups are also participating in finding solutions to the water shortage in Mexico City. A project worth mentioning is the collection of rainwater. Isla

Urbana is a project to install rainwater-harvesting systems in individual homes. More than a million households have installed water tanks to harvest water. It is estimated that water stored from the rainy season in the summer and fall can supply a household for up to six months. Isla Urbana's system requires the construction of gutters to channel rainwater as well as plumbing equipment and filters. However, these solutions are not comprehensive, since far-reaching policy changes and political consensus among stakeholders are needed to produce effective outcomes.

If Mexico's past is any indicator of the future, it is envisaged that the city will continue to face major challenges such as water scarcity, water pollution, and wastewater disposal problems, which are likely to aggravate social and economic inequality.

Andrea Arzaba

REFERENCES AND FURTHER READING

Conagua. "2030 Water Agenda." 2013. http://www.conagua.gob.mx/english07/publications/2030_water_agenda.pdf.

Fausset, Richard. "Aquifer Discovered That Could Spell Relief for Mexico City Residents." *Seattle Times*, January 21, 2013. http://www.seattletimes.com/nation-world/aquifer-discovered-that-could-spell-relief-for-mexico-city-residents/.

Geo-Mexico. "Round-up of Recent Developments in Mexico's Drug War." July 14, 2011. http://geo-mexico.com/?p=4496.

Malkin, Elisabeth. "Once a Vision of Water, Mexico's Capital Now Thirsts for It." *New York Times*, March 16, 2006. http://www.nytimes.com/2006/03/16/international/americas/16mexico.html.

Notimex. "El DF es el 'peor ejemplo' en el tratamiento de aguas residuales." May 19, 2011. http://www.excelsior.com.mx/2011/05/19/comunidad/738131.

Rios, Fernando. "Desperdicio de agua potable Ilega al 40%." *El Sol de Mexico*, February 4, 2016. http://elsoldemexico.com.mx/metropoli/118835-desperdicio-de-agua-potable-llega-al-40.

United Nations. "International Decade for Action 'Water for Life' 2005–2015." http://www.un.org/waterforlifedecade/waterforlifevoices/cases_arp_mexico.shtml.

United Nations University. "The Basin of Mexico." 1992. http://archive.unu.edu/unupress/unupbooks/uu14re/uu14re0s.htm.

Case Study 3: Water Stress in African Cities

Africa is currently undergoing rapid urbanization. Over 320 million people, about 37 percent of the continent, live in towns and cities (Jacobsen et al. 2013, 16). Slightly over 40 percent of all houses and yards are connected to a pipe, whether an individual or communal standpipe. Less than 20 percent have sewer connections (Jacobsen et al. 2013, 16). By 2030 it is expected that 50 percent of Africans, or approximately 654 million people, will be city dwellers (Jacobsen et al. 2013, 16). This translates to a huge demand for water and sanitation services.

Between 2000 and 2010, around 83 million urban Africans accessed improved water supplies, and 42 million received improved sanitation services (Jacobsen et al. 2013, 19). However, population growth and the rise of the urban population dampened these achievements, so that the huge gap in services still remains. Cities are unable to fulfill the urban population's current needs for water, let alone keep up with future growth.

However, not all African cities suffer the same water supply problems. Some, such as Johannesburg in South Africa and Windhoek in Namibia, are naturally deprived of freshwater; others, such as Kinshasa in the Democratic Republic of Congo, have abundant natural water supplies. Nairobi, located in a semiarid region in Kenya, must rely on water coming from a distance. Dar es Salaam in Tanzania depends on artisan groundwater deposits. Khartoum, while located in an arid area of the Sudan, is situated at the confluence of the White Nile and the Blue Nile, which provide the city with abundant surface water (Jacobsen et al. 2013).

Overall, the rise of urban population reflects Africa's economic growth and rapid changes in its economic structure. More people are moving to the cities, and industries and services are rapidly expanding. Increased water demand and the growth of a solid middle class are reflected in the expansion of suburbs and increased water consumption.

At least seventeen cities in Africa are growing at an annual rate of over 3.7 percent per year, which will double the urban population time in only 13 years. Five cities—Abuja in Nigeria, Kumasi in Ghana, Luanda in Angola, Ouagadougou in Burkina Faso, and Yaounde in Cameroon—have growth rates of over 5.3 percent per year, which is expected to lead to a doubling of the population in about 10 years (Jacobsen et al. 2013, 74). This highlights the need for urban planners to address water supply problems to cover the needs of residents, including wastewater recycling and greater water-efficient technologies.

Shrinking Water Supply

The dilemma in African urban environments is that while demand is growing, water supply is shrinking, and water quality is declining. The expansion of built-up areas in the city and the spread of suburbs in African cities means that urban centers such as Lagos in Nigeria, Accra in Ghana, and Nairobi in Kenya are forced to increasingly tap water sources from distant rural areas—at a higher cost. At the same time, groundwater or surface water sources in rural areas are under pressure due to the water demand for agriculture and hydropower. There is fierce competition for the same finite water resources everywhere.

Deforestation and Runoff

To complicate matters, deforestation and soil erosion of the watersheds of major African rivers have caused land use changes upstream and affected seasonal patterns of water runoff downstream. Due to forest depletion, there is more flooding in the rainy season and drought in the dry season. Overpumping of aquifers for irrigation and the growing contamination of groundwater through sewage and pollutants reduce groundwater sources available for urban (and rural) dwellers. The already observed impacts of climate change such as flooding of low-lying cities and salt intrusion of the water table are contributing further uncertainties to the future of African water resources.

Selected Examples of City Problems

To highlight some examples of technical and institutional challenges and opportunities in water management for urban areas, this section will focus on case studies of selected African cities: Arua in Uganda, Douala in Cameroon, Windhoek in Namibia, and Naivasha in Kenya.

Arua, Uganda

Arua has a population of about 70,000, and is located on the Enyau River in the western part of Uganda (Jacobsen et al. 2013, 97). Settlements upstream of the town have cut down trees on the river banks, and are also using the river for subsistence farming, small scale irrigation, livestock herding, brick making and other purposes, reducing the town's available water supply. Farmers have diverted large quantities of water from the Enyau River by digging small water canals for farming and brick making, leading to large sediment flows, as much as 70 percent to the river (Jacobsen et al. 2013, 97). This situation causes potable water to run turbid for city dwellers.

The different users in the Arua area have legitimate rights to access and use the water from the Enyau River, but there is no coordination to resolve competing uses and ensure water security and cleanliness for all. The existing water treatment plant in Arua is overloaded by the turbidity of the water coming from upstream. Filtering costs are high, since high sediment loads damage pumps and impede the functioning of settling tanks and clarifiers.

Constant damage to pumps means that water supplies to urban dwellers are costly and usually intermittent. In addition, the dumping of sewage and solid waste into boreholes and streams, and the cross contamination of water pipes by sewers are also bringing about severe problems of water quality in the city. Most municipalities in Arua are already showing signs of pollution and negative public health impacts.

Solutions to the problem under consideration range from the possible use of other rivers from far away, including the Nile River, 40 kilometers away from the city, and the drilling of new groundwater aquifers. There is no guarantee that the same problem would not occur with new river or groundwater sources, unless structural changes were made in water management and consensus among all stakeholders. The success of any measure will need to address destruction of watersheds further upstream, reduction of source pollution, and participation of all affected groups in the city and outside of the city.

Douala, Cameroon

Water management challenges in Douala, Cameroon, are also formidable. Douala has a population of over 2 million—approximately 50 percent receive water coverage (Jacobsen et al. 2013, 105). The city is growing at a fast rate, and is unable to cover its current water needs, let alone future ones.

The poorest residents in Douala live in low-lying areas, which are subject to flooding and contamination of wells. Increased construction activity in the wake of city growth has caused runoff of sediments, while deforestation in the areas of higher elevation surrounding the city has increased flooding.

This situation is compounded by high tides, which consistently overwhelm the lowest areas of the city. Contaminants are formed from overflowing latrines; and the unregulated dumping of solid, commercial, animal, and industrial wastes pollutes the water supply and blocks urban storm drains. In Douala, cholera is endemic. The only sewage treatment plant in the city no longer works due to technical failure.

To resolve this problem Douala needs a sound water policy and regulations on water protection, waste disposal, and public sanitation programs. However, due to institutional weaknesses, financial constraints, and top-down decision making in city administration, few measures are in place to involve communities towards sustainable solutions.

Poor recovery of property taxes and other municipal costs means that Douala does not have sufficient funds for public sanitation, water, and waste management investments as well as public hygiene and education campaigns. It is doubtful that the city is in a position to protect its residents and resolve current and future needs.

While the above examples point to challenges, the following examples from Windhoek, Namibia, and Naivasha, Kenya underscore opportunities that may help resolve some water issues facing African cities.

Windhoek, Namibia

Windhoek is the capital of Namibia. It has severe water challenges, since it is not located on a river, has low annual rainfall, little access to rainwater, and high surface evaporation rates. There are few potable water resources within 300 miles (Jacobsen et al. 2013, 53). In this arid environment, there are only three main sources of water: dams, groundwater boreholes, and reclaimed water. Windhoek is one of the few cities in the world that uses state-of-the art technology to safely recycle treated wastewater for use as drinking water. This experience provides valuable lessons for other African cities.

Naivasha, Kenya

In Naivasha, wastewater from toilets and washrooms of the bus terminal is currently used to generate and transmit biogas for cooking (Jacobsen et al. 2013, 59). This solution addresses the problem of wastewater disposal and also creates energy. Treated effluents from the facility, built next to the city's bus terminal in 2008, are discharged to a public sewer, and the resulting sludge is used as fertilizer. This example demonstrates that integrated and innovative approaches such as combined lavatory, wastewater treatment, and biogas plant facilities can satisfy the needs of stakeholders in the urban and rural environment.

Camille Gaskin-Reyes

REFERENCES AND FURTHER READING

Jacobsen, Michael, Michael Webster, and Kalanithy Vairavamoorthy. *The Future of Water in African Cities: Why Waste Water?* Washington, DC: World Bank Publications, 2013. Available online at https://openknowledge.worldbank.org/handle/10986/11964.

Mogaka, Hezron, Samuel Gichere, Richard Davis, and Rafik Hirji. *Climate Variability and Water Resources Degradation in Kenya: Improving Water Resources Development and Management*. World Bank Working Paper Series no. 69. Washington, DC: World Bank Publications, 2006.

ANNOTATED DOCUMENT

UN-Habitat: Global Water Operators' Partnership Alliance (GWOPA): GWOPA Strategy 2013–2017

Background

The Global Water Operators' Partnerships Alliance (GWOPA) is an international network created by UN-Habitat in 2009 to support Water Operator Partnerships (WOPs). UN-Habitat is a United Nations specialized agency, headquartered in Nairobi, Kenya, with the mandate to support sustainable urban development policies and programs in member countries.

The UN-Habitat/GWOPA developed a strategy for 2013 to 2017 to support water operators, since these are the institutions in charge of delivering water and urban sanitation services in cities across the world. Operators include public sector utilities, private companies, water managers, and local and community-based organizations that manage or allocate water supplies. The alliance promotes sharing of information and exchange of best practices.

Developing countries face a huge problem in the urban water sector due to the inadequate capacity of operators to manage water, collect water payments, build or maintain water systems, and deliver water in a timely, equitable, and efficient manner. Since most of the growth in demand for water services will come from urban areas in the next decades, it is important that water operators in these countries be prepared for the enormous challenge of supplying the existing and future population with water for drinking and other purposes.

The United Nations and its specialized agencies, such as UN-Habitat, have led global discussions in the last decades on people's rights to water, sanitation, wastewater collection, and solid waste disposal. UN-Habitat played an important role in incorporating water and sanitation in urban areas into the UN Sustainable Development Goals. In preparation for sponsoring the Partnerships Alliance, UN-Habitat conducted a number of studies on the success or failure factors of urban water operators, and the linkages among water operators, development institutions, communities, water users, and consumers.

Summary

UN-Habitat set up a GWOPA secretariat to coordinate the work of all members of the alliance, and to extend membership among water operators and associations, all types of utilities, engineering groups, knowledge and research centers, and funding institutions in cities and towns of all sizes. Building on the studies and inputs of many stakeholders in the water sector, the GWOPA developed a 2013–2017 strategy, which proposed 10 areas for priority action; they are included in the excerpts of the document outlined below.

These areas encompass improvements in water billing and water payments; reduction of water wastage; improved governance structures and capacity; attention to sanitation systems, wastewater collection, disposal, and recycling programs; energy efficiency in the operation of utilities and the reduction of greenhouse gases; climate change adaptation measures; water quality management and reduction of pollution; capacity building of water operators; and increased access of the urban poor to water and sanitation.

These priority areas are based on the document's statistics, which show increasing demand for urban water and sanitation services, and inequitable access of the poor to such services. Despite progress on the provision of water and sanitation services to the world's population, 780 million people in the world still lack access to safe drinking water, and 1.8 million people do not have toilets. Half of the population lives in urban centers, and more than one-third of the world's urban growth is concentrated in urban slums. In Africa, this percentage is even greater, representing more than 80 percent of urban growth on the continent. The highest levels of urbanization are occurring in developing countries.

Urban growth, whether in megacities or smaller towns, shows the same pervasive patterns throughout the world: pressures on water supply, problems of sanitation and garbage disposal services, and overstretched infrastructure and transportation networks. According to the document, people using piped connections rose from 32 percent in 1990 to 46 percent in 2010, but many municipal governments in poor countries do not afford priority to water and waste services, or lack funds to provide services. Many public utilities are financially insolvent, and their institutional capacity is poor. Water tariffs are woefully inadequate, and collection rates are too low to cover the costs of investments and maintenance in new water infrastructure and respond to growing urban demands.

The GWOPA therefore combines two important themes: urban growth and water. It is intended to address the common problems related to the mismanagement of urban water and wastewater services, and ongoing issues of quantity and quality, all of which have major impacts on urban ecology, preservation of water sources, and watershed protection. According to the strategy document, about 4,000 children in the world, mostly in developing countries, die daily from waterborne diseases, and it is therefore important to address the problem of unsafe water supply in urban cities and reduce waterborne diseases by shifting to piped water connections.

Camille Gaskin-Reyes

Excerpts of the Document

Water Operators' Partnerships (WOPs)

WOPs make use of the fact that while many operators lack capacity, others have it in abundance, and are willing to share it on a solidarity basis. A WOP is a peer-support exchange between two or more water or sanitation operators, carried out on a not-for-profit basis with the objective of strengthening capacity, enhancing performance and enabling the water operator to provide a better service to more people, especially the poor.

WOPs have existed in one form or another for decades and vary greatly in their objectives, approach and outcomes. Nevertheless, they are always carried out by and for utilities, and according to a number of guiding principles, notably not-for-profit and integrity.

WOPs propose to increase the ability of water operators to meet the needs of the people they are meant to serve, by enabling them to improve and extend their services. WOPs work by harnessing the skills, know-how, and goodwill within a strong 'mentor' utility in order to sustainably build the capacity of another utility—the 'mentee'—that needs assistance.

Through mentorship, WOPs progressively strengthen and empower the mentee operator at management, financial, and technical levels to implement changes that will lead to better performance and service.

Some WOPs focus on a particular aspect of service provision or seek to change particular processes, while others are more comprehensive. Most work by changing processes that will result in increased efficiency, leading to greater financial sustainability and the eventual ability to improve and extend services.

Other WOPs help the mentee extend their services directly. They may focus on transfer of expertise around pro-poor service delivery, extension into informal settlements, fair tariff setting, and so on. Given the dual need of urgently expanding provision and ensuring the capacity to maintain that service over the long term, WOPs ideally pair an explicit focus on service extension with long-term efforts to ensure sustainability.

Box 2: Some of the Areas Where WOPs Can Help

Non-revenue water (NRW)

NRW, water that is "lost" (through leaks, theft, illegal usage or legal usage with no payment being made) is a major threat to the viability of water operators. WOPs can help in reducing losses though leakage reduction, better system operation and maintenance and/or improved commercial practices.

Billing and collection

Utility managers know that efficient billing and collection systems are critical to the financial health of a utility, but they should also support environmental and equity objectives. WOPs can bring in diagnostic skills and knowledge of robust systems that enhance billing accuracy, solve customers' billing queries and improve collections from overdue accounts. In addition, specific strategies are needed to address non-payment from government agencies.

Governance

WOPs can help operators improve governance and share experience of processes in which operators are more accountable to users and more aware of customer concerns.

Sanitation

Operators can play a role in getting more users connected to centralized sewer systems, or in enabling better decentralized sanitation systems. Introducing sanitation into un-served urban areas requires operators to have a range of social, financial, and technical skills, which makes the support of experienced operators invaluable.

Wastewater

Neglected until recently, wastewater is now becoming a priority for cities and development partners. Without proper wastewater collection and treatment, progress in drinking water coverage won't achieve its full impact on people's heath and dignity. A growing understanding of the water, energy, and food nexus is also highlighting the hidden value of 'waste' water as a potential resource. WOPs can help demonstrate the options and develop the capacity of staff to manage.

Energy efficiency

Energy commonly represents upwards of 30% of a utility's total operation and maintenance costs, and is a major contributor of greenhouse gases. But power is among the largest controllable costs of providing water and wastewater services. WOPs can help utilities to sustainably manage and reduce energy costs by helping operators analyze their current energy usage, implement energy audits to identify opportunities to improve their efficiency, and ultimately develop energy management programs.

Climate resilience

Climate change impacts pose acute challenges to water utilities: extreme weather events, sea-level rise, temperature changes, and shifting precipitation and runoff patterns may result in changes to water quality and availability and severe impacts on utility assets. Mentor and mentee utilities can together identify risks and adopt climate change adaptation practices, for example, in asset management, water supply and demand planning, and in security and emergency preparedness.

Water quality management, water resources protection and pollution prevention

Water quality, water pressure, continuity of service, protection of water resources and prevention of pollution are key topics in service management which can benefit from capacity building and experience exchange with more experienced operators. Water Safety Planning has been a common topic for short-term WOPs.

Human resources development and labour relations
Lack of skilled staff is the most common problem identified by water operators in developing countries and labour-management conflicts can be an obstacle to effective service provision. Water and sanitation work is often carried out in the absence of health and safety regulations. WOPs can provide training and mentorship to help both management and labour be more effective, and develop the capacity of staff to manage.

Expanding service access to the poor
Serving the poor, often living as tenants in slums, in informal settlements or highly congested areas, dealing with large-scale illegal connections, or managing small-scale vendor operations requires specific approaches, skills, and experience which many busy utilities do not have. WOPs can help build skills and offer tested approaches to extending access.

Source: Global Water Operators' Partnerships Alliance. GWOPA Strategy, 2013–2017. Copyright United Nations Human Settlements Programme [UN-HABITAT], 2013. Available online at http://unhabitat.org/global-water-operators-partnership-alliance-strategy-2013-2017/.

PERSPECTIVES

What Steps Should California Take to Deal with Its Ongoing Water Management Issues?

On November 30, 2007, officials in Orange County, California, opened a new water recycling plant. The plant subjects sewer water—that is, water from sinks, showers, and flushed toilets—to filtering, screening, scrubbing, and blasts of ultraviolet light before returning it to the ecosystem. "Gray water," as it is known, is not sent directly to homes; rather, it is pumped into the ground to replace groundwater that has been pumped out for irrigation or other purposes. The process is officially known as "indirect potable water reuse," but critics have nicknamed it "toilet to tap."

There are several water recycling plants in the United States, none of which recycle water for direct use by consumers as drinking water. However, as the name critics use might imply, some municipalities are considering making the transition to recycled potable water supplies: toilet to tap in truth. Such communities are mainly in the dry regions of the American West: Texas, Arizona, and California. This region contains only a small number of rivers and lakes and receives little rainfall. As a result, water availability has always been a problem. Water recycling is just one way to approach the chronic water shortage in these areas, which has only been compounded by the newer problem of population growth.

California: A History of Water Scarcity

California has been faced with water difficulties since the 18th century, when Spanish colonists first arrived, and the problem has endured through U.S. settlement to the present day. The question is not simply who has rights to the water in certain areas, but whether water should be moved from one place for use in another. And even if water is moved to where it is needed, the question of whose needs take precedence inevitably arises. Such questions tend to create contests for water rights that have many players.

City populations are one of the most important competitors for water. An elaborate system of dams and aqueducts has been built over the last century-and-a-half to keep large, southern metropolitan areas such as Los Angeles and San Diego supplied with water, often diverting flows hundreds of miles and potentially depriving other communities. Further, the issue of water management acquires an international as well as a regional component once we recall that Southern California abuts Mexico, and the two often share water resources like the Colorado River. In 1944, the two nations signed a binational water treaty, but controversies have flared across the border from time to time.

Much of California is prime agricultural land and the state exports the majority of the nation's fresh produce. Vast amounts of water are required for farming, and this fact often brings the agriculture industry into conflict with city populations that need drinking water.

Since the 1970s, another actor has entered the water rights stage: the environmental movement. Methods used to move water from place to place can often be very damaging to the ecosystem. Reduced lake water levels can wreak havoc on local wildlife, while damming can interrupt migration and spawning among fish populations. Many environmental groups, therefore, have attempted to halt or even reverse water management approaches that do not take the needs of nature into account.

Geography and Population

Water has affected California's history and its people in many ways. Two mountain ranges, the Cascade and the Sierra Nevada, run north and south through the state; California's two longest rivers, the Sacramento and San Joaquin, drain from these mountains and flow through the northern half of the state. Thus, the huge southern portion of the state has a limited natural water supply.

Most parts of California have only two well-marked seasons—a short rainy season and a longer dry season, with most precipitation coming during the winter months. The state's northern third gets 70 percent of the precipitation, with annual precipitation declining dramatically toward the south. Los Angeles receives only 15 inches of rainfall annually, and San Diego averages 10 inches. Most of the water needed to sustain the state does not come from rainfall, but rather from the runoff of melting snow in the High Sierras and the Rockies. The state's southeastern region, containing the Mojave Desert, receives even less rainfall.

California's population patterns conflict sharply with its naturally available water resources. For example, Southern California contains half the state's population, but has only 2 percent of its water resources. Even farther north, in the state's agricultural midsection, water demand for agriculture far outstrips available natural resources. Without water, fertile lands such as California's highly productive Imperial Valley might lie uncultivated. In addition, cities such as Los Angeles and San Diego could not grow or support industry without municipal water supplies that far exceed what nature offers.

The Politics of Water Scarcity and Management

Political leaders have long recognized the importance of water to California's development. As early as 1902, President Theodore Roosevelt approved a law creating programs to reclaim and manage water supplies for California and other arid regions. Two California-area reclamation projects—the All-American Canal, which brought water to farms in the Imperial Valley, and the Hoover Dam, which quenched the thirst of Los Angeles—diverted water from the Colorado River to Southern California. Though these water programs have had a detrimental effect on the environment and continue to impact Southwest ecosystems, Southern California's demand for water continues to grow. Numerous other water management projects, such as the California State Water Project and the Central Valley Project, carry water to far-off destinations through systems of aqueducts and pipelines that link reservoirs behind dams with towns and cities.

Water management has allowed California's communities and industries to explode. As a result, water managers face several problems. They must meet the state's water needs in a way that minimally impacts the natural environment, either at the water source or along the route. They must counteract the natural salinization process occurring in the irrigated agricultural areas, in which crops draw water from the soil but leave behind salt. They must develop new water sources without incurring costs that consumers (farmers, for example) cannot sustain. Californians have explored many solutions, including water reuse, in which wastewater is reclaimed for industrial purposes, and desalination, in which salts are removed from saltwater.

Water Scarcity: A Global Perspective

California's water troubles are not unique; in fact, they are part of a global problem of water access. Some areas lack water due to population pressures, poor management, or environmental problems. In addition, many scientists believe that global warming may shift wind and tide currents, which, in turn, will likely affect both the patterns and amounts of rainfall around the globe. This means that many areas that are already marginal in terms of their water supplies, such as sub-Saharan Africa, will receive even less rain than they do currently.

Widespread drought in such areas could cause great hardship as people are forced to struggle over scarce resources. Competition could lead to warfare. The combination of scarcity and conflict could combine and create masses of refugees, leading to even more suffering and social fragmentation. On the flip side, areas that begin to receive increased rainfall might have to wrestle with the problems that this can bring: widespread flood damage, waterlogged crops and inundated farmland, pollution from spilled sewage, and epidemic waterborne diseases. In light of these potential disasters, many eyes are turned toward the American West, and to California in particular, for clues as to what to do about the issue of water scarcity.

ABC-CLIO

Perspective 1: Sharing the Burden of Limits

When California governor Jerry Brown declared a drought state of emergency in 2014, it harkened back to when Arnold Schwarzenegger, the former governor of California, referred to water resource management as a "crisis" in 2007. Many of these same issues are being encountered today, nearly a decade after Schwarzenegger addressed the beginnings of what was soon to become one of the worst droughts in the state's long history of water scarcity. In this essay, we will briefly explore what these and other political actors may be called upon to determine: what steps should California take to address ongoing water management issues?

Call for Stronger Water Management

If California is serious about getting control of its water management in order to ensure a healthy environment and a strong economic future, then radical measures need to be undertaken. The hodgepodge of rules, regulations, laws, and governmental authority that make up the existing California water management regime needs to be abandoned in order to meet future environmental and economic demands.

Water management in California, if it can be called that, is not well-suited to meeting these future demands. It is fragmented and decentralized. In a nutshell, what California needs is a water czar or board with the authority to make the tough allocation decisions that will be necessary to protect the state's economy and environment. One needs only to examine other Western states to see that many of them have a state engineer or state director of water resources. The individual in this position has the authority to weigh competing uses and make decisions to issue permits for water based on established priorities. There is already a basis in water law to modify current water management practices should the state decide to meet the associated political challenges in order to do so.

California only has to look at its history to recognize repeating the past is not a viable solution for the future. At the turn of the last century, the city of Los Angeles constructed water works that ultimately destroyed the agricultural productivity—and some would say the community—of the Owens Valley in eastern California.

Currently, water is pumped and transported away from the Sacramento-San Joaquin Delta, commonly referred to as the "backbone" of the California water supply, in order to meet the steadily increasing thirst of Southern California. Delta water serves two-thirds of the California population with drinking water, provides irrigation to farmland statewide, and preserves important wildlife habitat. Pumping in the delta system is already beset with problems, including shortages, pollution, and the need to protect wildlife. Neither the mismanagement of state water resources nor a repeated history of Owens Valley should be allowed to destroy the Sacramento-San Joaquin Delta region.

Effective Water Resource Management

There is little, if any, doubt that California's future water use will involve a transition away from agricultural use to urban and municipal use, but the transition needs to be as smooth and nonconfrontational as possible. The creation of water markets is the best way to accomplish what undoubtedly will be a difficult process. Water rights in California are primarily use rights associated with land ownership. Although the state owns the water, it has little or no control over the amount of water used. Water management often defaults to private water districts and adjudication processes, that is, court settlements. Water markets will allow current holders of water rights being used for agriculture to transfer their rights to the state, cities, and other entities for nonagricultural use.

To date, a complex and nested hierarchy of federal, state, local, and regional institutions and organizations shows little evidence of successful water resource management. Citizens need to seriously consider opportunities afforded them in future elections. Their elected officials are accountable for ensuring a continued source of water to meet ever- increasing requirements. This may require severe measures such as the power of eminent domain that could jeopardize reelection, especially if the public is not sufficiently informed about options to offset the water crisis.

The scarcity and competition over a limited water resource is not unique to California. The state has demonstrated its ability to develop and implement important and precedent-setting public policies. It now has an opportunity to develop a water management system useful to other states and even nations, especially given the very real possibility of continued global warming. If state control of state water resources is not on the horizon, at a minimum, serious conservation measures need to be undertaken, particularly in the southern part of the state. The task is increasingly complex and will require, perhaps above all else, political will and extraordinary levels of cooperation.

Zachary Smith and Jane Whitmire

Perspective 2: California's Water: An Environmental Perspective

California is not alone when it comes to difficult questions about water management. This topic is a global concern and recent events illustrate formidable challenges across the United States. From an environmental perspective, California should do the following to address ongoing water management issues.

First, ensure existing habitat is protected, and restore habitat in developed areas using techniques that soften shorelines, mitigate wetland losses, and treat storm water runoff. Second, clean up existing groundwater contamination and prevent it from occurring in the future. Third, broadly educate citizens about all threats to water, including the importance of conserving both surface and groundwater to prepare for droughts that will inevitably occur. Fourth, support and fund water efficiency and recycling programs to meet future needs at a fraction of the cost of other drastic proposals, such as constructing new dams that are not needed. Finally, the California Department of Water Resources has a water plan that is updated every five years. Use it!

Water Interests and the Need for Environmental Consciousness

California includes some of the fastest growing cities in the nation and is home to more than 38 million people. Three key political groups represent California's water interests: urban, agricultural, and environmental. Urban issues cover domestic use and demands from business and commercial concerns. Agriculture is the primary industry, with prolific vegetable, fruit, dairy, and wine production. Obviously, water is crucial to the success of the state. But why is an environmental perspective relevant?

The environment provides humans with everything necessary for life—air, water, food sources, raw materials, and aesthetics. Protecting the environment protects humanity. Therefore, as humanity progresses, it is essential that we remember environmental concerns. After protecting the air we breathe, protecting water is the next priority. Common survival tenets include the concept that humans cannot survive beyond three minutes without air, three days without water, or three weeks without food. Long-term survival is impossible without clean, accessible water supplies.

Sources of Water

Earth has two kinds of water: saltwater and freshwater. Saltwater ecosystems are fundamental to the health of the planet, and the ocean is a big part of California living. However, the focus of this question is freshwater, which also comes to us in two ways: as surface water or groundwater. Surface water includes lakes, rivers, and streams; groundwater includes water from aquifers, or water bubbling to the surface from seeps and springs. In addition to urban and agricultural uses, it is essential for California to use water to protect natural assets.

For example, the Sacramento-San Joaquin Delta and the Colorado River are the main sources of surface water for urban and agricultural requirements, but they must also provide water for environmental protection. Why? If the water in these rivers were only used to meet agricultural and urban needs, and basic environmental protections were not in place, what would happen? To help answer this question, we use the example of a fish that is well known from the court decision *Natural Resource Defense Council v. Kempthorne et al.* (2009).

The Case of the Delta Smelt

The Delta smelt is a small fish found only in the Sacramento-San Joaquin Delta. These fish are special not only because they cannot be found anywhere else, but also because they serve as an "indicator species" for the delta. When they are doing well, it indicates that the overall health of the delta is good.

The Center for Biological Diversity explains that 29 species of fish originally lived in the delta, but today many have either become extinct, are threatened with extinction, or are rapidly declining. If this continues, the delta food web will collapse, meaning that all wildlife, bird, and human uses that depend upon a healthy delta would suffer greatly. Protecting the habitat of the smelt creates the ripple effect of protecting the delta food web. Unfortunately, increasing water diversions from these rivers, pollution from agricultural and urban contaminants, and invasions of nonnative species all combine to threaten the smelt, and thus, the overall health of the system.

Water Management and Urbanization

Land use questions are also prominent. The Water Education Foundation notes that California is the nation's most urbanized state. This means the natural landscape has been replaced on a massive scale with buildings, parking lots, driveways, and roads—impervious surfaces that water cannot penetrate. Precipitation falling on these places is not absorbed by the ground, resulting in runoff called storm water. It does not seep into the water table to replenish groundwater and it is not slowly channeled into streams. It does go somewhere, though, and if urbanized development is not engineered correctly, runoff can intensify in paved channels and result in serious flooding before it reaches a water source.

Storm water also carries whatever is deposited on impervious surfaces—oil and gas, fertilizers, chemicals, and so on. When it finally does reach a water supply, it creates a vicious "nonpoint source" water pollution; as opposed to "source point" pollution, which can be traced to a specific cause, nonpoint source pollution is not easy to identify. In addition, impervious surfaces mean loss of wetland habitat that controls flooding, purifies water, and preserves wildlife.

Unfortunately, land use decisions also create concerns about groundwater: examples include leaking underground storage tanks, agricultural runoff containing pesticides and chemicals, landfills where disposal of waste leaches into the water table, septage escaping from inadequate septic systems, and industrial point sources of pollution. These and other things make California groundwater vulnerable, which is extremely important because the Groundwater Protection Council notes that half of the state's population depends on groundwater for drinking water supplies.

Drought is another important factor. New things must now be considered when we think about drought, which is a recurring fact of life in California. Today we face the prospect of global climate change and all the uncertainty that comes with it. California understands the concept of living with drought, but it has never had to plan for the potentially severe conditions climate change could trigger, coupled with a much larger—and growing—population. What will this mean? To best prepare, California should implement all five recommendations offered from the environmental perspective noted above.

Grenetta Thomassey

REFERENCES AND FURTHER READING

California Department of Water Resources. "California Water Plan." http://www.waterplan.water.ca.gov/. Accessed October 9, 2015.

California Urban Water Conservation Council. https://www.cuwcc.org/.

Carle, David. *Introduction to Water in California*. 2nd ed. Berkeley, CA: University of California Press, 2004.

Center for Biological Diversity. "Saving the Delta Smelt." http://www.biologicaldiversity.org/species/fish/Delta_smelt/. Accessed October 9, 2015.

Hanak, Ellen, et al. *Managing California's Water: From Conflict to Reconciliation.* San Francisco: Public Policy Institute of California, 2011.

Hundley, Norris. *Great Thirst: Californians and Water—A History.* Rev. ed. Oakland: University of California Press, 2001.

Lassiter, Allison. *Sustainable Water: Challenges and Solutions from California.* Oakland: University of California Press, 2015.

The Nature Conservancy. http://www.nature.org/.

Northern California Water Association. http://www.norcalwater.org/.

State Water Resources Control Board. California Environmental Protection Agency. http://www.swrcb.ca.gov/.

Water Education Foundation. http://www.watereducation.org/.

Water Education Foundation. "Aquafornia: Water News You Need to Know." http://www.watereducation.org/aquafornia. Accessed October 9, 2015.

8 WATER, OCEAN EXPLORATION, AND LAND GRABS

OVERVIEW

In the same way that traffic flow today is enhanced by modern road networks and GPS technologies, in the past people used maritime routes, ocean currents, navigation instruments, maps, and ocean charts to embark on expeditions, attacks, and land grabs. Travel into the unknown intensified with human advancement in boat building techniques, navigational instruments, weaponry, and clearer understanding of the risks and rewards of venturing on the high seas.

This chapter outlines the significance of oceanic waterways for the rise of seafaring nations, the expansion of colonies, and growth in maritime trade on the high seas. The case studies review the importance of navigational devices, and the significance of Muslim and Chinese seafaring for the advancement of ocean exploration and maritime networks. The perspectives section discusses the question of how the Polynesians managed to travel across the Pacific Ocean using their own navigational skills, and without the aid of formal maps and charts.

Riding the Oceans

One can imagine that an explorer's, sailor's, or warrior's curiosity might have arisen while standing on a shore and looking out to the horizon. There was a time when sailors believed that if they sailed to the horizon, they would fall off a flat earth; some thought sea monsters lurking below the water would eat them. The quest to discover what lay beyond the beyond propelled people to hone boat-building and sailing skills. There were many factors at play in ocean exploration: greed, the quest for new trading partners and products, the capture of slaves, plunder, establishment of new settlements, and flight from invaders. For early settlements, location on maritime routes was a distinct advantage.

Early Maritime Trading Routes

Prior to the development of known sea routes, overland trade between Europe, Persia, and China existed and flourished on the Silk Road from 200 BCE to 200 CE. Huge quantities of silk, lacquerware, horses, glass, spices, gold, silver, and slaves were traded on these land routes.

In the first century CE, maritime trade started to grow between southern India and southern Arabia, based on increasing knowledge of the monsoon winds. Sailing routes soon emerged as initial alternatives to land trade, extending from Egyptian Red Sea ports to East Africa and the Ganges Delta in India. Merchants used these sea-lanes to trade in bronze, glass, ceramics, ivory, silk, spices, and other products.

Early Egyptians explored the seaways around the Red Sea, Arabia, and the east coast of Africa. Hieroglyphs dating back to 1500 BCE record that Queen Hatshepsut sent a fleet to the Land of Punt, thought to lie between the Red Sea in Arabia and the east coast of Africa. Records indicate that this foray brought back wood for boatbuilding, incense, slaves, and metals.

In the eighth century BCE, the Phoenicians, sailing out of Lebanese ports, started to dominate Mediterranean Sea routes. They established colonies along northern Africa, Sicily, and Spain, and in the fifth century BCE, they navigated the straits of Gibraltar and set up posts across the Mediterranean Sea to trade with the Greeks.

Following the decline of the Phoenicians, Greek colonization, under Alexander the Great, expanded throughout the Mediterranean Basin, first to countries such as Italy and Sicily, and later to North Africa, using the acquired knowledge of Mediterranean routes. In 490 BCE, the Persian empire entered the Mediterranean arena. The Persian king, Darius, sent two maritime expeditions to the Mediterranean to invade Greece. His maneuver forced the separate city-states in Greece to join together and pay Athens tribute to defend them through constructing of a large fleet.

The construction of the expansive Greek fleet employed many citizens of the Greek empire as boat builders, oarsmen, and weapon specialists, which further developed the seafaring skills of the Greeks. The tributes and loot obtained through

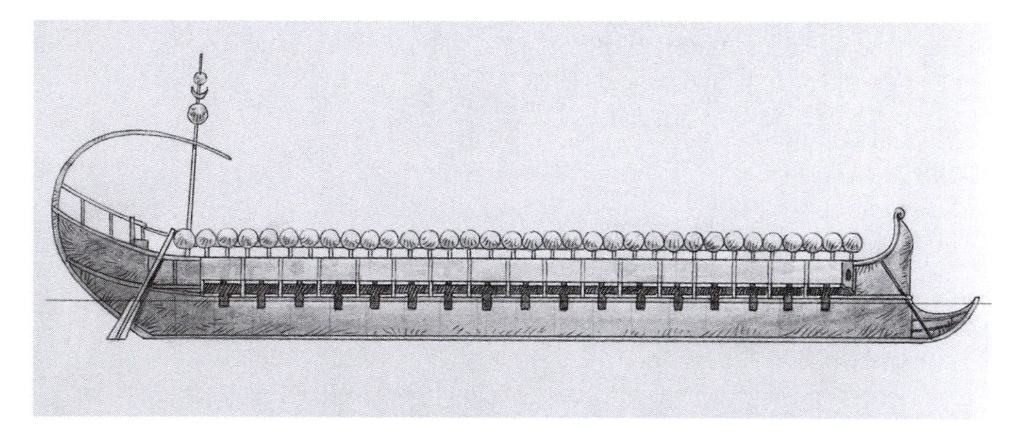

The rise of the Phoenicians in seafaring and sea trade is related to their shipbuilding skills that enabled huge loads and long hauls along the Mediterranean. (DeAgostini/Getty Images)

successful Greek naval campaigns against the Persians were used to fund the great monuments in Athens and provide offerings for the Gods.

Sailing Expertise and Ocean Domination

Using their sailing expertise, the Romans soon emerged as an important power in the Mediterranean. They defeated Carthage in naval battles, and after driving the Carthaginians out of Sicily, Corsica, and Sardinia, the Romans sent naval expeditions to take on Greece and defeat Macedonia, a Greek province. This victory provided Rome with abundant spoils of war such as slaves to work in their territory and looted Greek works of art. It also led to the expansion of Greek literature throughout Rome.

Critical sailing skills were also developing in Viking territory in Northern Europe. The Vikings were proficient seafarers, warriors, and traders in the ninth and 10th centuries, who dominated travel on the waters of Northern Europe. Their sphere of influence largely remained in the north. They traded goods, raided lands, and established settlements in England, France, Ireland, Scotland, Iceland, Greenland, and Newfoundland, even traversing the great rivers of Russia up to Kiev and Novgorod to trade with the Persians.

The Vikings' success in colonization, warfare, and the trade of valuable commodities, particularly slaves, was directly linked to their shipbuilding and navigation competencies. Through trial and error, they constantly improved their boats by building sturdier keels, masts, and oars. These innovations enabled the boats to sit flexibly in water and ride the waves without breaking up. Improving the size and structure of Viking longboats allowed Vikings to venture further—across the North Atlantic and Baltic Seas and eventually into North America.

While the European navigational horizon was generally limited to the Mediterranean and the seas of Northern Europe, Arab traders and explorers were expanding into the vast waters of the Indian Ocean from the 13th century onwards. In the 10th century Arabs used the astrolabe, a navigational instrument, to calculate the positions of celestial bodies; in the 12th century they became important mapmakers, and in the 13th century, Arab explorers, shipowners, and traders were already undertaking lengthy voyages across the ocean.

Manuscripts from this period describe Arab ships with square sails riding monsoon winds far away from their home ports, dedicated to the transport of passengers and cargo in large holds. In 1325 the Arab traveler Ibn Battutah left Morocco, and in the course of 25 years he visited 45 countries on land and sea routes in the Mediterranean, Arabia, Africa, Europe, and Asia.

In the past, India and China also forged important long-range overland and sea trading routes across the Indian and Pacific Oceans. Indian traders were knowledgeable about the Eurasian world and European demand for spices and other commodities. Indians also benefitted from overland and sea voyages throughout vast areas of the Indian subcontinent. In the eighth century, the Chinese made a big leap in technology through the invention of the magnetic compass; in 1000, they adapted it for navigation, 300 years before the Europeans would deploy it. In 1320, Chinese map makers were already showing the southern coast of Africa on Chinese maps, a full century before Portuguese rounded the Cape of Good Hope in southern Africa.

In 1405, the Ming Dynasty in China launched a series of ocean expeditions to Indochina, Malaysia, Indonesia, Southern India, the Persian Gulf, and Madagascar in East Africa, using large boats called junks that carried up to 27,000 people in one expedition. At this time, ships from Java (Indonesia) were also making sea journeys to Madagascar to trade in the seas of the Malay Archipelago and with the Philippines. However, up to that period, no seafarer had yet crossed the entire expanse of the central Pacific Ocean.

Maritime traders and colonists from the Moluccas area of Indonesia became adept at embarking on marine voyages. From the second millennia BCE onwards, they succeeded in completing perilous canoe journeys across the North Pacific. Over a period of several hundred years, Polynesian seafarers used their knowledge of the stars, winds, and ocean currents to progressively settle Melanesia, western Polynesia, Australia, New Zealand, and the Hawaiian and Marquesas Islands, as well as Easter Island off the Pacific coast of South America.

They completed their island-hopping voyages over many generations, carrying crops, livestock, and other essentials in small double and outrigger canoes over enormous expanses of water. In 1768, Captain Cook and other Europeans arrived in New Zealand and established contact with these groups. Following Captain Cook, the arrival of more European settlers to the country would eventually lead to violent confrontations between the settlers and locals, and the progressive breakdown of traditional Maori culture.

In 1770 British navigator and explorer James Cook discovered and charted New Zealand and Australia's Great Barrier Reef on the ship HMB *Endeavor*. (Ridpath, John Clark, *Ridpath's History of the World,* 1901)

Navigational Advances and European Land Grabs

Up to the 15th century, Europe was not what one might consider technologically advanced in comparison to other cultures. Important achievements such as the compass, astrolabe, gunpowder, and algebra—already prevalent in the Islamic and Chinese worlds—had not yet entered full-scale European usage. Up to the 15th century, ocean exploration in Europe had largely been confined to known trading routes among European countries and between Europe and some parts of Asia.

Up to this point, the Ming empire of China had completely overshadowed Europe in terms of navigational prowess and cartographic know how. In the 15th century, China had a population of over 100 million (more than all the European countries combined), 2,000 years of unbroken civilization, and detailed nautical and cartographic knowledge of the South China Sea, the Bay of Bengal, and the Arabian Sea. At the same time, the Ottoman Empire in the East had also gained political importance, and had closed off European overland trade routes to Asia.

Since Europe's growth depended on warfare, slave trading, the settlement of new areas, and the overland trade with Asia, the Ottoman Empire's sealing of the Asia route disrupted trade along the Silk Road with serious economic consequences for Europe. This break in trade increased the cost of Asian imports to Europe and forced Europeans to find an alternative way to the prized spices, teas, and silks from the East. Europeans looked to the oceans.

In 1498, the Portuguese explorer Vasco da Gama discovered a continuous ocean route from Europe to India around the tip of Africa, but it was considered costly and perilous. Da Gama's path involved sailing down the length of the African coast, rounding the treacherous Cape of Good Hope, and entering the Indian Ocean. Due to the risks of this journey, European traders were desperate to find a safer, alternative sea path to Asia, which they speculated would lie westward via the Atlantic Ocean. At that time, they were not aware that an entire continent, South America, lay in between.

This urge to sail west to get to India set the stage for heightened interest in 15th-century expeditions across the Atlantic. The Portuguese had a clear advantage. They were already seasoned navigators and makers of maps and ocean charts, which were guarded as state secrets. Henry the Navigator, a Portuguese prince, was a key figure in this quest. He had participated in the Portuguese conquest of Ceuta, Morocco, in 1415, and had dedicated most of his life to Portuguese colonization and slave-trading expeditions.

Early Portuguese forays along the African coast to acquire African slaves from Mauretania and Upper Guinea for sale in Europe had already established Portugal's hegemony in parts of Africa. In 1442, an official document from the pope proclaimed that the trading of African slaves fell within the limits of a just war; in 1452, the Pope extended this proclamation, giving the Portuguese the right to enslave captives from such expeditions.

The Africa expeditions also provided the Portuguese with advanced knowledge of high seas sailing, expertise in plotting ocean routes, and competencies in ocean warfare, navigation tools, and slave trading. These skills would prove valuable in the global race for ocean exploration after 1500, and lead to Portugal's later conquest of Brazil, the enslavement of indigenous peoples in the new world of the Americas, and the trading of African slaves.

Getting to the Americas

The Spanish, not wanting to be outmaneuvered by the Portuguese, were eager to jump onto the transoceanic exploration bandwagon across the Atlantic towards India.

Queen Isabel and King Ferdinand of Spain were at a critical juncture in the 1490s. They had defeated the Muslims and driven them out of Spain, and they had expelled the Jewish population from Spain in a process called the Inquisition. Having consolidated Catholicism and the Spanish empire, these monarchs were anxious to enter the race for the alternative routes to Asia via the Atlantic.

At that time Spain already possessed ships that could handle deep-sea voyages. It had also acquired gunpowder, the compass, and the astrolabe from the Chinese and the Arabs. The Spanish were able to improve the weaponry of their ships by adding gunpowder in forged iron–cast ship cannons. Desperate to find the new route across the Atlantic to the Indies before the Portuguese did, the Spanish monarchs backed Christopher Columbus's expedition with three small ships, a crew of 90, funds, and provisions.

In those days, undertaking such a voyage into the unknown would have been comparable to a trip to Mars in modern times. The sponsorship of the Spanish crown was a leap of faith. Explorers had sailing ships, navigational skills, instruments, and maps, but no one knew exactly what lay beyond the oceans to the west. An expedition's success was very dependent on political backing and funding from the reigning authorities. When news of Columbus's arrival in the Americas reached Portugal (Spain's rival), the Portuguese crown moved swiftly to launch its own expeditions to safeguard its interests across the Atlantic.

Direct intervention of the pope became necessary to resolve the ensuing feud between Spain and Portugal. In the Treaty of Tordesillas, signed in Spain in 1494, the pope drew an imaginary line through the Atlantic Ocean, east of the Azores Islands, which the Portuguese had already settled. The treaty declared that Portugal could claim rights to territories south and east of the line, while Spain would be entitled to lands north and west of it. Following the treaty, a Portuguese explorer, Pedro Alvares Cabral, claimed Brazil in 1500. The Portuguese also stepped up other explorations and staked claims to lands on the African coast at the same time.

Spain and Portugal became the two principal European players on the Atlantic, poised to divvy up the rewards of new routes to the Indies. However, as they soon discovered, neither country had yet found the alternative route to Asia through the Atlantic as originally envisaged. Both countries had to settle for the spoils of the Americas: Portugal took over Brazil, while Spain conquered Mesoamerica and South America. In 1521, it would fall to the Portuguese explorer, Ferdinand Magellan, to find the westward route all the way to the Indies.

Magellan started out from Spain in 1519 and first sailed westward across the Atlantic to the tip of South America. His expedition successfully traversed the Strait of Magellan, between South America and Tierra del Fuego, and then entered the Pacific Ocean, sailing on to Guam in the Pacific. Magellan did not fully complete the entire voyage around the world because local communities killed him in the Philippines. Nonetheless, one of his five original ships and 18 survivors of his original crew of 270 completed the quest for circumnavigation around the globe.

The Spanish consolidated their domination of Central and South America, while the Portuguese—in addition to ruling Brazil—focused on Asia (Malabar, Malacca, the Moluccas, and Macao) and Africa in the 15th and 16th centuries. Spain was instrumental in setting up networks between Acapulco in Mexico and Manila in the Philippines in the Pacific. Both countries benefitted from their Atlantic-Pacific connections, and cemented their dominion over oceanic global trading networks for over a century. This strategy prevented other nations from using these sea routes, except when privateers tried to mount temporary raiding forays on the high seas.

The Age of Colonization and Hegemony

After the initial period of Spanish and Portuguese hegemony in the Americas, the Dutch, French, and British started undertaking their own exploration voyages. They, too, conquered new lands, opened up further trading routes, and participated in the already developing, lucrative trans-Atlantic slave trade as well. Between 1487 and 1616, explorers and agents from Holland, France, and Britain completed 29 voyages to South Asia and China; between 1492 and 1519, they undertook six voyages to the Caribbean, and between 1772 and 1780, six trips to the Pacific.

From 1632 onward, the British were instrumental in settling North America. Further British expeditions to the Pacific opened up Australia and New Zealand for British settlers. Through their Caribbean expeditions, the French and the Dutch progressively challenged the supremacy of the Spanish and Portuguese in South America and the Caribbean, and established their own colonies.

The discovery of the landmass of the Americas, combined with Europe's gradual settlement of other regions in Asia, Australia, and Africa, helped generate economic profits for Europe; exploration and conquest of new lands transformed Europe from a relatively undeveloped, self-contained region into a continent of individual, global powers. The period between 1500 and 1815 marked the largest expansion of Europeans throughout the New World. The takeover of new territories, the slave trade, and the enormous returns generated by the plantation economy provided the keys to unlock power for these relatively small European countries.

From the 16th to the 19th centuries, Europe finally emerged from its relative isolation and surged forward on the global stage. Building on the foundation of pillage, colonial settlements, subjugation of the indigenous population, slavery, and the lucrative trade in new products, European countries controlled all vital maritime trading routes. They administered and controlled colonies in the conquered lands, funneled profits back to Europe, and wielded robust political, military, and naval power to shape pivotal events around the world. They also waged numerous wars among themselves to protect their interests.

Infamous Slave Triangle across the Ocean

Motivated by the quest for trade, personal enrichment, national wealth, global power, and expansion, Europeans had, by 1780, mapped almost all of the world's seas and the outlines of all its continents. They expanded the infamous slave trade in a triangular pattern connecting Europe, Africa, and the Americas, dominated by the Portuguese and the Spanish in the early phases. However, in the 1650s, French, Dutch, and British companies joined the slave trade.

Trading companies such as the British East India Company, the Royal African Company, the South Sea Company, the Guinea Company and the Dutch West India Company—under contract with European governments—dominated the peak years of the slave trade from the 16th to the 18th centuries. Private merchants operated these slave ships, often as joint enterprises backed by state capital.

Ships involved in the trans-Atlantic slave trade followed a triangular route between Europe, Africa, and the Americas in three phases, or legs. The first leg of the triangle was the voyage from a European homeport to a West African port. These ships carried manufactured goods, beads, firearms, and other commodities to be traded for slaves. This leg took about a month.

The second leg involved the notorious Middle Passage across water. Slaves were taken from African villages to African ports, jam-packed into ships, and transported across the Atlantic to the Americas or the Caribbean. In peak years, British slave traders carried about 50,000 slaves annually in ships specially outfitted with copper hulls, lower decks, and holds to cram in as many people as possible. Shackled at the wrists, passengers could not stand up straight during the voyage. Crews meted out particularly harsh punishments to rebellious slaves. To escape severe conditions, some slaves committed suicide through hunger strikes or by jumping overboard.

On the third leg of the triangle, ships discharged their human cargo in the Americas. Slaves who had survived the crossing (about one third did not) were auctioned off to plantation owners. The same ships were then loaded up with tropical products (sugar, rice, cotton, tobacco) and sent to European homeports. Upon arrival in Europe, after unloading the cargo and sharing the sale proceeds, ship owners, sponsors, and captains outfitted the ships for the next slave voyage to Africa, thus continuing the infamous journey. The three legs of the voyage took about a year to complete. Enormous profits from the slave trade led to the accumulation of wealth in British ports such as London, Liverpool, and Bristol and the growth of these ports and cities.

Portugal shipped slaves from its African outposts to its settlements in Brazil. Spain also transported African slaves, after the sharp reduction of indigenous people in the Americas due to slavery and illnesses, to its territories in Latin America. The French, who had established territories in the Caribbean such as Guadalupe and Martinique, traded slaves to work on plantations in these islands, as did the Dutch, who possessed colonies in Suriname, Aruba, Bonaire, and Curacao.

Sugar production was the basis for exploitation of all these territories. The Dutch, in fierce competition with Portugal for sugar, seized Portugal's sugar plantations in Brazil from 1630 to 1654, and in the 1630s, the British established colonies in the West Indies, e.g., Saint Kitts, Barbados, Nevis, Montserrat, and Antigua. These were sugar-producing areas, which required large quantities of slave labor to satisfy the rising demand for this bittersweet product in Europe.

Pirates Ahoy!

The multitude of ships crisscrossing the oceans with all types of valuable cargo attracted another kind of oceanfarer, pirates, who ambushed trading vessels on the high seas and took their loot to secret hideouts. Bandits were also rampant in the

East China Sea. Dutch, Spanish, and Portuguese pirates set up bases on some islands in that area, from which they would venture out to attack Chinese trading vessels.

In the Atlantic, piracy was directly aimed at Spanish ships laden with gold and silver coming from Spanish colonies (today Peru and Bolivia). Spaniards first shipped the metals up the Pacific coast to the Isthmus of Panama, where they transferred the loot to mules along the shortest route between the Pacific and the Atlantic. They then repacked the load onto galleons in Panamanian Atlantic ports bound for mainland Spain. These galleons naturally became sitting ducks for pirates waiting to ambush them after they left Panama.

Engraving of slaves harvesting sugar cane as an overseer looks on. The sugar industry was made possible by the infamous slave trade across the Atlantic, the plantation system, and slave labor. (Duncan Walker/iStockphoto)

In the 16th century, hostility between Spain and England and the fight for control of the waves became very intense; skirmishes and cat-and-mouse games between the two fleets on the high seas were constant occurrences. England, which had been unable to make headway into Spanish territories in the Americas to obtain precious metals, turned to sponsoring English privateers (pirates) to attack Spanish ships at sea.

One such privateer was Sir Francis Drake. In 1588, he became vice admiral in command of the English fleet, which conquered the Spanish fleet, or armada, as it was about to invade England. Two other well-known English privateers were Captain Edward Mansfield, who attacked Spanish possessions off the coast of Nicaragua, and Captain Henry Morgan, who occupied the Spanish city of Portobello in Panama in 1668 and ransacked the capital of Panama in 1671.

The abolition of slavery in the 19th century led to the demise of privateers and trading companies, which had trafficked human cargo. Following the end of slavery, new waves of trade and communications networks emerged across the globe. In fact, trade between the Americas and Europe intensified after the independence of Spanish, British, French, and Dutch colonies. These newly independent countries were free to trade with anyone else in the world.

United States and the Panama Canal: Path through the Oceans

The turn of the 19th century saw the rise of the United States as a key player in global oceanic traffic and geopolitics. The United States became a propelling force for the development of the Panama Canal between the Atlantic and the Pacific Oceans through the waist of the Americas (replacing the mules of the Spaniards in colonial times). The gold rush in the United States was an important factor, since the Panama passage replaced the long, overland U.S. route to gold. Strategic, geopolitical interests were also in play.

The rise of international transportation, investment, and production networks, as well as global corporations in the 20th century, increased the shipping of global consumer goods and primary products worldwide. Globalization and trade liberalization were key factors propelling trade and travel. The use of oil tankers, shipping containers, research ships, leisure vessels, and fishing and diving expeditions exploded, as new forms of oceanic traffic emerged.

Key factors for this development were postwar peace; the resurgence of Japan's economy; the dynamic growth of Hong Kong, South Korea, Singapore, and Taiwan (the Asian tigers); the growth of China as a major player in the global economy; and the rise of affluent consumers in North America and Europe. The benefits of increased world trade were accompanied by the negative impacts associated with the expansion of ocean transportation: increased stress on maritime resources, ocean pollution, the use of oceans and seas as pathways for human trafficking, illicit arms trade, and transportation of endangered species.

Camille Gaskin-Reyes

REFERENCES AND FURTHER READING

Bergreen, Lawrence. *Magellan's Terrifying Circumnavigation of the World: Over the Edge of the World*. New York: Harper Collins, 2003.

Lansing, Alfred. *Endurance: Shackleton's Incredible Voyage*. New York: Carroll & Graf Publishers, 2006.

Mann, Charles. *1493: Uncovering the New World Columbus Created*. New York: Random House, 2012.

McCullough, David. *The Path Between the Seas: The Creation of the Panama Canal, 1870-1914*. New York: Simon & Schuster, 1977.

Morgan, Kenneth. *Slavery and the British Empire: From Africa to America*. New York: Oxford University Press, 2007.

Novaresio, Paolo. *The Explorers: From the Ancient World to the Present*. New York: U.S. Media Holdings, 1996.

Overey, Richard, ed. *Hammond Atlas of World History*. Maplewood, NJ: Times Books, 1999.

Rediker, Marcus. *The Slave Ship: A Human History*. New York: Penguin Group, 2007.

Smallwood, Stephanie. *Saltwater Slavery: A Middle Passage from Africa to American Diaspora*. Cambridge, MA: Harvard University Press, 2007.

St. Clair, William. *The Door of No Return: The History of Cape Coast Castle and the Atlantic Slave Trade*. New York: Bluebridge, 2007.

CASE STUDIES

The following case studies highlight the role of shipbuilding and sailing technologies as enabling factors for many groups to explore the high seas for trading, raiding, and the spread of religion. The first case study explores the development of new navigation devices such as the sandglass, compass, the astrolabe, and charts in medieval times, which also facilitated exploration across the high seas. The second case study examines the feats of Muslim explorers across the Mediterranean Sea and Indian Oceans, trips that were mainly motivated at first by the desire to spread Islam. The third case study documents the case of the early Chinese explorers and their competency in shipbuilding and navigation, which allowed them to successfully explore, colonize, and trade far beyond their homelands.

Case Study 1: Medieval Navigation Devices

The magnetic compass was one of the outstanding discoveries of medieval times. It led to the European journeys of discovery that are a major part of postmedieval history. Without a compass, a ship's captain needed the sun or stars to identify north. He could not sail away from land with confidence because he could not know his direction on cloudy nights. With the compass, the development of specialized maps for the sea, and the introduction of the sandglass and astrolabe, sailing navigation became a precision craft that allowed ships to sail year-round and away from land.

Compass

The compass began with the Chinese discovery of magnetism. Their earliest compass was a piece of lodestone, a naturally occurring magnetic iron ore (magnetite). They learned to rub an iron needle on lodestone to magnetize it and then either hang the needle from a strand of silk or put it through a straw or a bit of wood in a bowl of water. Free-floating, the needle would swing to a north-south position. This basic compass almost certainly made its way from China to Europe via the Silk Road, perhaps because of its usefulness in travel or because it may also have had astrological purposes.

Several medieval books and poems make references to compasses in use in the Mediterranean during the 12th and 13th centuries. The earliest 12th-century reference by the English scholar Alexander Neckam described a compass that was still in the Chinese form, a needle floating on a wood chip in water. This compass could be made as needed with a lodestone, a bowl, and a needle, but it was not completely useful since its accuracy depended on keeping the bowl of water steady. The next development of a true compass is credited to the sailors of Amalfi, particularly to Flavio Gioia. When a monk, Peter the Pilgrim, described the compass in use in southern Italy in 1269, he described a true instrument in a box. By the end of the 13th century, compass use was routine and widespread.

The medieval compass was a circular card marked with 32 directional points. The card was placed below a free-swinging magnetized needle attached to a dry pivot and housed in a wooden box. The compass rose that points to north, south, east, and west was developed by the sailors of Amalfi. Europe had not yet adopted the 360-degree directional convention.

Navigators knew the needle pointed to the north magnetic pole, not to true north, and made adjustments for that fact. Use of the compass first became common in the Mediterranean Sea, which in many places was too deep for sailors to determine their position by sounding, the traditional method of determining the water's depth. The compass was less used in the shallower northern waters, such as the Baltic Sea and the North Sea, where sounding continued to be the main navigational tool.

Other Instruments

The hourglass, or sand glass, was the ship's only clock; there are records of its use in the 14th century, although it could have been used earlier. On a ship, the hourglass was turned every 30 minutes. A crew member (often the cabin boy) was assigned to keep careful watch and to strike a bell when he turned the glass over. The number of bells told the time. An hourglass also was used in calculating a ship's speed to determine its location. A piece of wood was attached to a rope that had knots tied in it at measured intervals. This wood was let down to the water to trail from the back of the ship while a ship's officer timed the seconds between knots and calculated the speed. (This is why speeds at sea are spoken of in knots.)

The hourglass had to be carefully calibrated by its maker with exactly the right amount of dry sand or pulverized eggshell. The sand had to maintain a steady rate of flow and had to be fine enough not to erode the glass opening it passed through. The glass bulbs had to have the right angle to keep the sand flowing evenly. For accuracy, the hourglass was set on a flat, even surface. Because a ship at sea is constantly heaving up and down on the waves, the hourglass was often hung by a cord in a holder that allowed the hourglass to be easily turned over.

The astrolabe, borrowed from astronomy, became a way to find position at sea some time during the 13th century, but it is unclear how widely it was used until the 15th century. The form developed for use at sea was both simpler and more practical than the astronomical tool. It had a heavy brass ring with an alidade for sighting a star or the sun. It was a ring, not a plate, because when a sailor held it by its top ring, the wind at sea blew it so it was hard to use; a ring offered less surface for the wind than a plate. Because it was so difficult, it was easier for a ship to use it for determining the latitude of an island when they were at anchor or on land.

An even simpler instrument came into use in the middle of the 15th century. The quadrant became the primary tool for determining a star's altitude. It had an arc marked off in degrees and an alidade-sighting tool along one side. A plumb bob (a string with a pointed lead weight) hung from the other side. As one sailor held the quadrant steady, lined up between his eye and the star, another read the degree where the lead pointer hung.

The cross-staff may have been invented by a Dutch sailor in the 13th century. It measured the angle between the horizon and a star, most usefully the North Star. It was a simple shaft held up to eye level and a moving crossbar called the transom. When the bottom of the transom was at the horizon and the top at the target star, the distance mark along the staff could tell the viewer's position with the help of a chart made for the cross-staff.

Soundings, Buoys, and Sea Charts

In ancient and medieval times, sounding was done by dropping a weighted rope until it touched bottom and then measuring that depth. Soundings were taken to see whether the ship was about to run onto rocks and also to help establish the position of the ship. Soundings were usually taken by using a rope coated with tallow. A big wad of tallow on the end of the rope could bring up sand or gravel to show what the sea bottom was like and help the captain estimate the position of the ship.

Sounding was useful in the Baltic Sea, where the water was not very deep and the coastlines were shallow, but less useful in the much deeper Atlantic Ocean and Mediterranean Sea. The Baltic region was also an early adopter of channel markers. Many of the flat shorelines had few landmarks visible from a ship, which had to navigate through shallow water and identify the correct river mouth to enter. There were established sea routes between the cities in the Hanseatic League; cities established several kinds of markers to help navigators identify their location.

Archaeologists believe buoys were used as channel markers as early as 1066 on the River Weser. Buoys marked the entrance to the Zuider Zee bay beginning in 1323, a practice that spread in following years. Early buoys may have been small, watertight barrels. Sometimes buoys were marked to show whether a channel was going upstream or downstream. The marker was a besom, a bundle of twigs attached to a handle, which was placed with its point upward for one direction, downward for the other.

The Mediterranean Sea had known lighthouses during Greek and Roman times, but many had fallen into disrepair. In Northern Europe, there was no early lighthouse tradition except the lighting of beacons in bad weather in some places. The first known European lighthouse was in use in 1202, with more lighthouses established later in that century. Large wooden beacons with distinctive top-marks also were placed to identify localities. By 1280, lighted beacons marked the location of some rivers. Lighthouses also were established along the shores of the Strait of Dover, the earliest being at Winchelsea in 1261 and on the Isle of Wight in 1314.

The portolan was essentially a port-finding chart or map. For several centuries, Mediterranean seamen wrote down information about ports, tides, winds, and dangerous coastlines. Gradually, this information was written into pilot guides (in Greek they were called *peripli*, and in Italian, *portolani*). Portolan sea charts were mapped versions of these guides.

A portolan showed highly detailed coastlines marked with ports, sources of water, and hazards such as reefs or pirates. The names of the ports, capes, and so on were written at right angles to the coast, on the inland side so as not to obscure the coastline. There was no attempt to show scale of distances accurately or to be true to how maps were made, and there was no up or down to the portolan.

The most distinguishing feature is a network of rhumb lines (lines of a specific geographic direction). These are straight lines for navigation; 16 lines radiate from a central point. The lines were often color coded for the main directions (north, south, east, and west) and the intermediate directions (northeast, southeast, northwest, and southwest). The lines ran through 16 intersecting compass stars, giving a navigator a continuum of straight navigation lines he could use to work his way to the desired port by using his mariner's compass for navigating by dead reckoning. Other aids were log lines to estimate distance and an hourglass for telling time.

Ruth A. Johnston

REFERENCES AND FURTHER READING

Gies, Frances, and Joseph Gies. *Cathedral, Forge, and Waterwheel*. New York: Harper Collins, 1994.

Launer, Donald. *Navigation through the Ages*. Dobbs Ferry, NY: Sheridan House, 2009.

Woodman, Richard. *The History of the Ship: The Comprehensive Story of Seafaring from the Earliest Times to the Present Day*. Guilford, CT: Lyons Press, 2002.

Case Study 2: Muslim Exploration

Beginning with ancient times, Arab civilizations engaged in maritime exploration. Egyptian seafarers were thought to have built ships as long ago as 2400 BCE, and Phoenician records describe early explorations of the Red Sea. Ancient Arab rulers held colonies in East Africa, and the last ruling dynasty of the Persian empire—the Sassanians—established several faraway trading ports and encouraged their people to explore by sea in order to increase trade. When Muslims conquered the Persian empire and began to unite the Arabian Peninsula under Islam in the mid-seventh century, such seafaring activities took on heightened importance.

After the Muslim conquests, the Mediterranean Sea and the Indian Ocean were both partially controlled by the Islamic empire. The Islamic rulers required naval power to maintain dominance across their widespread territories and to combat possible invaders. Because of this, wood resources from Syria were combined with maritime knowledge from ancient Egypt to create a large and imposing Islamic navy.

However, protection was not the only reason that Muslim sailors took to the seas. A thriving trade between the Muslim empire and China necessitated navigable water routes to China. By the mid-ninth century, such routes had been established and Muslim traders were regularly traveling to China by sea.

Religion also played an important role in Muslim exploration. According to the Koran, the sacred text of Islam, Muslims should travel the world and look for signs of God everywhere they go. Islam also asks its followers to make a pilgrimage to the holy site of Mecca, in modern-day Saudi Arabia, at least once in their lifetime. Additionally, travel was seen as a way to spread the Islamic faith to other lands and peoples. Together, these factors helped make the Muslims some of the most powerful traders and marine navigators on the Indian Ocean and beyond.

In addition to China, Muslim traders and travelers sailed to India, the East Indies, and Africa. There is evidence that Muslim explorers may have even possibly crossed the Atlantic Ocean and reached the Americas as early as the late ninth century. Perhaps the most renowned Muslim explorer was Ibn Battuta, who is sometimes known as "the Arab Marco Polo." During the 14th century, Battuta traveled throughout both the Near East and Far East, as well as into Muslim Spain and to the African empire of Mali. All told, Battuta spent more than 30 years traveling the world. He documented his experiences in a *rihla*, an Islamic travel book with a religious slant.

Hasan Ali Ibn al-Husain al-Masudi is another Muslim explorer who wrote about his travels. Al-Masudi traveled through Persia and India, as well as into Madagascar and Indochina, in the early part of the 10th century.

Sulaiman al-Mahiri is the Muslim explorer who is thought by some to have reached the Americas. He wrote of his experiences sailing the Indian, Pacific, and Atlantic Oceans and is known to have at least reached the Bering Strait during his adventures.

Other notable Muslim explorers include Ibn Fadlan, who was made an ambassador to Bulgaria in 921 and traveled extensively through the Russian region at the time, and Abu Rehan Beruni, who recorded detailed information about the geography of India in his 11th-century book, *Kitab-al-Hiplcl*.

Muslim scholars and sailors greatly contributed to modern navigational and ship technologies. They used the lateen sail, a triangular sail that could take wind on either side and allowed a ship to tack into the wind instead of having to wait for it to come to them. Although it was not necessarily an Islamic invention, successful use of the sail by Muslim explorers brought it to the attention of others who traveled the Mediterranean Sea and beyond. Soon, it became a regular fixture on oceangoing vessels.

Originally invented by the Greeks, a sophisticated astronomical calculator called the astrolabe was also fine-tuned in the Islamic empire. This tool allowed astronomers to compute the positions of objects in the sky and to find the time of sunrise and sunset, among other things. A smaller version of the tool, later called a mariner's astrolabe, could be used to help navigate ships at sea. While it was not as accurate due to its smaller size, it was still an enormous advancement from earlier navigation systems. Although the earliest documented use of this tool was in 1481, experts suspect the tool was in use long before then.

Additionally, navigating dangerous waters and currents off of trading ports in Africa meant that Muslim sailors needed to have an understanding of latitude in order to plot an accurate course. Out of this necessity, early Muslim seafarers helped develop a tool called a quadrant that used the stars and sun to triangulate their position in the water.

Tamar Burris

REFERENCES AND FURTHER READING

Alavi, S. M. Ziauddin. *Arab Geography in the Ninth and Tenth Centuries*. Alighar, India: Alighar University Press, 1965.

Battuta, Ibn. *The Travels of Ibn Battutah*. Edited by Tim Mackintosh-Smith. London: Picador, 2002.

Case Study 3: Chinese Exploration

It is thought that Asian exploration began with the people of what is currently southern China some 50,000 years ago. Traveling via bamboo rafts, they began populating nearby islands in the Pacific. Evidence suggests that sometime between 14,000 BCE and 4000 BCE, another wave of southeastern Asians took these travels further, colonizing Indonesia and Polynesia. It is even suspected by some scholars and experts that early Chinese explorers sailed across the Indian Ocean, and may even have ventured as far as the Americas between 3000 BCE and 2500 BCE. What is certain is that by the mid-12th century, the Chinese had a fleet of technologically advanced, oceangoing trade ships that were able to cross the China Sea and far beyond.

The earliest known emperors to fund explorers were the Qin ruler Shi Huangdi and Han leader Wudi in about 219 BCE and some hundred years later, respectively. These men were driven in part by the search for an elixir of life that would give the gift of immortality. Xu Fu was arguably the most famous Asian explorer of this time. Shi Huangdi sponsored two of his voyages in search of the magic immortality herb in around 219 BCE and 210 BCE, and he is known to have ventured into Japan before his death. Not all emperors supported exploration expeditions, however, as some distrusted outsiders and preferred to keep their people within their own borders. Therefore, the timeline of exploration was often interrupted. A seafaring wreck found off the coast of Quanzhou in the 1970s has been dated to the Song dynasty, which ruled from about 960 to 1279.

The Song developed the magnetic navigational compass as early as 1125, as confirmed in a book written at that time. This invention was essentially a magnetic needle floating in a small bowl of water, but it did the trick. The Song also invented the sternpost rudder, which provided greater control over the direction of vessels, and movable sails, which meant that sailors no longer had to wait for the wind to blow in the right direction in order to move their ships. Along with all this, the Song also developed separate, watertight compartments in which to store goods; this made it so that if one area sprang a leak, only the cargo in that particular hold would be damaged. When investigating the wreck off of Quanzhou, scientists found traces of exotic spices, woods, and shells from East Africa, suggesting that these early Chinese were already trading with civilizations across the Indian Ocean.

The Mongol emperors of the Yuan dynasty were the first to commission an imperial fleet during their reign between the late 11th century and the mid-12th century. Marco Polo described their ships in his book, *The Travels of Marco Polo*, noting that they had four mastheads, cabins for 60 traveling merchants, and room for a crew of 300. With their ability to navigate the seas, the Mongols established trading posts in such places as Sumatra, Ceylon, and southern India. When the Ming dynasty was established after the fall of the Mongols around 1368, the new rulers inherited the advanced fleet and extensive trade network. After taking down his ruling uncle in 1402, Emperor Zhu Di commissioned more than 300 new ships to be built and ordered nearly 200 more to be retrofitted for longer oceangoing voyages. This was the beginning of what came to be known as the treasure fleet, ushering in a golden age of Asian exploration. Between 1404 and 1407, around 1,500 new ships were built. The largest wooden ships on record at the time, the galleons had hulls that stretched about 500 feet in length.

Eventually, the magnificent Ming navy had a crew that numbered in the tens of thousands. Grand Eunuch Zheng He commanded the fleet, and his seven maritime voyages took the Chinese as far as Mecca in Arabia and Mogadishu in Africa. Zheng He's adventures were commemorated on a stone pillar found in the Fujian Province in the early 20th century. According to what is written there, he and his armada battled pirates, encountered an enormous hurricane, and brought home such exotic goods as "dragon saliva" and "celestial horses" (zebra), before Zheng He died at sea in 1433. Around this time, conservative Confucianism gained in popularity. As part of this growing culture, it was viewed as improper to travel far distances while one's parents still lived—a notion that put a damper on young men's desires to be mariners. In addition, criticism against the emperor's lavish spending on the treasure fleet grew, and conservatives railed against the idea that interactions with distant, "barbaric" civilizations held any value for the Chinese. The threat of another Mongol invasion along China's northern borders took interest away from the naval fleet and funneled military funds towards land protections.

By the early years of the 16th century, the treasure fleet had downsized by more than 1,000 ships. Then, in 1525, the government dealt the final blow to exploration by ordering that all oceangoing vessels be destroyed. With that, the country entered into an extended period of xenophobic isolationism that predominated until the 20th century.

Tamar Burris

REFERENCES AND FURTHER READING

Asia for Educators. "The Ming Voyages." Columbia University. http://afe.easia.columbia.edu/special/china_1000ce_mingvoyages.htm.

Grice, Elizabeth. "Explorer from China Who 'Beat Columbus to America.'" March 4, 2002. http://www.telegraph.co.uk/news/worldnews/asia/china/1386655/Explorer-from-China-who-beat-Columbus-to-America.html.

Hadingham, Evan. "Ancient Chinese Explorers." NOVA. PBS. January 16, 2001. http://www.pbs.org/wgbh/nova/ancient/ancient-chinese-explorers.html.

Mariner's Museum. "Chinese Exploration." http://ageofex.marinersmuseum.org/index.php?type=webpage&id=8.

NOVA. "China's Age of Invention." PBS. February 29, 2000. http://www.pbs.org/wgbh/nova/ancient/song-dynasty.html.

ANNOTATED DOCUMENT

Diego Gutiérrez's 1562 Map of America from the Library of Congress, Washington, D.C.

Background

The Library of Congress was established by an act of the U.S. Congress in 1800. President John Adams signed a bill transferring the seat of government from Philadelphia to the new capital city of Washington. The legislation contained provisions for a reference library for Congress, called the Library of Congress.

The Library of Congress contains the largest and most comprehensive cartographic collection in the world. Its collections encompass over 5.5 million maps, 80,000 atlases, 6,000 reference works, over 500 globes, 3,000 raised relief models, and a wide range of cartographic materials in other formats. The Library of Congress also has online map collections, which represent a small fraction of the maps that the Library has converted to digital form.

Summary

The 1562 Map of America, described in this section, was created by Diego Gutiérrez, a Spanish cartographer, and Hieronymus Cock, an engraver from Antwerp. It belongs to the Library of Congress's map collection, which documents discovery and explorations in manuscripts and published maps. Many of these maps reflect the European age of discovery, which ranged from the late 15th century to the 17th century.

In the 16th century, Europeans were very interested in determining the outline of the continents, and they explored and mapped coastal areas and major waterways to draw the outlines as accurately as possible. These mapping exercises became significant tools with which to support European explorations and expeditions across the Atlantic Ocean.

Christopher Columbus's Exploration

The arrival of Christopher Columbus in the Bahamas opened the way for a slew of expeditions to the Americas, primarily by Spain and Portugal, which were vying for the discovery of an alternative route to India westward through the Atlantic Ocean. We know now, but these powers did not know then, that Columbus had reached the Americas, a continent in its own right, and not part of Asia.

Even getting to that point was a tremendous feat for Columbus, as well as other Spanish and Portuguese explorers and state-sponsored navigators, who would follow Columbus's epic landfall. These voyages into the Americas eventually added to the information regarding major rivers, coasts, contours, and boundaries of the new continent, and helped extend the parameters of European knowledge about the ocean and what lay beyond.

In 16th-century Europe, feedback from explorers was fed into the ongoing work on maps and charts of official cosmographers and cartographers of the Spanish and Portuguese court. This information was vital to support and guide the takeover and exploitation of new lands and the formation of Spanish and Portuguese colonies in the Americas.

At that time, Europe's sphere of influence was shifting from the arena of the Mediterranean Sea to settlement and takeover of lands and the acquisition of valuable metals in the Atlantic, and consolidation of naval power on the high seas. The increasing knowledge of new geographical areas helped forge new ideas and debunk old myths or beliefs about the shape of the world and the composition of its territories.

The Map

Against this backdrop, in 1562, Diego Gutiérrez, a Spanish cartographer, and Hieronymus Cock, a cartographer or map maker from Antwerp in Belgium, played a significant role in the joint preparation of an ornate map of what was then referred to as

America, the fourth part of the world. Their map was the largest engraved map of the continent at the time.

The map shows the eastern coast of North America, all parts of central and South America, and some parts of the western coasts of Europe and Africa. It covers an area in longitude from 0° and 115° west of Greenwich in England, and extends up to 57° north and 70° south in latitude. It depicts the Equator, and the Tropics of Cancer and Capricorn in a north and south direction. These boundaries were the limits of knowledge of the area surrounding the Atlantic Ocean at the time. Its title was "America," which was identified as the fourth part of the world. It is speculated that this title was deliberately placed by the map makers to bolster Spain's claim to the Americas, and to stave off claims of any other European power at the time for what Spain considered to be its lands.

The map identifies major river basins in the Americas such as the Amazon and other rivers in South America, Lake Titicaca (in today's Bolivia), and other important centers such as Potosi (for mining), Mexico City (huge indigenous center), Florida, and the southeastern part of the United States. The coastal contours and features of Central America and the Caribbean are also included in the map, which up until the 18th century continued to be the largest Spanish map of America in print.

While the map might have had the objective of Spain's staking a claim to Atlantic Ocean territories, it might have also stoked claims of other European powers such as France to new lands around what are now South Carolina and Florida after 1562. It has been pointed out that the map, as official as it was, did not contain the famous line of demarcation in the Atlantic Ocean, which the pope had decreed to be the dividing boundary between feuding Spanish and Portuguese crowns over colonial territories.

It is speculated that Spain was not ready to concede land to Portugal until it had settled further territory in the Americas, or that Spain had not accepted the pope's authority as arbitrator, or might have been planning to encroach upon Portugal's claims. It is impossible to know the reason for Spain's omission of this boundary.

The only discernible line on the map was the Tropic of Cancer, an important line for exploration because it ran through channels called the Straits of Florida, which was at the time one of the safest entry points to Cuban and Caribbean waters and the Gulf of Mexico. For Spain, this was a strategically significant point to enter and leave the Americas, since it was bringing out tons of gold and silver from the Americas (Mexico, Peru, Bolivia) through the Caribbean.

Ships of any European nation, particularly France or England, that posed a threat to Spain's hegemony in this area had to be intercepted and beaten off before entering the area around the Tropic of Cancer. It was important for Spain to stake its claim to overseas territories and their access points through highlighting these positions on the America map, while downplaying the Portuguese territories on the same map.

Strategic Considerations

The purpose of the America map was not to serve as an exact navigational or scientific tool, even though it was drawn in a spectacular fashion and was the largest map of the new continent for a century. It looked more like a ceremonial instrument, and might

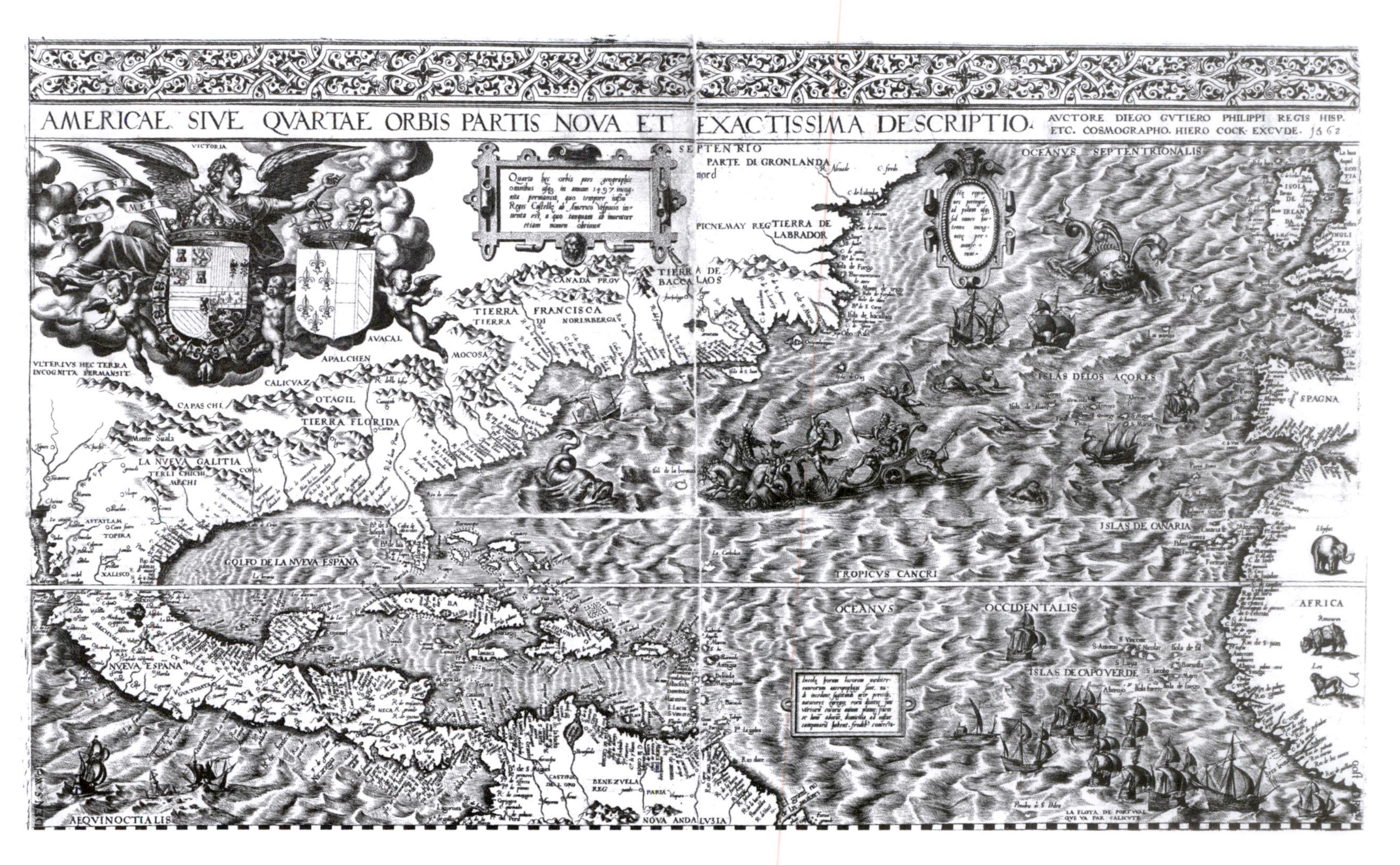
AMERICAE SIVE QVARTAE ORBIS PARTIS NOVA ET EXACTISSIMA DESCRIPTIO.
AVCTORE DIEGO GVTIERO PHILIPPI REGIS HISP. ETC. COSMOGRAPHO. HIERO COCK. EXCVDE. 1562
VICTORIA
SEPTENTRIO
PARTE DI GRONLANDA
TIERRA DE LABRADOR
TIERRA DE BACCALAOS
CANADA PROV.
TIERRA FRANCISCA
TIERRA DI NORIMBERGA
MOCOSA
AVACAL
APALCHEN
CALICVAZ
OTAGIL
TIERRA FLORIDA
CAPAS CHI
Monte Suala
LA NVEVA GALITIA
TERLI CHICHI
MECHI
TOPIRA
XALISCO
VLTERIVS HEC TERRA INCOGNITA PERMANSIT
GOLFO DE LA NVEVA ESPAÑA
NVEVA ESPAÑA
CVBA
BENEZVELA REG.
PARIA
NOVA ANDALVSIA
AEQVINOCTIALIS
OCEANVS SEPTENTRIONALIS
ISLAS DELOS AÇORES
SPAGNA
ISLAS DE CANARIA
TROPICVS CANCRI
OCEANVS OCCIDENTALIS
ISLAS DE CAPOVERDE
AFRICA
LA FLOTA DE PORTVGAL QVE VA PAR CALICVT

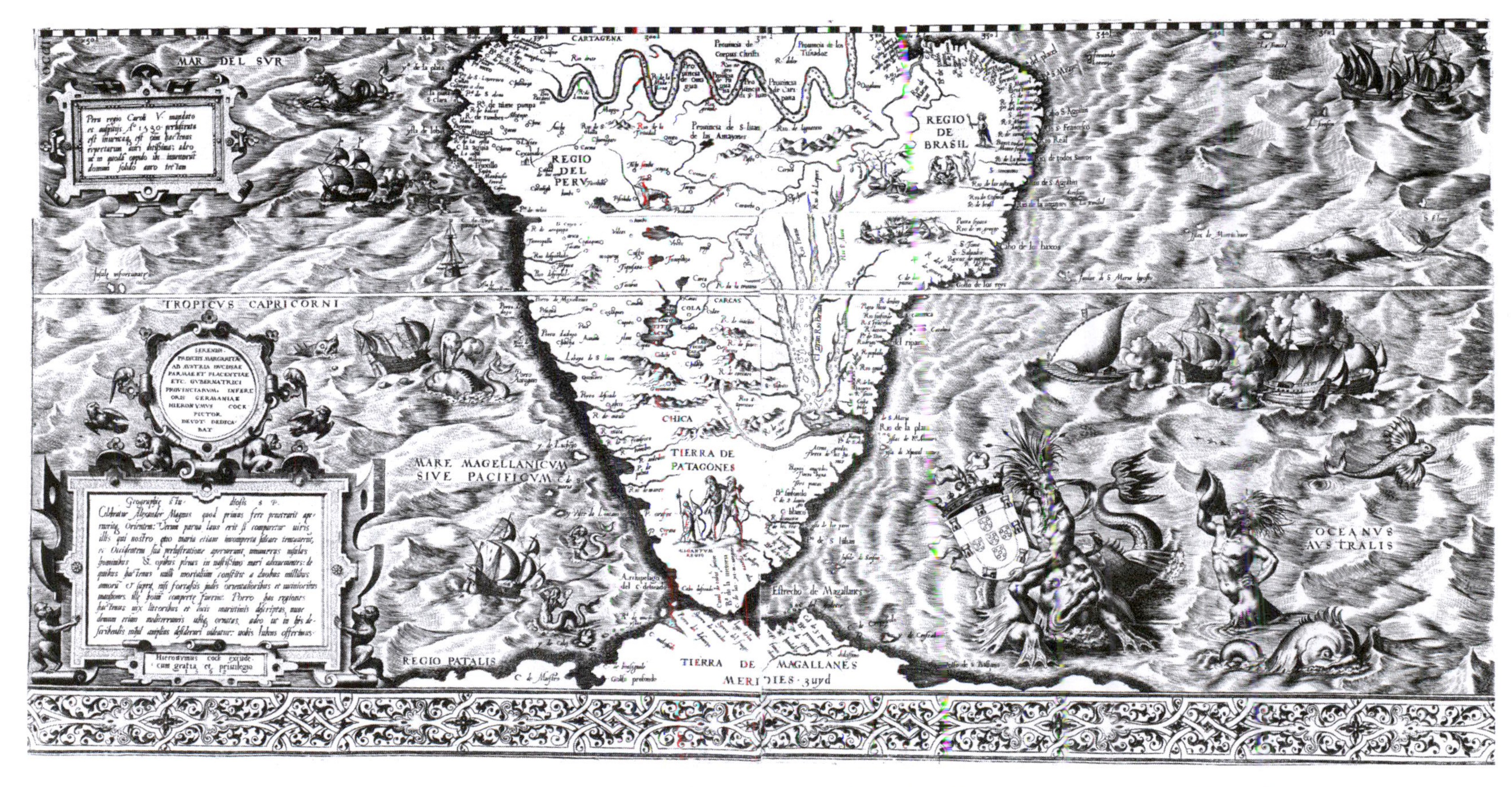

Source: Library of Congress. https://www.loc.gov/collections/discovery-and-exploration/articles-and-essays/the-1562-map-of-america/%20-%20the-1562-map-of-america

have been a diplomatic ploy of the Spanish Court to showcase its claims and to send a message to competing European powers that Spain was in charge of the areas depicted on the map. It portrayed the Crown's official seal, the Spanish coat of arms, on the map, which lent an air of official Crown ownership and signaled that Spain was willing to go to any lengths to defend the Crown's possessions.

Even though this cartographic showpiece achieved obscurity after the 18th century, it was the predominant authoritative map of America for a century, at a time when the conquest of new lands and European settlement had reached a frenzy of competition for the shipping of new crops and products (including the slave trade) from the Americas. European powers were interested in military dominion on the high seas to secure safe passage of such products, and the map would have been valuable. Four hundred years after production of the map, there is still some uncertainty about its true authors and where it was printed.

The lack of exact knowledge of its background is complicated by the fact that there are only two copies of this map available in the world: one in the Library of Congress and the other in the British Museum in London. It is still considered a cartographic treasure that helped pave the way for further European exploration of the Americas at a crucial period.

Camille Gaskin-Reyes

PERSPECTIVES

Did Early Polynesians Deliberately Travel Great Distances to Colonize Pacific Islands?

Overview

Polynesians initially peopled one-third of the vast Pacific Ocean over 2,000 years ago. They accomplished this remarkable feat by sailing in skillfully crafted, oceangoing canoes, some measuring over 100 feet in length. Once they had settled these newly found islands and atolls, the sea continued to be part of their daily lives simply because the Polynesians had to live near the shore, and they relied upon its resources to provide food in their daily lives. Also, the ocean was the easiest means of travel and communication between the various districts on their island or between other islands in the archipelago; the canoe was the facilitator of that travel.

Polynesian canoes ranged in size from the small outrigger canoe of perhaps 10 to 15 feet in length to the gigantic sea-voyaging, double-hulled canoes that could measure up to 100 feet in length and accommodate 100 people or more. Most Polynesian canoes were of three types. The first was a small dugout constructed from a single felled tree, outfitted with a smaller outrigger support for stability, and used primarily for inshore or lagoon fishing and for transportation around the shoreline of the island. It certainly would not have exceeded more than 20 feet in length. The second type was a larger dugout with built-up sides, outfitted more often with another dugout of equal size, lashed together, with high upward-curving sterns. These canoes could measure 70 feet in length, and various types of platforms constructed between them could accommodate numerous people. The third type was a double-hulled, composite-plank keeled canoe made of internal ribs over which were placed wooden planks fastened together by stitching or lashing with sennit (coconut-fiber rope). The seams and any holes were plugged by using caulking made from the sap of the breadfruit tree. These double-hulled "catamarans" were then finished by constructing various wooden platforms between them where the passengers could sit, cook, and sleep. Often protective canopies and/or small huts, made of palm fronds, were also built on top of the platform. Ships this size often averaged 30 to 100 feet in length and 35 feet in width. Outfitted with food and other provisions, this type of canoe could sail the open ocean for 30 days before needing to restock its supplies. Most likely, it is this type of canoe that made most of the long-distance voyages between the various island groups of Polynesia.

Most of the small canoes in Polynesia were utilitarian and did not warrant decorations of any kind, but the larger, more formal canoes sported various types of decorations. The prows of the double-hulled canoes, for example, almost always extended to a great height (some 26 feet high) and were decorated with streamers made of sennit and feathers. The Maoris of New Zealand hand-carved enormous prows and sterns with intricate geometric designs and humanlike figures. Sometimes additional carvings were added down the top narrow beading on each of the outer hulls. The prows, sterns, and beading were generally painted black with splashes of white and decorated with white feathers.

All Polynesian canoes could be propelled either by human paddlers or by sail. A small, one-man dugout would not normally have a sail; its forward movement came completely from paddling. Even the larger canoes, including the great war canoes, could be maneuvered by the use of paddlers, but on most other occasions, the canoes sped forward in the water by the use of sails. Sails were made of fine woven mats in the form of an inverted triangle or in the form of a crab's claw and lashed to masts rising from the centers of the hulls. Polynesians were experts in steering their ships by means of the stars, the sun, the swells of the ocean waves, the winds, the birds, and the clouds, and a seagoing vessel could easily make 100 to 150 miles a day. Round-trips of over 2,000 miles were recorded in ancient Polynesian legends and stories.

Constructing a canoe was a sacred ritual in which prayers and sacrifices were required. Most of the timber for canoe building came from forests that grew only on the upper slopes of the mountains. Timber such as the tall kamani (*Colophyllum inophyllum*), ironwood, breadfruit, mango, coconut palm, or even the hibiscus could be used. Once it had been decided to fell a tree or gather wood for a canoe, certain individuals would be designated for the task, and they would be held sacred or taboo (*tapu*) while they were at their task. Canoe builders were usually male, but sometimes women helped with the less skillful tasks. Prayers would be invoked to the various gods—Kane (Tane), Taere, Tangaloa (Taaroa), Ku, and so forth—and then the workers would set out to find the proper trees. Once a tree had been found, it was often observed for days before any cutting began. A priest would observe to see if any flycatcher bird would light upon the tree and peck at certain portions of it. If it did, then that tree was determined to be hollow or possibly rotten. Once a tree was approved, other sacred rituals and prayers followed, and then the tree was felled. Most often, trees were dragged down the mountains to shore, where they were customarily finished. If the tree was huge, however, much of the hollowing out or making timber planks was done on site. The single most important tool of the canoe builders was the stone adze, for metal was unknown to the Polynesians, and specific prayers were invoked to the gods to bless the builders' tools during construction.

Usually, the canoe builders constructed sheds at the seashore to protect the workers and the emerging canoe. Once the canoe was completed, a great feast would be held to name the canoe. It was decked out with fine tapa cloth, mats, and feathers, and then the villagers would make their way to the temple, where prayers and sacrifices were offered for the proper dedication of the canoe. In some instances, a human sacrifice was made, and in others, the sacrifice was a pig. Afterwards, the canoe was pushed onto rollers and launched into the sea. Priests were on hand to offer whatever particular chants and prayers were needed.

Robert D. Craig

Perspective 1: There Was Organized Seafaring in the Settlement of Polynesia

The exploration and colonization of the islands in the Pacific Ocean have puzzled generations of scholars and have spurred wide-scale speculation among a broader public. Over a span of more than 200 years, the debate about how the Pacific was settled has produced an enormous amount of literature, ranging from observations about the lives and customs of Polynesian islanders in voyagers' logbooks to scientific investigations about ancient seafaring, archaeology, and population genetics.

The geographic distances in Oceania are enormous. It is some 6,000 nautical miles (or 10,000 kilometers, respectively) from Australia or New Guinea in the west to Easter Island in the east and some 5,000 nautical miles (about 8,500 kilometers) from the Hawaiian archipelago in the north to the South Island of New Zealand.

Seafaring in the western Pacific and on the Pacific Rim started at a very early time. Those who crossed the strait that separated Southeast Asia from New Guinea during the Ice Age were the first seafaring humans. They undertook their voyages in canoes or on rafts more than 62,000 years ago. Those early seafarers did not have any technical devices for their navigation, but they certainly observed phenomena in their natural surroundings very carefully and learned how to respond to the impact of environmental factors on their living conditions.

The colonization of the Pacific islands unfolded in a long-lasting process of several migratory waves, with intervals of stabilization of newly explored sea routes. It was mainly Austronesian-speaking people who explored and colonized Oceania. The migration of Austronesian populations began with the second wave, starting around 1500 BCE.

Those migrants profited from the seafaring experiences of their predecessors, who were not of the same ethnic stock as the Austronesians. The technology of building vessels capable of covering long nautical distances improved over generations, as did the methods of navigation, and it eventually enabled the Polynesians to explore the far-distant peripheries of Oceania (i.e., Easter Island, the Hawaiian Islands, and New Zealand).

The first migration (before ca. 40,000 BCE), coming from the mainland of Southeast Asia, to reach New Guinea more than 40,000 years ago were the ancestors of the modern Papuans, whose languages are unrelated to the Austronesian phylum. The first wave of migration proceeded as far as the Bismarck Archipelago and the Solomon Islands. The second migration (between 1500 and 1000 BCE) were Austronesian populations. The Austronesian migrants reached the Carolines in the north, the Samoa group in the east, and New Caledonia in the south.

The Fiji Islands are located in the central part of this area (i.e., Melanesia) that was colonized during the second migration. The Melanesian cultural complex, centered on Fiji, provides the foundation for the subsequently developing Polynesian culture, which proliferated into a wide array of local cultures during later migrations. The third migration, starting ca. 200 BCE with the Fiji group, was directed to the east.

The settlements on the Society Islands, with Tahiti as their cultural center, date to the time of the third migration. The fourth migration (starting ca. 300 CE) saw the greatest distances in the Pacific Ocean traversed by migrants from Tahiti. The distant Easter Island was settled around 300. The sea route to the north of Tahiti brought settlers to the Hawaiian Islands. The fifth migration (10th century) involved the exploration of the sea route to the southwest of Tahiti, and migrants reached the islands of New Zealand toward the end of the 10th century.

It has been demonstrated by means of computer simulations of the migratory movements in space and time that the peopling of the Pacific Islands cannot have occurred as the result of accidental drifts. Such accidental drifts would have produced random patterns, with densely populated islands contrasting with widely unpopulated areas. The only explanation for the regularities and the emergence of the dense web of local settlements is that the migrations were carefully planned.

Polynesian demography is the outcome of intentional seafaring. In time, the seafaring endeavors became more frequent and better organized. This was not only the result of farsighted planning of voyages but also the accumulation of knowledge about climatic conditions and the advancement of technological development. Each wave of migrants could profit from the experiences of previous generations to improve their know how of seafaring.

Many of the sea routes that had been explored in the course of colonization were already regularly frequented for trading more than 2,000 years ago. The migrants from the second wave onward who set out on their voyages were agriculturalists, and they transferred their knowledge of food production to newly explored islands. To establish new settlements at a distance from the existing ones, it was necessary to transport plants and livestock (i.e., pigs, fowls, and dogs). To safeguard the continuity of new settlements over generations, the migrant groups had to include both men and women.

The Polynesians did not have any technical devices to facilitate their navigation, and they did not possess systems of notation to draw charts of currents or of star configurations. What they definitely instrumentalized for their orientation at sea was their refined perception of natural phenomena. Over time, a vast amount of specialized knowledge about seafaring accumulated, and this knowledge was transferred, by means of oral memory, from master to disciple in the professional domains of specialized handicraft. The builders of canoes and catamarans profited from the experiences of seafarers to make their vessels apt for maneuvering against high waves and gusty winds. And the settlers who took livestock and provisions on their journeys learned from the stories that earlier migrants told to make improvements for their selection.

A crucial question that has puzzled many scholars is the proportional relationship between one-way voyages (go and no return) and two-way voyages (go and return). Trade and cultural ties can only be kept up under the condition that voyagers explored new terrain for settlement and returned to their base to report about their discoveries. There are areas in Oceania that were settled in one-way voyages and remained isolated. This is true for Easter Island and for New Zealand.

Geographic distance, though, was not a decisive condition for either isolation or permanent contact with other islands. The Hawaiian Islands are located on the extreme northern periphery of Oceania, thousands of miles away from other major island groups. Nevertheless, a trade route existed between Hawaii and Tahiti. An experienced seafarer would certainly explore the chances for a safe return, especially when sailing out into unknown space. One method of safeguarding return was sailing upwind. Computer simulations of far-distant voyages, undertaken by groups of migrants in several canoes, and their chances of two-way success show that the ratio for the loss of canoes and their voyagers was fairly low.

An astounding level of sophistication of craftsmanship is revealed in the Polynesian technology of constructing vessels apt for long-distance sea traveling and in the sailing equipment of those vessels. These vessels were not built by unskilled craftsmen, and they were not built for random travel. They reflect the accumulated know how of many generations of seafaring people. The big canoes for seafaring were outrigged in Micronesia and double hulled in Polynesia, with a length of up to 22 meters. They could sail at eight knots and cover up to 150 nautical miles in a day. The aptness of the canoes for seafaring is manifested in every detail of the technical equipment. In such vessels, provisions could be stored from one to three months. In a month's voyage, these canoes could traverse distances of several thousands of nautical miles.

A major factor used extensively by the Polynesians and essential for successful navigation under natural conditions was the knowledge of the orientation at the position of major stars and star configurations in the night sky. The Polynesian seafarers were well acquainted with the star configurations of the Southern Hemisphere. Practically every major group of islands in Oceania has its own particular star configuration. Using the position of single bright stars and these configurations for navigation was not as easy as looking at the sky, because the constellations, forming sections in the bigger mosaic of visible heavenly bodies, may shift their relative position depending on the seasons. At times, certain configurations are invisible. The seafarers had to learn from the experience of earlier generations to what stars and configurations to orient, and where and when.

Once sea routes had been explored, these were used not only for the transport of settlers and their livestock but also for the transfer of trade goods. The earliest evidence for overseas trading activities comes from Melanesia. The most important merchandise to be traded in early times was obsidian as raw material. In the course of time, a greater variety of goods were traded. Since sea routes for trading had already been explored in Melanesia before the colonization of the Polynesian islands, the history of seaborne trading is older than the Polynesian settlements, and the Melanesian tradition was inherited by the Polynesians.

The evidence documenting the existence of ancient trade routes—for short-distance as well as for long-distance trading—speaks in favor of organized seafaring. Seafaring in terms of frequenting certain sea routes in regular intervals without navigational skills is unthinkable. The simple opening of a trade route requires knowledge about the conditions of two-way voyaging. Seaborne trading grew in magnitude and importance for the interconnection of scattered settlements in Melanesia and throughout Polynesia.

A modern reconstruction of the Hawaiian two-sailed canoe was used in 1976 to make a voyage from Maui in the Hawaiian islands to Tahiti. This voyage was made without the help of modern instruments of navigation, with star configurations and the observation of currents and winds as the only means of orientation. The voyage was successful and may serve as a demonstration to illustrate the possibilities of seafaring over long distances. Another voyage with the same vessel was undertaken from central Polynesia to New Zealand several years later, and this endeavor was also successfully concluded.

Harald Haarmann

Perspective 2: The Polynesians Drifted Accidently into New Territories

The debate over Polynesian migrations is a long-running one. Early European voyagers in the Pacific were perplexed by the existence of people who were obviously culturally related but who inhabited the widely dispersed islands of the Pacific. European navigators were particularly confused, as their own technology had only just allowed them to voyage to these islands, and yet upon landing they seemed to be inevitably confronted by people using Stone Age technology, who had reached these small and widely scattered pieces of land well before them. Among the explanations proposed was that of accidental voyaging.

The islands of Polynesia are generally described as being contained within a triangle, with the corners comprised of New Zealand, Hawaii, and Easter Island. The corners of the triangle are distant from everywhere. Hawaii is approximately 2,000 miles from the North American continent. It is 2,400 miles from Tahiti, and 3,800 miles from New Zealand. Similarly, New Zealand is about 1,300 miles distant from its nearest continent (Australia) and only slightly closer to Fiji, the nearest island group. Easter Island is often described as the most isolated place on Earth, as it is 2,400 miles west of South America and 1,200 miles east of Pitcairn, the nearest island.

The debate over the navigational practices of Polynesians continues into the present because the existence of a recognizably single culture across this area of the world is an astonishing feat. The difficulty of settling Polynesia meant that it was the last region on Earth to receive humans. Voyagers had to set out without being able to see their destination and travel without land in sight for long periods of time. Easter Island was only reached about 1,700 years ago, Hawaii, 1,600 years ago, and New Zealand, possibly as recently as 700 years ago. Whether settlement of this region involved deliberate voyaging or contained elements of chance remains an open question.

The idea of regular deliberate voyaging between the far reaches of Polynesia has its problems. Certainly such voyages were not occurring when Europeans entered the Pacific, and the possible place of chance in Polynesian settlement should not be ignored. While the idea of Polynesian navigation across the vast Pacific is presently fashionable, older ideas about the role of drift voyaging in Pacific settlement deserve consideration. The drift voyage thesis maintains that the Pacific is a difficult environment and that some Polynesian settlement can be explained by voyages of chance when fishing parties or groups of refugees stumbled across islands when blown there by unexpected winds or pushed there by unknown currents.

As Europeans explored the Pacific they were fascinated by the people who lived there, recording observations of them and information gained from talking to them. Part of early European interest in Polynesian navigation was pragmatic; Europeans wanted help in navigating the vast Pacific and they were eager for information that could be gained from Polynesians, Tahitians in particular.

Thus, the famous British navigator Captain James Cook eagerly compared information about the location of Pacific islands with his passenger, the Tahitian navigator Tupaia, who joined Cook's *Endeavour* voyage in 1769. While Tupaia was able to add 80 islands to Cook's charts, he did not impart any information about lands as distant as New Zealand or Hawaii, indicating that voyaging between Tahiti and these places did not occur at that time and had not occurred for a long time, if at all.

This experience of a lack of knowledge of the far corners of Polynesia was not limited to Tahiti, but occurred throughout the islands of central Polynesia. While in Tonga, Cook collected information about other Pacific islands from his hosts. That list included 156 islands known to his Tongan informants, but despite this wealth of knowledge, islands that required open-water journeys of more than 30 miles from Tonga were not included. Thus large and significant island groups in the vicinity but more than 30 miles distant were not among those described to Cook.

It would seem that Tongan navigation covered only limited legs of the ocean, although island hopping along chains of islands meant that Tongans were aware of many islands. However, the information collected by Cook indicated no knowledge in Tonga of the existence of Hawaii, Easter Island, or New Zealand, and again clearly indicated that regular voyages to any of those places did not occur at the time of contact and had not occurred in the recent past.

In this early period of contact there is no evidence that Tahitian navigators visited, or were even aware of, islands as far distant as Hawaii. Deliberate navigation took place, but it occurred only within limited regions, such as within the island groups of Tahiti, Hawaii, and Tonga/Samoa. Travel between these discrete groups, and between island Polynesia and New Zealand, was not observed and seemed not to have occurred at all recently. In Cook's observations the only indication of the existence of more distant islands were Polynesian stories of distant origins, but no voyages to the corners of the Polynesian triangle occurred.

As a result of the lack of any clear evidence of Polynesian long-distance navigation at contact, the first theorist to suggest that Polynesian navigators reached their islands by drift voyages was Cook himself. Cook regularly observed Polynesian canoes during his visits to island Polynesia, and he was impressed by the size, speed, and maneuverability of those that he observed in Tahiti and Tonga. However, Cook argued that Tahitian navigation did not deal with very long voyages and did not have the tools to cope with long periods at sea out of sight of land.

In addition to his observations of the limits of Tupaia's knowledge of Polynesia, Cook also observed evidence of drift voyages occurring across long distances. While in the southern Cook Islands, he observed survivors of a drift voyage from Tahiti. A canoe had been blown off course, and although 15 members of the crew had died, five had survived and landed in Tahiti. Cook argued that such chance events accounted for the settlement of Polynesia, as people in canoes were blown away from known land, and some were fortunate enough to land on previously unknown islands and establish societies. No evidence of more deliberate long-range voyaging was found by Cook.

In 1956, Andrew Sharp published the book *Ancient Voyagers in the Pacific*, challenging the notion that ancient Polynesians navigated the vast expanse of the Pacific easily and often. Sharp did not argue that all long-distance voyages in Polynesia were solely the result of aimless drifting, but he did argue that the notion of repeated navigations between the center and far corners of Polynesia was not clearly supported by tangible evidence.

Sharp's attack on the romantic notion of a vast Pacific highway did not deny the possibility of Polynesian navigation in all its forms, but rather it questioned the regularity and control that Polynesian navigators could exercise over voyages more than about 300 miles in length out of sight of land—especially those thought to connect the corners of the Polynesian triangle to the center.

Sharp raised useful questions that challenged unexamined assumptions about Polynesian long-distance navigation. He argued that the materials available to islanders when building their craft could not construct a craft capable of reliably sailing long distances. In particular, Sharp argued that the rope materials available within Polynesia were vulnerable to strain on long voyages and could not be relied on in rough conditions or when sailing against the wind.

Sharp also questioned the assumption that Polynesian canoe building and navigational technology had declined by the time of Cook's first visit, arguing against the assumption that knowledge had previously existed and somehow been lost. And Sharp was able to produce evidence of well-documented cases of unintentional and unnavigated voyages across the Pacific in historic times.

The most famous intervention in the argument about deliberate voyaging in the settlement of the Pacific involved the Norwegian Thor Heyerdahl. In 1947 Heyerdahl set off across the Pacific on the balsa-wood raft, the *Kon-Tiki*. The raft was towed about 50 miles offshore from Peru and then drifted and sailed across the Pacific before reaching the Tuamotu Archipelago, a distance of approximately 4,300 miles, in 101 days. The voyage seized the public's imagination and demonstrated that, despite the vast size of the Pacific, it was possible to run into land by chance rather than design.

Heyerdahl's expedition was intended to deal with questions of Polynesian origins rather than their means of migration, and it drew on the older idea that, because the prevailing winds in the Pacific blow from the east, people were blown into the Pacific from South America. Heyerdahl's South American origin theory for Polynesians has generally been rejected, but his expedition indicated that long-distance voyages guided by luck rather than design could succeed, and so produced evidence to support the possibility of significant drift voyages in the settlement of the Pacific.

Current theories of the way in which Polynesians settled the Pacific and of long-distance navigation tend to reflect contemporary concerns. At present it is unfashionable to argue that drift played a prominent part in Polynesian settlement of the Pacific. However, the continued process of drift voyages across seemingly impossible distances cannot be dismissed, and the role of accident and drift in the settlement of Polynesia requires consideration. Polynesians were fine mariners and efficient navigators within island groups, but the Pacific is a vast ocean, and Polynesian canoes were certainly at times directed by wind and currents rather than human will.

The colonization of islands throughout the vast area of the Pacific Ocean called Polynesia is widely considered one of humanity's greatest achievements of exploration and navigation. Using well-crafted canoes, the largest of which were over 100 feet long and could carry as many as 100 people, Polynesian seafarers could travel great distances with enough provisions to last several weeks. Whether most of the long-distance travel was the result of planning or was instead the result of unintentional drifting, however, remains a matter of debate.

Claire Brennan

REFERENCES AND FURTHER READING

Barthel, Thomas S. *The Eighth Land: The Polynesian Discovery and Settlement of Easter Island.* Translated by Anneliese Martin. Honolulu: University of Hawai'i Press, 1978.

Bellwood, Peter S. *Man's Conquest of the Pacific: The Prehistory of Southeast Asia and Oceania.* London: Collins, 1978.

Goldman, Irving. *Ancient Polynesian Society.* Chicago: University of Chicago Press, 1970.

Heyerdahl, Thor. *Kon-Tiki Expedition.* London: Allen & Unwin, 1950.

Jennings, Jesse D., ed. *The Prehistory of Polynesia.* Cambridge, MA: Harvard University Press, 1979.

Lewis, David. *From Maui to Cook: The Discovery and Settlement of the Pacific.* Sydney: Doubleday Australia, 1977.

Lewis, David. *The Voyaging Stars: Secrets of the Pacific Island Navigators.* New York: W. W. Norton, 1978.

Lewis, David. *We, the Navigators.* Honolulu: University of Hawai'i Press, 1994.

Oliver, Douglas. *Ancient Tahitian Society.* 3 vols. Honolulu: University of Hawai'i Press, 1974.

Oliver, Douglas. *Oceania: The Native Cultures of Australia and the Pacific Islands.* 2 vols. Honolulu: University of Hawai'i Press, 1988.

Sharp, Andrew. *Ancient Voyagers in the Pacific.* Baltimore: Penguin Books Inc., 1957.

Thorne, Alan, and Robert Raymond. *Man on the Rim: The Peopling of the Pacific.* Sydney and London: Angus & Robertson, 1989.

9 WATER AND ENERGY

OVERVIEW

This chapter reviews the use of water for energy generation and the environmental and other impacts related to hydroelectric dams and oil exploration and drilling in coastal environments. The case studies explore the controversy of hydroelectric dams in Chiapas, Mexico, the role of geothermal energy, and the robustness of environmental impact assessments in hydroelectric dams. The perspectives section examines the regional conflicts around construction of the Grand Ethiopian Renaissance Dam.

Water Is Power

People have been harnessing water for energy for thousands of years. Up to the early 20th century, water mills were used to grind grain. Later, the use of water for energy evolved to the construction of hydroelectric dams, the cooling of nuclear power plants, steam generation through geothermal energy, and the harnessing of ocean tides and waves.

In the last 100 years, hydroelectric power has become a major source of water-based energy in many countries. The word *hydro* comes from the Greek word for water; hydroelectric energy is therefore energy generated from moving water. Hydropower involves damming rivers to create reservoirs and channeling the flow of water to turn turbines for energy. Tidal and wave energy uses the movement of water, i.e., the surge of oceans and the rise and fall of tides to turn turbines to produce energy. Geysers and steam are a source of geothermal energy coming from below the earth and harnessed for energy.

All of these water-based sources, whether old or new, are renewable, low–fossil fuel ways to generate energy. This is a positive aspect, given the problems associated

The Grand Renaissance Dam is located along the Blue Nile and within the Nile Basin, which is shared with Egypt and Sudan. (DigitalGlobe via Getty Images)

with the emission of greenhouse gases from fossil fuel use. However, hydroelectric dams and tidal energy plants can also produce environmental and social impacts, as discussed later.

Turning Watermills

It is believed that the Greeks invented watermills in the first century BCE by placing huge wheels in the middle of rivers, where the current was usually strongest. The concept was simple but effective. The flow of the river current turned the wheel to move simple machinery to grind corn and wheat. Prior to water mills, most people had to pound or mill grain by hand; hence the origin of the word "miller."

Before connecting the mill to the source of waterpower, the river, farmers sometimes used animal power to turn millstones to grind corn. The watermill was an important technological advance in its time. After a while, in addition to grinding grain, the watermill was used for small-scale artisan production, e.g., flattening metal pieces and sawing wood and marble. Larger industries used watermills to produce paper, cotton, oils, and spokes for horse carriages. Early manufacturing in Europe emerged along rivers since industries needed waterpower to run mills. In the Caribbean, plantation owners expanded use of water mills to drive presses to crush sugar cane for sugar production.

China, India, and a number of Islamic countries made widespread use and adaptation of watermill technology. Some users mounted water mills to bridges instead of river banks or river beds to achieve better flow; others attached them to ships moored on rivers or added complex systems of gears and sluices to control water flow, eventually using steam engines to augment water flow to the wheels used for grinding.

Steam Engines Replace Watermills

After the water mill had developed to its highest efficiency and widespread use in the 19th century, it became obsolete in the early 20th century. It was displaced by the steam engine and the increasing availability of cheap and more convenient electrical energy. Many developing countries continued to use working watermills up to the 20th century, but the harnessing of rivers for hydroelectric power and expanding use of fossil fuels eventually displaced watermills in those countries as well. In recent years, a new take on watermills has emerged which does not require the physical construction of large wheels, but rather the placement of turbines in fast-flowing rivers and transport of the resulting energy to an electric grid on land.

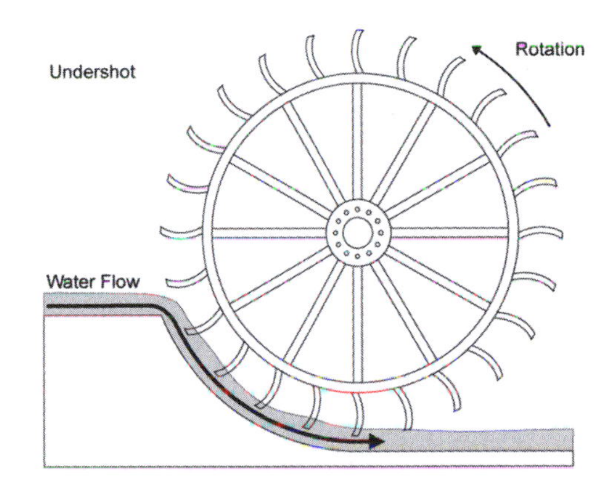

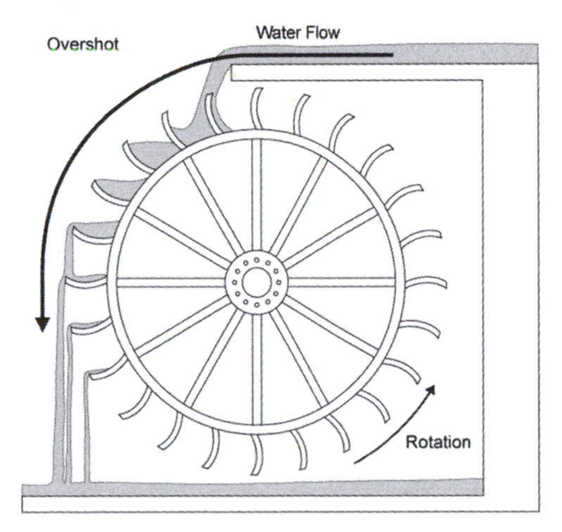

Early waterwheel designs enabled the use of water for energy to mill grain and develop small industries. (Jeff Dixon after an illustration by Robert Karl Koslowsky)

Growth of Hydroelectric Dams

In the last 50 years, numerous countries, especially those with extensive river systems, have intensified use of hydroelectric dams for energy generation, flood control, and water storage for irrigation. How does it work? First, water is stored in a reservoir, usually through damming of a river or lake. This impoundment interrupts the normal flow of the river and stores the impounded water before channeling it through tunnels. The process that regulates this flow is called the intake system. Water flowing from a height and at high speed turns the turbines and moves generators, the machines that produce energy.

An important feature of dam construction is that dam engineers and operators can control the amount of water flow or intake going through the tunnels according to demand for energy and the supply of water from the reservoir. In periods of less energy demand, operators can close some tunnels and reduce the flow of water turning the turbines. In periods of high demand, they can open all tunnels and let out millions of gallons of water to increase energy supply.

Low levels of water in reservoirs negatively affect intake and the functioning of the dams as well as the production of energy. In recent years, drought, rising temperatures, and high evaporation levels have compromised the functioning of many reservoirs such as Lake Mead in the United States, and reduced the capacity of a number of dams to generate energy consistently. There are also concerns about safety hazards and the obsolete nature of many old dams that now have limited life span. It is estimated that about 26,000 federal dams in the United States out of a total of 80,000 pose such hazards or no longer serve their purpose because their reservoirs are too low, or they have degraded watersheds, endangered salmon populations, or produced greenhouse gas emissions that outweigh their benefits (Chouinard 2014). Many are being taken down.

Niagara Falls provide a spectacular border between the U.S. and Canada and an important source of energy. (Camille Gaskin-Reyes)

Construction of hydroelectric dams expanded at the turn of the 20th century; in the period following World War II there was an explosion of dam building, big dam building in particular. Some hydroelectric plants were built close to waterfalls, e.g., Niagara Falls on the U.S./Canada border, to unleash energy from the natural flow of huge volumes of water plunging off the escarpment. In others, levels of reservoirs were raised to create higher downflows. Postwar construction of dams intensified in tandem with a peacetime spurt of worldwide economic development, rising industrialization and globalization, and higher demand for energy, as well as for irrigation for food production.

More recent dams emphasize sustainable aspects of construction. Building techniques have evolved over the years and include safeguards to avoid overflowing of water stored in reservoirs. New dams have a structure, called a spillway, to cope with excessive rainfall, floods, landslides, earthquakes, or other natural disasters, which may cause water to overtop a dam. Spillways also help prevent damage to surrounding communities by sending dam overflow water directly to rivers or to other impoundments below the dam, bypassing tunnels, turbines, and generators. However, this mechanism is not foolproof. In 2010, the Tehri Dam on the Ganges River overflowed due to heavy rains, causing 60 deaths and the evacuation of villages close to the dam (Piper 2014, 135).

Rise of Megadams

In 1950, about 5,000 hydroelectric dams existed worldwide, 10 of which were very large, or megadams. By 2000, there were over 40,000 big dams, including 300 megadams with higher energy capacity (International Rivers). Dam construction rose in the 2000s in at least 140 different countries around the globe—with increasing size and cost. Over 90 percent of the world's major dams were built over the last 40 years, mostly in countries with escalating energy demand, i.e., Turkey, China, Colombia, India, South Korea, and Brazil.

The increase in size and capacity of hydroelectric dams was accompanied by a rise in dam height. For example, the British constructed the Aswan Low Dam on the Nile River in Egypt around 1900, but in the 1960s, this dam was subsequently raised to form the Aswan High Dam (Engineering Timelines).

The Hoover Dam and the Tennessee Valley Authority (TVA) in the 1940s, as well as the large Shasta Dam (on the Sacramento River) and the Grand Coulee Dam (on the Columbia River), all in the United States, are examples of this earlier trend towards larger and higher dams. Modern megadams today are larger-scale, higher-cost structures of over 300 feet tall with reservoirs holding up to 800 million gallons of water (Black 2004, 116). As the size and height of dams has risen, concerns have grown about their social and environmental impacts, as outlined further in this chapter.

Climate Change and Dams

An issue affecting some dams is climate change–related melting of ice caps and glaciers. Too-rapid melting and runoff disrupts the flow of melted ice into rivers and affects the stable volume of water in reservoirs. Combined with high evaporation, the rapid runoff of water reduces reservoir levels even further and undermines the dependability or consistency of electricity generation. Smaller hydroelectric plants—which use the natural speed of river currents and the natural drop in river gradients—have a less invasive impact on river ecosystems. However, they produce less power and suffer more from high fluctuations in energy output due to drought, seasonal rainfall, and climate change.

Dams for Development

Many megadams in developing and transitioning countries in Eastern Europe (after the Cold War) were financed after the creation of the World Bank and regional

Water crashes into the canyon walls during testing of the jet flow gates at the Hoover Dam. The gates control water flow through tunnels that were originally used to divert water around the dam construction site. Today the tunnels act as spillway outlets. (U.S. Department of the Interior, Bureau of Reclamation)

development banks in the 1950s. The World Bank played a leading role in lending billions of dollars to countries to propel economic takeoff and the growing demand for energy.

Funding of dam construction was part of economic development strategies to provide infrastructure and energy for industries, agriculture, services, and telecommunications. Development finance and aid agencies also funded river basin development programs to protect watersheds surrounding dams from deforestation, and construction of water and sanitation projects to address water supply gaps, public health, and irrigation needs.

The rapid expansion of private investors, credit and funding agencies, energy companies, and transnational corporations from the 1960s onward further enabled the spread of hydroelectric dams around the world. Most developing countries lacked financial and technical resources to plan and construct such projects. They thus had to borrow huge sums of money to construct hydroelectric works and to contract engineers, consultants, builders, and construction companies for dam development, increasing their indebtedness.

While dams improve countries' ability to address rising demand for energy and water for irrigation, they have associated social and environmental costs. Reservoir construction often leads to the resettlement of entire communities from watershed areas that are flooded to impound water. More than 90 percent of the world's renewable electricity comes from hydroelectric plants with such reservoirs. This process has in many cases disrupted communities' lives, livelihoods, and cultural traditions, especially in areas with indigenous communities. Examples of such disruptions are found in Canada, India, and Brazil. In addition, flooded forests cause methane release.

Protests against Dams

Local communities and grassroots organizations often claim that dam builders, financiers, and government agencies do not adequately consult with affected communities (particularly indigenous groups), do not obtain informed consent, and do not fairly or adequately compensate affected people for the loss of flooded homes and crops or ancestral, cultural, and religious assets. They also argue that dam proponents do not sufficiently consider alternative sources of energy before giving the green light to costly dam projects. Dam financiers, sponsors, or governments dealing with energy bottlenecks continue to hail the overall economic benefits of energy production and irrigation at large, and point out increasing efforts to consult with and compensate affected people.

Nongovernmental organizations (NGOs), community leaders, and village elders have waged public protest campaigns to stop dam construction in many countries, and in some instances physically brought construction to a halt, even if only temporarily. In many cases, disputes between proponents (usually powerful energy sponsors and government planners) and opponents of dams (local communities and environmental and indigenous groups) have spilled over into the courts, the public media, and the international arena.

Local protestors often ally themselves with international environmental organizations, while industrial and agricultural interests that stand to benefit from dams usually back dam proponents. The controversy on energy alternatives and longer-term energy benefits versus shorter-term, localized social and environmental costs continues to rage. Sections of this chapter discuss specific examples of public mobilization against dams by groups.

Environmental Impacts of Dams

Hydroelectric energy is largely considered a clean, renewable energy source compared to fossil fuels such as coal and oil. However, there are known environmental impacts. In North America, for example, dams have affected the reproductive cycles of salmon, which need to swim upstream to spawn. To mitigate this problem, dam builders construct fish ladders on the sides of rivers and dams to assist salmon to climb upward. Dams also discharge water from mechanical systems at high temperatures, altering dissolved oxygen and nitrogen levels in downriver waters. They inundate land and wildlife habitats, reengineer the flow of rivers, disrupt fluvial or riverine ecosystems, and affect migratory paths of animals. Reservoirs that flood forests release methane.

A fish ladder at Bonneville Dam in Oregon is intended to help fish swim upstream to spawning grounds. (Lori Howard/Dreamstime.com)

Upstream Dams, Downstream Problems

In addition, when a country dams a river upstream and impounds water for a reservoir, it can deprive a downstream region or neighboring country of regular flows of water and create conflicts among areas and bordering states. Such disputes occur regularly and can be mitigated by advance planning and coordination. China's plans to construct multiple dams on the Brahmaputra River before it flows into its neighboring country, India, has raised concerns in India about the potential downstream effects of these dams on its water supply. These and other examples are explored in this chapter.

Similar areas of discontent are occurring in Central Asia between Uzbekistan and its neighbors, Tajikistan and Kyrgyzstan. These two countries contend that they need to build dams upstream because they have few alternatives to generate energy. Uzbekistan, on the other hand, has threatened military reprisals if Tajikistan goes ahead with plans for hydropower projects on the Vakhsh River.

Conflicts abound around Nile waters. In the Nile River basin, Egypt—located downstream from Sudan and Ethiopia—has always maintained a historical right to

Nile waters. Egypt was the first country in the region to impound waters from the Nile River to build the Aswan Low Dam in 1902 and the Aswan High Dam in the 1970s. In 1959 Egypt signed a treaty with Sudan, which established Egypt's acquired rights to billions of gallons of Nile water each year and granted Sudan rights to a significantly smaller amount. At that time, the accord did not specify any share of water reserved for Ethiopia, another upstream country that contributes high volumes of water to the Blue Nile.

Ethiopia, faced with a growing demand for energy and agricultural production, is currently completing the Renaissance Dam, Africa's largest dam upstream on the Nile. It is expected to triple the country's electricity generation and, at a cost of over $4 billion, will flood about 650 square miles of land and displace 3,700 households. Ethiopia has also unveiled plans to build more dams on the Nile and other rivers, prompting Egypt to lobby donors and international lenders against financing these dams.

Another upstream/downstream problem is brewing with Turkey's Southeastern Anatolia Project (abbreviated as "GAP" from the original Turkish). This dam covers nine of Turkey's provinces in the Tigris-Euphrates Basin and is a source of concern for neighboring countries such as Syria, Iraq, and Kurdistan, which share the river basins of these important rivers. The GAP was originally an irrigation project, which the Turkish government upgraded to a full-fledged mega-hydropower project, including irrigation and regulation of river flow.

The GAP dam has caused regional controversy. Turkey's neighboring countries contend that in such an arid area the GAP project would consume almost half of the

A man waters his land above the newly constructed Ataturk Dam in central Turkey. (Lynn Johnson/National Geographic/Getty Images)

flow from the Tigris and Euphrates Rivers and deprive them of water for their own development. Turkey counters that the dam is needed to develop the less-developed eastern part of the country and provide at least 2 percent of Turkey's electricity. The GAP is expected to contain 22 dams, 19 hydroelectric power stations, and an irrigation project of about 1.7 million hectares at a total cost of about 32 billion dollars. Once completed, it is slated to generate 27 billion kilowatts per hour (Sansal).

Opponents of the project claim that the dam would inundate protected areas, cause environmental damage and downstream pollution, displace local Kurdish communities, and disrupt the region's cultural heritage. Some villages have already been flooded; critics anticipate that many historic sites that once lay on the Silk Road will be inundated (Harte 2014). The wider area of the dam is already a political conflict zone; disputes over joint resources of the Tigris and Euphrates could further raise tensions. More details on environmental aspects of Turkey's dam projects are discussed in a later section.

In China, the Philippines, and India, dam construction is also a huge source of conflicts. A controversial megadam in China (the world's largest) called the Three Gorges Dam has sparked huge local and international protests due to the displacement of 1.7 million Chinese and significant environmental impacts (Chellaney 2013, 234). In the Philippines during the 1980s, protests of indigenous communities against a World Bank—funded dam project on the Chico River led to the cancellation of the project and ignited worldwide awareness of the impact of dams on affected peoples. In India, the largest and longest antidam movement was launched against the Sardar Sarovar hydroelectric project in the Narmada Valley. In 1985, local communities started campaigning against the project and its 47,000-mile canal system, causing the World Bank to withdraw from the funding consortium. While these protests delayed construction, they did not stop the dam altogether.

View of the Three Gorges Dam—the world's largest—on the Yangtze River in China. (iStockPhoto)

Dam Problems in Chile

Across Latin America, dam construction is fraught with conflicts. Chile is no exception. Its policy is to address some of its energy demand and acute electricity shortages through dam construction. In the 1980s, private energy companies conceived of a multibillion-dollar project (called HidroAysen) to create five hydroelectric dams on the Baker and Pascua Rivers in Patagonia, a relatively pristine area with stunning glaciers and lakes and thousands of nature visitors yearly.

This proposal sparked outrage in the local area and the country. Here, as in other places, project proponents pointed to Chile's pressing need for energy to fuel economic development and exports and prosperity for all; opponents argued that the strategy had high environmental and social risks for Patagonian communities. They viewed the analyses of environmental and social impacts as flawed and pointed to the lack of community consultation from the outset. They argued that flooding would cause damage to Patagonia's unique landscape and wetlands, disrupt residents' lives, damage local culture, and flood part of a protected national park.

A major concern voiced by locals was that the project would force them to bear the environmental costs, while nonlocal energy interests would reap the economic benefits and profits from selling energy. Other issues raised were related to direct impacts such as construction dust, runoff of contaminants, noise pollution, the disturbance of Patagonian fauna (condors, guanos, and pumas) and social disruption of a small, closely knit community by such a megaproject. Opponents also claimed that the project would destroy the local ecotourism and ranching economy, displace locals, and cause the loss of land and livelihood through flooding of the reservoir area.

Fishing communities highlighted potential damage to fish spawning areas and wetlands in the dam watershed. Environmental NGOs pointed to potential risks of seismic activity in Patagonia and climate change variations in rain or snowmelt patterns that could affect functioning of the reservoir. Since the project also contemplated construction of a 1,200-mile transmission line across virtually untouched areas to transport energy from Patagonia to distant urban, mining, and production centers, there were concerns about further environmental impacts beyond the narrow project area.

For years leading up to this project, Chilean communities had been expressing growing dissatisfaction with hydropower proposals in southern Chile. Private projects to dam the River Bio-Bio, another pristine area in the South, were implemented in the 1990s, generating massive conflicts between developers and indigenous Mapuche communities fighting for land rights. The project on the Baker and Pascua Rivers in Patagonia, approved for design stage in 2011, further ignited this tinderbox of conflicts in the area.

Water Privatization Policies in Chile

Much of the citizen uproar in Chile over the Patagonian dams had to do with water privatization policies of the previous military regime of 1973, which laid the foundation for private control of water use and the emphasis on dams. Chile's 1981 Water Code had a strong private sector approach to water control, moving the concept far

Located in Argentina and Chile, the diverse Patagonian geographical region comprises the southernmost portion of the Andes mountains to the west and south, and low plains to the east. Due to the presence of rivers emerging from the highlands in sharp drops, the area has become an important but controversial source of hydroelectric energy. (Shutterstock)

away from the United Nations' definition of water as a basic human right. It viewed water as a marketable commodity that could be traded or leased by private concerns; opponents felt that this approach undermined their rights to water as a public good. Chile's Electricity Law of 1999 supported trading of water rights for nonconsumptive water use such as hydropower, thus opening the door further for profit-oriented interests in water projects such as those in Patagonia.

National groups opposing the Patagonia project argued that the possible use of alternative energy sources such as solar, geothermal, wind, and tidal wave power had been ignored in the conception of the project. They argued that the greatest demand for energy was located in the capital, Santiago, and mining centers further north in the country, areas close to Chile's Atacama Desert, a potential source of solar and wind energy. They also contended that the lengthy Chilean Pacific coast should be explored for possible development of tidal energy to help meet Chile's needs (Bodzin 2011).

Project proponents, on the other hand, counteracted that Chile desperately needed energy, and that hydropower was a clean, renewable energy resource and a good substitute for damaging fossil fuel use in the country. This position, however, ignored anticipated environmental impacts in Patagonia and the fact that dams are not emissions-free due to potential decomposition of flooded forests and the release of methane, a greenhouse gas.

In 2014, a new government of Chile reviewed the proposal and the controversies surrounding it, finally shelving the Baker/Pascua dam development due to environmental concerns. This ruling discontinued dam development in this part of Patagonia, and gave Chile an opportunity to reset its energy and environmental policies, and rethink water resource use in the interest of long-term sustainability.

Brazil Dam Building

In Brazil, big dam building has been an essential part of generating electricity to fuel the country's development strategy and its industrialization and commodity export economy. Brazil has built the most dams of any country in Latin America and was ranked among the top ten countries of the world in large dam construction in the 20th century. Between 1965 and 1985 alone, the World Bank funded projects for a value of 3 billion dollars in Brazil's energy sector (Khagram 2004, 143).

The increased construction of Brazilian dams in areas close to indigenous communities also fueled local and international protests against anticipated social and environmental impacts. The construction of megadams such as Itaipu (jointly with neighboring Paraguay), Tucurui, and Balbina led to community opposition and fierce protests.

Citizen outcry continued with the Sobradinho Dam in the São Francisco Valley, which displaced over 70,000 people and inundated more than 1,544 square miles of land, and with the Itaparica Dam, which displaced about 44,000 people in the watershed area (Khagram 2004, 145). Another dam, the Belo Monte Dam, received public criticism for its 14-billion-dollar price tag and potential social and environmental impacts, including the flooding of forests and potential methane gas release (Altamira 2013).

Broad-based protests eventually led to the improvement of environmental standards for dam construction in Brazil and the practice of public consultations with affected communities. These changes were also, in part, a response to international lenders' more stringent requirements for the funding of dams. In spite of this progress, mobilization campaigns against dam construction have continued in Brazil. As in many other countries, protestors try to force hydroelectric dam lenders and sponsors to halt construction outright, or at least to modify design and construction rules to minimize social and environmental impacts on local and indigenous groups.

Water for Cooling of Nuclear Reactors

Water is also important for the cooling of nuclear power reactors, am additional source of energy. The entire nuclear fuel production cycle is water-intensive. For example, the Indian Point nuclear power plant in New York State extracts millions of gallons of water per day from the Hudson River. Areas of concern relate to hot water outflows pumped back into the river, which can potentially damage plant life and fisheries. The location of nuclear power stations in close proximity to water poses problems in natural disasters. The 2011 reactor meltdown at the Japanese Fukushima

Daiichi plant after a major earthquake and related tsunami demonstrated this vulnerability.

Water for Solar Energy

Solar energy generation using concentrated solar power plants requires large quantities of water for cooling, since they focus sunlight on water, generating steam to power turbines. In Arizona alone, solar panels use about 53 billion gallons of water a year (Cain 2010). However, given water shortages in drier, sun-rich areas, there is a push to use solar photovoltaic power that converts sunlight directly to energy without needing cooling water.

Water and Fracking

Shale oil and gas extraction can be achieved through a fracking or hydraulic fracturing process. This involves sinking drills into open rock spaces two miles deep, turning them to move horizontally through rock, and then shooting water into cracks at high pressure to pulverize or fracture the rock and extract the shale oil and gas. It takes about four barrels of freshwater to produce about one barrel of oil. In the North Dakota shale oil boom, the demand for water has been burdening aquifers. There has been a steep rise in applications for permits to tap wells for fracking, causing concern in other water use areas such as farming, ranching, and fisheries. Demand for water in North Dakota is expected to increase to 1 billion gallons per year as long as the boom continues (Kellman and Schneider 2010).

Water for Tidal and Wave Power

A newer source of water-based energy is tidal and wave power. The harnessing of tidal power is still in its infancy worldwide. It is a good source of clean, non-fossil-fuel-producing energy that capitalizes on the surge of ocean waters during the rise and fall of tides. Differences in intertidal height and water volumes create opportunities to deploy turbines and generators and convert tidal energy into electricity. Tidal plants do not emit greenhouse gases, but there are concerns that they could produce some environmental impacts on marine organisms and mammals. Conflicts may arise through competition of such plants with fishing, shipping, boating, and other activities in the same marine space.

The technology of harnessing tidal energy developed during the 20th century. The first plant was located in France. In North America, the Bay of Fundy tidal plant on Canada's east coast is more well known. This plant has operated for a number of years, and there are plans for its upgrade. It takes advantage of a huge differential between the heights of low and high tides on Fundy Bay to turn turbines. Other plants in operation or development are the Sihwa Lake Tidal Power Station in South Korea, the Tidal Lagoon Swansea Bay in the United Kingdom, and the MeyGen tidal energy project in Scotland. Further projects are planned for the southern coast of Australia (power-technology.com 2014).

Wave-generated energy within the oceans is expected to rise with ongoing research and development of innovative technologies. Wave farms use wave movement patterns to harness energy through the construction of snake- or carpet-like energy converters anchored offshore in the oceans. There are proposals to use such converters to also power desalination of ocean water as a means of producing freshwater. A 1.5-megawatt wave energy park is already in operation two-and-a-half miles away from the Oregon coast. It uses floating buoys that move up and down coasts with ocean swells and drive electrical generators to transmit electricity to shore via subsea cables.

Research is ongoing into the use of floating wave-powered pumps to push high-pressured ocean water through underwater pipelines and move on-shore turbines. Such projects under study or implementation are likely to change the way people look at the oceans' potential for energy production and freshwater generation, but they require further environmental impact analyses.

Camille Gaskin-Reyes

REFERENCES AND FURTHER READING

Altamira. "The Rights and Wrongs of Belo Monte." *The Economist*, May 4, 2013. http://www.economist.com/news/americas/21577073-having-spent-heavily-make-worlds-third-biggest-hydroelectric-project-greener-brazil.

Barrionuevo, Alexei. "Chilean Court Blocks Plan for Patagonia Dam Project." *New York Times*, June 20, 2011. http://www.nytimes.com/2011/06/21/world/americas/21chile.html?_r=0.

Barrionuevo, Alexei. "Plan for Hydroelectric Dam in Patagonia Outrages Chileans." *New York Times*, June 16, 2011. http://www.nytimes.com/2011/06/17/world/americas/17chile.html.

Black, Maggie. *The No-Nonsense Guide to Water*. Oxford, UK: New Internationalist Publications, 2004.

Bodzin, Steven. "Chileans Protest Government Approval of Five Patagonia Dams." *Christian Science Monitor*, May 10, 2011. http://www.csmonitor.com/World/Americas/2011/0510/Chileans-protest-government-approval-of-five-Patagonia-dams.

Cain, Nicholas L. "In Solar Power Lies Path to Reducing Water Use for Energy." *Circle of Blue*. August 31, 2010. http://www.circleofblue.org/2010/world/in-solar-power-lies-path-to-reducing-water-use-for-energy/.

CBC News Nova Scotia. "Bay of Fundy Tidal Power Gets 2 New Entrants and $4M." March 28, 2014. http://www.cbc.ca/news/canada/nova-scotia/bay-of-fundy-tidal-power-gets-2-new-entrants-and-4m-1.2590233.

CBC News North. "Yukon First Nation Leaders Worried About Impact of Hydro-Dams." January 30, 2015. http://www.cbc.ca/news/canada/north/yukon-first-nation-leaders-worried-about-impact-of-hydro-dams-1.2937666.

Chellaney, Brahma. *Water, Peace, and War: Confronting the Global Water Crisis*. Lanham, MD: Rowman & Littlefield, 2013.

Chouinard, Yvon. "Tear Down 'Deadbeat' Dams." *New York Times*, May 7, 2014. http://www.nytimes.com/2014/05/08/opinion/tear-down-deadbeat-dams.html.

Engineering Timelines. "Benjamin Baker." http://www.engineering-timelines.com/who/Baker_B/bakerBenjamin9.asp.

Handwerk, Brian. "Five Striking Concepts for Harnessing the Sea's Power." *National Geographic*, February 21, 2014. http://news.nationalgeographic.com/news/energy/2014/02/140220-five-striking-wave-and-tidal-energy-concepts/.

Harte, Julia. "New Dam in Turkey Threatens to Flood Ancient City and Archaeological Sites." *National Geographic*, February 21, 2014. http://news.nationalgeographic.com/news/2014/02/140221-tigris-river-dam-hasankeyf-turkey-iraq-water/.

International Rivers. "Questions and Answers about Large Dams." https://www.internationalrivers.org/questions-and-answers-about-large-dams.

Kellman, Steve, and Keith Schneider. "Water Demand Is Flash Point in Dakota Oil Boom." *Circle of Blue*. September 15, 2010. http://www.circleofblue.org/2010/world/scarce-water-is-no-limit-yet-to-north-dakota-oil-shale-boom/.

Khagram, Sanjeev. *Dams and Development*. Ithaca, NY: Cornell University Press, 2004.

Piper, Karen. *The Price of Thirst: Global Water Inequality and the Coming Chaos*. Minneapolis: University of Minnesota Press, 2014.

Ponce, Roberto D., Felipe Vasquez, Alejandra Stehr, Patrick Debels, and Carlos Orihuela. "Estimating the Economic Value of Landscape Losses Due to Flooding by Hydropower Plants in the Chilean Patagonia." *Water Resources Management* 25, no. 10 (2011): 2449-2466.

power-technology.com. "Tidal Giants—The World's Five Biggest Tidal Plants." April 11, 2014. http://www.power-technology.com/features/featuretidal-giants---the-worlds-five-biggest-tidal-power-plants-4211218/.

Rothfeder, Jeffrey. *Every Drop for Sale: Our Desperate Battle Over Water in a World about to Run Out*. New York: Putnam, 2001.

Sansal, Burak. "GAP Project." All about Turkey. http://www.allaboutturkey.com/gap.htm.

Shiva, Vandana. *Water Wars: Privatization, Pollution and Profit*. Cambridge, MA: South End Press, 2002.

Vince, Gaia. "Dams for Patagonia." *Science* 329 (July 23, 2010): 382-385. www.sciencemag.org.

Willms & Shier Environment Aboriginal Energy Law. "First Nation Challenges to Hydroelectric Development, A Tale of Two Provinces." November 25, 2014. Accessed at http://www.willmsshier.com/docs/default-source/articles/2014-11-25_first-nation-challenges-to-hydroelectric-development_article_julie-abouchar_nicole-petersen-(energy-law_aboriginal-law_first-nations_hydroelectric_site-c).pdf?sfvrsn=2.

World Bank. http://www.worldbank.org.

World Bank. "Resettlement and Development: The Bankwide Review of Projects Involving Involuntary Resettlement 1986–1993." Washington, DC: World Bank Publications. http://documents.worldbank.org/curated/en/1996/03/696707/resettlement-development-bankwide-review-projects-involving-involuntary-resettlement-1986-1993.

CASE STUDIES

The first case study in this section highlights the controversy of hydroelectric dam building in Chiapas, one of the poorest states in Mexico, with its accompanying problems of social and cultural displacement and persistent inequities. The second case study provides details on the development of geothermal energy and its benefits as a clean source of energy using steam, and the third case study presents a juxtaposition of theory and practice of environmental impact assessment methodologies and their application in projects on the ground.

Case Study 1: The Clean Energy Poverty Market: The Case of Chiapas, Mexico

Chiapas, a southern state of Mexico, is home to the country's best flora and fauna and is considered an ecological wonder. It contains a historical heritage of impressive

Mayan sites and Mexico's largest indigenous-speaking communities. The main crops for export are coffee and maize. Chiapas is naturally resource-rich in water as a result of heavy rainfall. Due to the abundance of rainfall and rivers, Chiapas produces 50 percent of Mexico's energy. Hydroelectric dams are built along its two major rivers, the Usumacinta and the Grijalva. However, despite its resources and agricultural production, Chiapas remains one of the poorest states in Mexico in terms of per capita income.

Water-Rich Chiapas

However, even with so much water in Chiapas, access to potable water for daily consumption and crops is difficult for indigenous communities. This problem is particularly pervasive for indigenous communities living downstream in the Grijalva River basin, home to four hydroelectric dams: La Angostura (Belisario Dominguez), Chicoasén (Manuel M. Torres), Malpaso (Netzahualcóyotl), and Peñitas (Ángel Albino Corzo).

Hydroelectric Dams

Hydroelectric dams are considered important infrastructural works to help improve people's daily lives, particularly in developing countries. Dams are also responsible for job creation, meeting energy and water needs, irrigation control for agricultural production, and flood management. However, communities living in the catchment area are greatly impacted by hydroelectric dams and the regulation of water distribution, and are subject to vested interests and political disputes. These political disputes and differences in power are referred to as hydro-hegemony, i.e., control of water resources in river basins as the result of asymmetrical power relationships between the state and poor communities living in the catchment (Zeitoun and Warner 2006).

The social and environmental problems associated with hydroelectric dams pose threats to the goal of achieving sustainable development. Water that remains sedentary in the reservoir areas of hydroelectric dams can release methane, a greenhouse gas. Mercury can also be a by-product in the water. Hydroelectric dams also disrupt aquatic life and change migratory patterns for wildlife.

As in the case of Chiapas, communities living in the catchment area are the most affected by dams, allocation and distribution rights to irrigation, and access to clean water. Nonetheless, hydroelectric dams in Chiapas have been a useful source of clean, low-cost energy for commercial ventures. In 2013, the Mexican government initiated an energy reform plan, which included the proposal to construct five new dams in Chiapas. Protests of local indigenous communities in 2010 against the construction of this second stage of dam construction, Chicoasén II, were not taken into account.

Conflicts in Chiapas

In 2010, the Mexican Federal Electricity Commission (CFE) announced plans for the Chicoasén II hydroelectric project. It had the goals of providing inexpensive clean energy for the rural indigenous communities of Chicoasén, a municipality in Chiapas.

Local residents protested that the CFE had yet to repay the resettlement costs incurred from the first Chicoasén hydroelectric dam, in 1974. They complained that they had been displaced by the first project and still continued to pay high rates of electricity. They also made the point that in some cases they had not received any electricity from the original project, and that their access to clean, potable water had become extremely precarious.

The social implications of hydroelectric dams are unfortunately not a new phenomenon for indigenous communities in Chiapas. Looking back, experience shows that the construction of La Angostura hydroelectric dam in the 1970s adversely affected a land cooperative, called Venustiano Carranza. There, the indigenous farming community protested against landowners and the state due to flooding of their land during the construction of La Angostura.

At that time, remaining parcels of indigenous farmland were heavily guarded by federal troops and transferred to private landowners with political connections to the state (Harvey 1998). In response, indigenous people continued to organize protests, despite repression, imprisonment, and other abuses administered by the state, military, and local elites (Harvey 1998).

The case of Chiapas suggests that communities living in the catchment and surrounding areas of the dams bore adverse social and environmental impacts of dams. Yet these communities did not seem to be in a position to acquire the benefits of low energy costs in the same way as commercial enterprises; their access to potable water was diminished.

It seems that when lessons are not learned, the mistakes of the past are doomed to be repeated. Despite the negative experience and impacts of the dam, the Mexican government opted in 2013 under its energy reform program to build five new hydroelectric dams in Chiapas—including Chicoasén II.

In this regard, it would appear that serious thought has to be given to reconsidering the model of planning, designing, and implementing hydroelectric dams. At minimum, it has to be explored if alternative renewable means of energy production—beyond hydroelectric dams in socially and environmentally vulnerable regions—can be used to bring clean energy to consumers and commercial businesses without the associated negative impacts and community opposition.

Bernadette Hobson

REFERENCES AND FURTHER READING

Duflo, Esther, and Rohini Pande. "Dams." *Quarterly Journal of Economics* 122 no. 2 (2007): 601–646. doi: 10.1162/qjec.122.2.601.

Harvey, Neil. *The Chiapas Rebellion: The Struggle for Land and Democracy*. Durham, NC: Duke University Press, 1998.

Hobson, Bernadette. "Shared Development? A Case Study of How the Nixon Shock Impacted Small-Scale Farmers in Chiapas, México." *McNair Scholars Journal*, 2013.

World Commission on Dams. "Dams and Development: A New Framework for Decision-Making." Earthscan Publications, 2000. http://www.unep.org/dams/WCD/report/WCD_DAMS%20report.pdf.

Zeitoun, Mark, and Jeroen Warner. "Hydro-hegemony—A Framework for Analysis of Trans-Boundary Water Conflicts." *Water Policy* 8 (2006): 435–460. doi: 10.2166/wp.2006.054.

Case Study 2: Geothermal Energy

The word "geothermal" comes from Greek words *geo* (earth) and *therme* (heat). Geothermal energy is a renewable energy source because heated material located in the earth's interior is constantly being produced. It is extracted from such sources as steam, hot water, and hot dry rocks, and it can be used as power for the direct heating of buildings or for generating electricity. Geothermal energy works only where magma (melted rock) exists close to the earth's surface due to volcanic activity.

History of Use

The first use of geothermal resources in North America took place more than 10,000 years ago. According to historians, early humans used the minerals in hot springs as a source of healing. Some historians believe that Native Americans in the United States settled in many locations where hot springs were available.

In 1892, in Boise, Idaho, the world's first district heating system for the town was installed. Water was piped from nearby hot springs to town buildings and homes. Within a few years, the geothermal system served many homes and several downtown businesses. Even today, there are several district geothermal heating systems in Boise that provide heat to homes and businesses.

In 1921 the United States' first geothermal power plant went into operation at The Geysers, an area north of San Francisco. In that year, the geothermal plant produced 250 kilowatts of energy. This was enough power to provide electricity to light the local Geyser Resort Hotel and other buildings and street lights. But, according to the Department of Energy, the geothermal plant was not competitive with other energy sources of power and it was soon disused.

In 1948 Professor Carl Nielsen of Ohio State University developed the first geothermal heat pump for use at his residence. During the same time, J. D. Krocker, an engineer in Portland, Oregon, installed a geothermal heat pump to heat a local commercial building.

The Geysers made a comeback in the 1960s, becoming the country's first large-scale geothermal electricity-generating plant. The power plant's first turbine produced 11 megawatts of power.

In 1970 the Geothermal Resources Council was formed to encourage the development of geothermal resources worldwide. In 1997 the Lake County Southeast Geysers Effluent Pipeline project was the first of its kind to deliver treated water to The Geysers every day to be used as a geothermal resource. In 2000 The Geysers celebrated 40 years in operation with some 350 steam wells and about 80 miles of pipelines. In 2007, Calpine, a major U.S. power company, launched a $200 million program to enhance geothermal production by up to 80 megawatts.

Today, geothermal energy is the fourth-largest source of renewable energy in the United States. As of the end of 2012, the United States had about 3,180 megawatts of geothermal electricity connected to the grid, less than 1 percent of total U.S. electricity production and about 4 percent of renewable-based electricity. The United States continues to produce more geothermal electricity than any other country, representing about 30 percent of the world's total. According to the U.S. Energy Information Agency, geothermal has the potential to provide the United States with 49,000 megawatts by 2030.

Harnessing Geothermal Energy

If you are a geothermal engineer, the first place to start in your quest for geothermal resources is to look at a map showing the Pacific Ring of Fire. The Pacific Ring of Fire (or sometimes just Ring of Fire) is an area in the Pacific Ocean where large numbers of earthquakes and volcanic eruptions occur. Some of the countries along or near the Ring of Fire include the western United States (including Alaska), New Zealand, Indonesia, Philippines, Japan, and Russia. The Ring of Fire is associated with oceanic trenches, volcanic belts (earthquake areas, too), and with plate movements. In fact, the Pacific Ring of Fire includes a long list of hundreds of active and dormant volcanoes and a number of major earthquakes.

Geothermal energy can be harnessed from Earth's natural heat where there are active volcanoes and even inactive volcanoes that still produce interior heat. The internal fluids of the volcanoes can range from low-temperature to high-temperature heat. Steam from high-temperature fluids is powerful enough to drive turbines and generate electrical power. Lower temperature geothermal fluids can also be used to provide heat sources for homes, industrial uses, greenhouses, and hot or warm springs at resort spas.

In geothermal power plants, steam, heat, or hot water from geothermal reservoirs provides the power that rotates the turbine generators to produce electricity. Any excess geothermal water is then returned into the geothermal reservoir to be reheated naturally and recycled.

There are several commercial types of geothermal power plants, according to the Department of Energy, including the dry steam power plant, the binary cycle power plant, the single-flash steam power plant, and the double-flash power plant. There is also much interest in using hot dry rock systems, or enhanced geothermal systems, as a geothermal source of energy.

Benefits of Geothermal Energy

Geothermal energy has proven to be extremely reliable and flexible. Hydrothermal electric power plants operate very consistently approximately 97 percent of the time. That means that they are not shut down too often for maintenance.

Geothermal power plants, like wind and solar power plants, do not have to burn fossil fuels to manufacture steam to turn the turbines. Therefore, sulfur dioxide and nitrogen oxide emissions are much lower than at fossil power plants. Also, the land area needed for a geothermal power plant is smaller than that needed by most other types of power plants. Geothermal power plants are designed to operate 24 hours a day, all year. A geothermal power plant is built directly on top of its fuel source. There are no interruptions of power generation as a result of weather or natural disasters.

John F. Mongillo

REFERENCES AND FURTHER READING

Mongillo, John F. *A Student Guide to Energy*. Santa Barbara, CA: Greenwood, 2011.

Case Study 3: Environmental Impact Assessments: Theory and Practice and the Ilisu Dam and HEPP Hydroelectric Project in Turkey

Hydroelectric dams, particularly megadams, require environmental assessments (EA) to ensure that they are well designed. Dam planners must demonstrate they have taken precautions to mitigate social and environmental impacts. When such projects are submitted to development banks or private sector investors for financing, they must comply with stringent criteria to ensure that the construction is technically viable, and socially and environmentally sound. They also have to show that overall project benefits outweigh the social and environmental risks of dam and reservoir construction.

This case study explores the nature, scope, process, and content of such an environmental assessment (EA) and the in-depth analyses needed to comply with internally accepted standards. To illustrate the practical application of an EA, the case study uses a concrete example of an independent evaluation of a dam in Turkey (in the area of the GAP project). This independent evaluation uses World Bank environmental appraisal policies and standards as a benchmark to appraise the quality of the original analytical work on the environmental assessment of the dam project.

The Theory

The breadth, depth, and complexity of environmental assessments depend on the size and scale of a hydroelectric dam, its physical location and potential damage to the ecosystems around the proposed dam site. The scope of the upfront analytical work is also related to social and cultural aspects of the project, any impacts on local and indigenous groups, and potential displacement and relocation of communities in reservoir and flood areas.

Preliminary Screening

A preliminary environmental screening of the project determines the level of detail needed for an environmental assessment. This screening is necessary to classify the proposed hydroelectric dam into categories ranging from moderate to severe impact depending on the type and size of dam, location, and possible impacts. This initial scanning exercise gives first insight into the project and uncovers how risky, contentious, or potentially fraught with social, environmental, and regional conflicts it is likely to be.

The first technical scanning exercise, once completed, determines the type of further in-depth social and environmental analyses required to ensure strict adherence to social and environmental protection regulations. The findings of these detailed studies determine whether the project is worth pursuing or not, if these negative impacts can be adequately mitigated or not, and, ultimately, if the rewards or benefits of the entire project are estimated to be greater than the risk to the environment or to people's livelihoods.

Detailed Social and Environmental Analysis

A full-fledged social and environmental assessment involves the following:

1. A study of the geographic, environmental, social, and economic context of the proposed project, and estimates of the economic and financial costs of the investments for the hydroelectric dam. These estimates cover the anticipated costs associated with the reservoir, dam, turbines, pipes, access roads, power plant, transmission lines, water and wastewater infrastructure, and housing and construction equipment.
2. Stakeholder consultations with affected communities to determine their views and obtain their informed consent, including detailed studies on the extent to which residents' lands, fields and villages, and cultural assets would be flooded, and any impacts on indigenous peoples.
3. In-depth valuation of potential losses and required compensation measures, including resettlement plans and community rebuilding programs in the new location. The gold standard is that displaced communities should enjoy a better standard of living after relocation.
4. The collection of ecological and social baseline data, including an analysis of physical, biological, and socioeconomic conditions in the dam area before the beginning of the project.
5. Evaluation of the project's environmental impacts, including detailed reviews of possible destruction or potential changes in flora and fauna, fish stocks, and river and water flows, as well as any likely positive environmental impacts.
6. In-depth analyses of possible alternatives to the proposed project site or a comparison of different options that could generate the same quantity of energy; and the review of alternate construction design, technology, and operating rules, including the possibility of a "no project" option if environmental and social impacts are judged to be too severe.

Final Determination

Once the above-mentioned information is compiled, project developers, designers, and sponsors, as well as any independent review or audit groups as needed, would use the analyses to make a final determination on risks versus benefits of the project and pronounce a green or red light for the project.

If the environmental and social risks are deemed to be too high—if the ecosystems are deemed to be too fragile, if there are serious political, ethnic, or social conflicts, or if construction costs are too costly—then decisions may be taken to postpone, restructure, stop, or reposition the project.

If the project is given the green light to proceed to approval (if benefits are considered to be higher than costs), then ongoing stakeholder consultation activities and social and environmental management or mitigation plans need to be prepared, monitored, and implemented (with estimation of associated financial costs) throughout dam construction to address any previously identified negative impacts.

The setting up of monitoring and evaluation systems is essential to oversee and ensure the least possible disruption of ecosystems and minimum damage to the livelihoods and welfare of residents in the project area. In theory, the gold standard of dam design is to reduce pollution and greenhouse gas emissions as much as possible, minimize risks to local communities, and safeguard and/or improve the living standards of the affected population.

Experience shows that practice often deviates from theory. Many megadams have had inadequate environmental appraisal studies which gave insufficient attention to negative or often-devastating impacts on regional ecosystems and displaced populations. One key aspect of hydroelectric dam design and construction is the obligation of project planners and sponsors to consult with and inform, i.e., fully disclose, the project's impacts to affected groups prior to construction and monitor them during construction, but this requirement is not always followed.

The Practice

The social and environmental assessment requirements outlined above provide a solid technical, social, and environmental framework for analyzing and making decisions on complex and potentially harmful hydroelectric projects, if followed. The reality is that the growing demand for energy in many countries (especially in developing countries), combined with strong political, investor, energy, and construction interests behind dam projects, usually push hydroelectric dams at a very fast pace. Too often this means that preliminary screenings and environmental analyses are rushed or flawed. Construction plans and schedules often proceed faster than completion of the environmental and social advance work and implementation of required safeguards and measures to protect ecosystems and communities.

The following example illustrates this point. It outlines the findings of an independent evaluation of environmental studies conducted by project designers of the Ilisu Dam in Turkey. The studies are judged by the independent evaluation to contain a number of inadequacies. Some of these reported flaws are consistent with the findings of an in-depth study of the World Commission on Dams (WCD) in 2000 on the experience of thousands of large dams. The WCD study is also summarized in the annotated document section of this chapter. It is worth noting that recommendations of the WCD report were integrated into World Bank operational standards for environmental and social analyses of hydroprojects. The independent evaluator used the World Bank and other benchmarks to assess the quality of the Ilisu Dam studies.

Environmental Appraisal and Independent Evaluation of the Ilisu Dam and HEPP Project in Turkey

The construction of the hydroelectric dam in southeastern Turkey (called the Ilisu Dam and HEPP project, part of the wider GAP project area) involved the damming of the Tigris (or Dicle River, in Turkish) and flooding of a reservoir to create the Ilisu Dam for hydropower. The design of this dam entailed detailed environmental analyses for submission to international sponsors and financiers of the project in compliance with required standards.

Independent Evaluation

The independent review of the project's environmental assessment (Derneği 2006) pointed out shortcomings in the depth and breadth of the technical and environmental analyses conducted by the project developers prior to the project. It indicated gaps in the databases that supported the assessment, and inadequacies of the proposed mitigation plans.

The evaluation concluded that the assessment of the project was incomplete and did not cover all the elements needed—as outlined above in the section on theory—to enable a complete and comprehensive appraisal of the environmental and sociocultural impacts of the project. The independent analysis used as a benchmark key operational policies and state-of-the-art guidelines of the World Bank such as OP-4.04 on Natural Habitats, OP-401 on Environmental Assessment, OP-412 on Involuntary Resettlement, OP-411 on Cultural Property, and the World Bank Environmental Assessment Sourcebook.

The independent evaluation's findings indicate that the Ilisu project is likely to have major environmental impacts and lead to severe conversion or degradation of critical natural habitats. It argued that these potential impacts were not adequately covered in the original analyses. The impacts, judged as inadequately documented in the project's EA, are associated with the negative effects of dam construction and the flooding of the Dicle River valley, a unique and critical riverine and canyon ecosystem. The area is said to be irreplaceable once flooded, since it is a zone of high biodiversity, and contains rare and endangered bird species.

The evaluation mentioned serious gaps in baseline information, underaccounting and poor assessment of environmental risks related to the loss of biodiversity in the Dicle watershed area. It maintained that proposed environmental mitigation measures would not adequately compensate for such losses, that the analysis of project alternatives was weak, and the institutional capacity of project managers to coordinate environmental and social protection measures was incomplete.

In short, the environmental assessment was deemed to be less comprehensive than required, and an inadequate response to the potential reduction of key bird populations affected by the loss of 105 miles and flooding of the valley. The evaluation opined that there was little evidence of the baseline situation of the Dicle Valley before the project, and incomplete proposals to minimize the anticipated impact on reptiles, mammals, and birds.

It pointed out that a potential revision of the project's location was not considered, and concluded that the Environmental Action Plan (EAP) of the project did not comply with internationally recognized standards. It also pronounced that the environmental mitigation measures as proposed would be inadequate. This evaluation revealed the strong contrast between theory and practice. This situation of the design studies in the Ilisu Dam project was similar to the experiences of many large dams in the WCD report.

Camille Gaskin-Reyes

REFERENCES AND FURTHER READING

Derneği, Doğa. "Review of the Environmental Impact Assessment Report Submitted for the Ilisu Dam and Hydro-electric Power Project." 2006. Ankara, Turkey. http://www2.weed-online.org/uploads/DD_Review_IlisuEIAR.pdf.

World Bank. "Environmental Assessment Sourcebook and Updates." http://documents.worldbank.org/curated/en/1996/01/1047426/environmental-assessment-sourcebook-updates-1995-1996.

World Bank Operational Manual. "OP 4.01, Annex B—Content of an Environmental Assessment Report for a Category A Project." http://web.worldbank.org/WBSITE/EXTERNAL/PROJECTS/EXTPOLICIES/EXTOPMANUAL/0,,contentMDK:20065951~menuPK:64701637~pagePK:64709096~piPK:64709108~theSitePK:502184,00.html.

World Commission on Dams (WCD). "Dams and Development: A New Framework for Decision-Making. The Report of the World Commission on Dams." Earthscan Publications, 2000. http://www.unep.org/dams/WCD/report/WCD_DAMS%20report.pdf.

ANNOTATED DOCUMENT

The 2000 Report of the World Commission on Dams: "Dams and Development: A New Framework for Decision-Making"

Background

The World Commission on Dams (WCD) is a global multistakeholder entity, set up in 1998 by the World Bank and the World Conservation Union (IUCN) to respond to rising protests against large and megadam projects across the world. The Commission's mandate had a main focus: assess the development effectiveness of large dams and develop internationally acceptable standards and recommendations for the planning, construction, operation, and maintenance of such dams to ensure better protection of ecosystems, the rights of affected people, and more equitable distribution of benefits.

Preparation of the WCD report spanned two years. It commissioned 130 technical papers, and consulted with 1,400 stakeholders across the world. WCD also reviewed the specific performance of seven dams, conducted detailed studies of dams in three countries, assessed 125 dams in less detail, and reviewed 950 submissions from experts and the public.

At the end of the study period, the WCD review had covered experiences from 1,000 dams in 79 countries worldwide. The wide range of inputs and studies provided in-depth information on the adequacy of dam design, recommendations for improved construction and operation, and findings, key lessons learned, and proposed measures to reduce negative impacts. Out of these findings, many development agencies and dam builders such as the World Bank and other financing agents have adapted dam construction and operation rules and regulations.

Summary

The WCD's report (called the Report) contained 10 chapters, numerous statistical tables, and eight annexes, covering a wide range of topics. The most important subject areas include the role of water resources in dam development. These points cover technical, financial, economic, social, and environmental aspects; the impacts of dams on communities and ecosystems; decision-making, planning, and governance of dams; attention to human rights of affected population; environmental and social risks of dams; strategic priorities for a new policy framework to address problems; and guidelines to implement strategic priorities towards a future agenda for change.

The Report was critical of the performance of large hydroelectric dams on technical, financial, social, and environmental aspects. While it recognized the important contribution of large dams to economic development and the supply of energy, it lamented the social and environmental costs of dam construction, the inadequacy of measures to address the displacement of millions of people worldwide, and the significant or irreversible loss of biodiversity observed in the dams reviewed under the study.

The commission outlined seven areas of strategic priorities to improve the situation (included below in excerpts of the document), and the application of best practice standards and guidelines in all facets of dam building. The main points for action include measures to improve the mitigation of severe environmental impacts on watersheds and reduce social repercussions on communities. The WCD emphasized the importance of upfront planning to avoid major problems before, during, and after dam construction, and the imperative of ensuring public consultation and inputs throughout the entire process.

Follow-Up on Recommendations of the WCD Report

Many governments, institutions, energy sector agents, stakeholders, and communities around the world have endorsed the Report's recommendations, and accepted use of the implementation checklists linked to the strategic areas. Some have started the process of integrating WCD priorities into national policy frameworks or initiated a national or regional dialogue on dams and alternative energy generation methods. This discussion has also been sparked by reflections on climate change and the need to cut fossil fuel emissions.

Within the context of the recommendations on reform of policy frameworks and priority action areas, the commission emphasized that social and environmental factors should have the same weight as technical and financial parameters in the development of hydroelectric projects.

Action Items of WCD Report

Specific action items recommended by the WCD include focus on the participatory process in hydroelectric projects; the identification of options and alternatives to meet energy needs, particularly if there is public rejection of projects; maintenance of integrity of river and watershed ecosystems; sensitivity to community culture; repair of old, inefficient, or environmentally damaging dams; compliance of nations and project developers with regulations and enforcement; and peaceful use of rivers and watersheds for the good of society at large, particularly in border areas with shared water resources.

The Report warned against constructing dams on shared rivers without consensus of neighboring states and recommended that independent panels review and validate objections of stakeholders. The WCD also opined that decisions on dam construction affecting indigenous peoples should be taken with their free, prior, and informed consent. It underscored the principle of fair compensation to people displaced or regions and communities affected downstream by loss of water, destruction of biodiversity, cultural dislocation, and the impacts of reservoir infrastructure and transmission lines.

Citizen's Guide to the WCD

After publication of the commission's Report, several NGOs around the world took further action to support, follow up on, or interpret the commission's recommendations. For example, the International Rivers Network (IRN) developed a "Citizen's Guide to the WCD" and championed widespread dissemination of the WCD Report to communities and NGOs in all countries. It also called for translation of WCD guidelines into local languages, workshops for affected communities, national networks to raise awareness on the impacts of dams, and briefings for the media to discuss findings of the Report (Imhof et al. 2002).

The Citizen's Guide emphasized the need for informed consent and the right of communities to gain more information on those hydroelectric projects affecting their welfare. It also urged monitoring of compliance of dam construction with regulations and called for independent review panels of dams and expert opinions to better understand the impacts of planned dams or those in execution. In particular, it advocated for the incorporation of WCD recommendations into social and environmental policy and management frameworks of all nations in the hydroenergy sector.

Camille Gaskin-Reyes

Excerpts from the Executive Summary of the WCD Report: Dams and Development: A Framework for Decision-Making

Performance of large dams

The knowledge base indicates that shortfalls in technical, financial and economic performance have occurred and are compounded by significant social and environmental impacts, the costs of which are often disproportionately borne by poor people, indigenous peoples, and other vulnerable groups.

Given the large capital investment in large dams, the Commission was disturbed to find that substantive evaluations of completed projects are few in number, narrow in scope, poorly integrated across impact categories and scales, and inadequately linked to decisions on operations.

Large dams display a high degree of variability in delivering predicted water and electricity services—and related social benefits—with a considerable portion falling short of physical and economic targets, while others continue generating benefits after 30 to 40 years.

Large dams have demonstrated a marked tendency towards schedule delays and significant cost overruns.

Large dams designed to deliver irrigation services have typically fallen short of physical targets, did not recover their costs and have been less profitable in economic terms than expected.

Large hydropower dams tend to perform closer to, but still below, targets for power generation, generally meet their financial targets but demonstrate variable economic performance relative to targets, with a number of notable under-and over-performers.

Large dams generally have a range of extensive impacts on rivers, watersheds, and aquatic ecosystems—these impacts are more negative than positive and, in many cases, have led to irreversible loss of species and ecosystems.

Efforts to date to counter the ecosystem impacts of large dams have met with limited success owing to the lack of attention to anticipating and avoiding impacts, the poor quality and uncertainty of predictions, the difficulty of coping with all impacts, and the only partial implementation and success of mitigation measures.

Pervasive and systematic failure to assess the range of potential negative impacts and implement adequate mitigation, resettlement and development programmes for the displaced, and the failure to account for the consequences of large dams for downstream livelihoods have led to the impoverishment and suffering of millions, giving rise to growing opposition to dams by affected communities worldwide.

Since the environmental and social costs of large dams have been poorly accounted for in economic terms, the true profitability of these schemes remains elusive.

Perhaps of most significance is the fact that social groups bearing the social and environmental costs and risks of large dams, especially the poor, vulnerable, and future generations, are often not the same groups that receive the water and electricity services, nor the social and economic benefits from these.

Applying a 'balance-sheet' approach to assess the costs and benefits of large dams, where large inequities exist in the distribution of these costs and benefits, is seen as unacceptable given existing commitments to human rights and sustainable development.

Recommendations for a New Policy Framework

Researching and analyzing the history of water resources management, the emergence of large dams, their impacts and performance, and the resultant dams debate led the Commission to view the controversy surrounding dams within a broader normative framework. This framework, within which the dams' debate clearly resides, builds upon international recognition of human rights, the right to development, and the right to a healthy environment.

Within this framework the Commission has developed seven strategic priorities and related policy principles. It has translated these priorities and principles into a set of corresponding criteria and guidelines for key decision points in the planning and project cycles.

Together, they provide guidance on translating this framework into practice. They help us move from a traditional, top-down, technology-focused approach to advocate significant innovations in assessing options, managing existing dams—including processes for assessing reparations and environmental restoration, gaining public acceptance, and negotiating and sharing benefits.

The seven strategic priorities each supported by a set of policy principles, provide a principled and practical way forward for decision-making. Presented here as expressions of an achieved outcome, they summarize key principles and actions that the Commission proposes all actors should adopt and implement.

1. Gaining Public Acceptance

Public acceptance of key decisions is essential for equitable and sustainable water and energy resources development. Acceptance emerges from recognizing rights, addressing risks, and safeguarding the entitlements of all groups of affected people, particularly indigenous and tribal peoples, women, and other vulnerable groups. Decision-making processes and mechanisms are used that enable informed participation by all groups of people, and result in the demonstrable acceptance of key decisions. Where projects affect indigenous and tribal peoples, such processes are guided by their free, prior, and informed consent.

2. Comprehensive Options Assessment

Alternatives to dams do often exist. To explore these alternatives, needs for water, food, and energy are assessed and objectives clearly defined. The appropriate development response is identified from a range of possible options. The selection is based on a comprehensive and participatory assessment of the full range of policy, institutional, and technical options. In the assessment process social and environmental aspects have the same significance as economic and financial factors. The options assessment process continues through all stages of planning, project development, and operations.

3. Addressing Existing Dams

Opportunities exist to optimize benefits from many existing dams, address outstanding social issues, and strengthen environmental mitigation and restoration measures. Dams and the context in which they operate are not seen as static over time. Benefits and impacts may be transformed by changes in water use priorities, physical and land use changes in the river basin, technological developments, and changes in public policy expressed in environment, safety, economic, and technical regulations. Management and operation practices must adapt continuously to changing circumstances over the project's life and must address outstanding social issues.

4. Sustaining Rivers and Livelihoods

Rivers, watersheds, and aquatic ecosystems are the biological engines of the planet. They are the basis for life and the livelihoods of local communities. Dams transform landscapes and create risks of irreversible impacts. Understanding, protecting, and restoring ecosystems at river basin level is essential to foster equitable human development and the welfare of all species. Options assessment and decision-making around river development prioritizes the avoidance of impacts, followed by the minimization and mitigation of harm to the health and integrity of the river system. Avoiding impacts through good site selection and project design is a priority. Releasing tailor-made environmental flows can help maintain downstream ecosystems and the communities that depend on them.

5. Recognizing Entitlements and Sharing Benefits

Joint negotiations with adversely affected people result in mutually agreed and legally enforceable mitigation and development provisions. These recognize entitlements that improve livelihoods and quality of life, and affected people are beneficiaries of the project. Successful mitigation, resettlement, and development are fundamental commitments and responsibilities of the State and the developer. They bear the onus to satisfy all affected people that moving from their current context and resources will improve their livelihoods. Accountability of responsible parties to agreed mitigation, resettlement, and development provisions is ensured through legal means, such as contracts, and through accessible legal recourse at the national and international level.

6. Ensuring Compliance

Ensuring public trust and confidence requires that the governments, developers, regulators, and operators meet all commitments made for the planning, implementation, and operation of dams. Compliance with applicable regulations, criteria and guidelines, and project-specific negotiated agreements is secured at all critical stages in project planning and implementation. A set of mutually reinforcing incentives and mechanisms is required for social, environmental, and technical measures. These should involve an appropriate mix of regulatory and non-regulatory measures, incorporating incentives and sanctions. Regulatory and compliance frameworks use incentives and sanctions to ensure effectiveness where flexibility is needed to accommodate changing circumstances.

7. Sharing Rivers for Peace, Development and Security
Storage and diversion of water on trans-boundary rivers has been a source of considerable tension between countries and within countries. As specific interventions for diverting water, dams require constructive co-operation. Consequently, the use and management of resources increasingly becomes the subject of agreement between States to promote mutual self-interest for regional co-operation and peaceful collaboration. This leads to a shift in focus from the narrow approach of allocating a finite resource to the sharing of rivers and their associated benefits in which States are innovative in defining the scope of issues for discussion. External financing agencies support the principles of good faith negotiations between riparian States.

If we are to achieve equitable and sustainable outcomes, free of the divisive conflicts of the past, future decision-making about water and energy resource projects will need to reflect and integrate these strategic priorities and their associated policy principles in the planning and project cycles.

Source: World Commission on Dams (WCD). "Dams and Development: A New Framework for Decision-Making. The Report of the World Commission on Dams." London: Earthscan Publications, 2000. http://www.unep.org/dams/WCD/report/WCD_DAMS%20report.pdf. Aviva Imhof, Susanne Wong, and Peter Bosshard, *Citizens' Guide to the World Commission*. Berkeley: International Rivers Network, 2002. https://www.internationalrivers.org/files/attached-files/wcdguide.pdf.

PERSPECTIVES

How Will the Grand Ethiopian Renaissance Dam Affect Peace in the Region?

Overview

The Grand Ethiopian Renaissance Dam, formerly known as the Millennium Dam and sometimes referred to as Hidase Dam, is a gravity dam on the Blue Nile River in Ethiopia. As the project moves forward and nears completion, it has sparked a series of conflicts and intense diplomatic maneuvering, with some in Egypt threatening to sabotage it. Frustrated with the process, the Ethiopian government has more or less made the unilateral decision to build and operate the dam "no matter what," though in public the government states that it is cognizant of the downstream impacts, is willing to remain open to discussion, and will consider modifying the design if there are significant impacts. As the two essays in this section will show, there are arguments suggesting that the project promotes stability in the region as well as arguments that emphasize how it has destabilized political relations in an already unstable region.

The Nile River, found in North and East Africa, is said to be the longest river in the world, with a length exceeding 6,850 kilometers (4,255 miles). The river cuts a path through many different cultures, climates, and landscapes before it empties into the Mediterranean Sea. Though the Nile basin spans 11 countries, only Egypt and Sudan have extensively developed the river's water resources for human use for food and electricity. The other basin countries (Burundi, Democratic Republic of the Congo, Eritrea, Ethiopia, Kenya, Rwanda, South Sudan, Tanzania, and Uganda) are only marginally involved with Nile politics. With the exception of Egypt and Kenya, most of the countries are considered to be among the world's least developed nations. Until recently, economic poverty and/or political instability have prevented significant development projects on the river, but this is changing. The Grand Ethiopian Renaissance Dam being constructed in the Nile basin is the largest hydroelectric project under construction on the African continent. The dam has become a potent symbol of African modernization and development and, as such, is a lightning rod of conflict and negotiation—primarily involving Egypt, Ethiopia, and the Sudan.

Water resources are problematic in North and East Africa, as water is scarce and unavailable for much of the year due to the arid climate and highly variable precipitation patterns. The Nile basin includes many tributaries and two main river sections: the Blue Nile and the White Nile. The Blue Nile is where the Renaissance Dam is being constructed. That the Blue Nile accounts for over 80 percent of the water flow in the Nile River is an issue of concern for the downstream countries of Egypt and Sudan. The Blue Nile water primarily comes from precipitation that falls in the Ethiopian highlands during the summer monsoon season. There is already a hydropower dam on the Blue Nile, the Roseires, located in Sudan near the Ethiopian border. Otherwise, the river remains largely undeveloped and wild, subject to intense flooding in the monsoon summer rainy season and bouts of minimal flow in the dry winter season.

The Grand Ethiopian Renaissance Dam has a planned potential generating capacity of 6000 megawatts (MW). The dam's current design is for 16 turbines with 375 MW installed capacity. For comparison, the Hoover Dam has an installed generating capacity of 2080 MW, and the city of Toronto on an average Saturday consumes 3000 MW (CNW 2014). Ethiopia currently has less than 2000 MW in the grid from its existing power plants. The potential energy increase for Ethiopia and neighboring countries gives an unprecedented opportunity for economic development based on electricity availability. Given information provided by a U.S. Bureau of Reclamation survey conducted in the 1950s, the Blue Nile has the potential to provide about 10,000 MW of hydropower generation (Consulate General of Ethiopia 2013). Construction of the Renaissance Dam began in 2011 and it is planned to be completed by 2017. The cost exceeds 4 billion U.S. dollars, and as of 2015 the money has largely come from contributions and purchases of bonds by the Ethiopian people both living in Ethiopia and living in other countries.

That the Renaissance Dam is the first major development project on the Nile River to occur outside Sudan and Egypt is significant politically and economically. The existing international treaty that regulates water rights on the Nile River stipulates that Egypt and Sudan have exclusive rights to develop and use Nile water resources at the exclusion of the other upstream basin countries. Today, the treaty's validity is questioned because it was drafted before many of the countries in the Nile basin were independent, and because of this regional people feel the treaty does not give fair representation of their rights. The treaty was modified once to change water allocations in Egypt and Sudan in 1954, following the construction of the High Aswan Dam in Egypt. Today, the nine other countries in the Nile River basin have sought to amend or even completely abolish the existing treaty. Through diplomatic negotiations, the countries have endeavored to develop new terms for a treaty that would allow water rights for development outside the core countries. The Nile treaty itself is controversial for most, as it was created during the colonial era.

The dam will displace approximately 20,000 local people who subsist on the river and surrounding natural resources. The dam site is located in a remote location in Ethiopia 730 kilometers from Addis Ababa and 17 kilometers from the Sudan border in the Benishangul-Gumuz state. The Benishangul-Gumuz state borders Sudan, is considered remote, experiences food insecurity, is sparsely populated, and is one of the most impoverished states in Ethiopia. The local people, mostly from an ethnic minority group collectively called the Gumuz, subsist from the river, and do not have access to electricity or basic services. The Gumuz people pan for gold, practice flood-recession agriculture, and fish in the river. Due to the reservoir that will result from the Renaissance Dam, this population will no longer have access to the river resources listed above. What potential access to the water resources these 20,000 displaced will have in the future will be modified from how they use the river today. Historically, the Gumuz do not have good relations with other ethnic groups, particularly the highland Amhara, and are considered still today to be violent and hostile people. However, the Gumuz communities living in the shadow of the Renaissance Dam have not expressed any action or words to be interpreted as resistant to the project.

The Ethiopian national agenda is to improve the national economy and alleviate poverty. The country's population doubled from 40,000 million to 80,000 between 1990 and 2010. The Blue Nile accounts for close to 50 percent of all surface freshwater resources within Ethiopia. The Ethiopian Electric Power Corporation has already signed agreements for the electricity expected to generate from several hydropower projects across the country, including, most significantly, the Renaissance Dam. The countries that have signed some level of agreement to buy electricity from Ethiopia are Kenya, Sudan, Djibouti, and potentially South Sudan, before the civil war erupted. The Ethiopian government has a plan to take the revenue and apply it to further development projects throughout the country.

Jennifer C. Veilleux and Mark Troy Burnett

Perspective 1: The Contentious GERD: A Thorn in the Side of Nile Basin Politics

Despite its promise, the Grand Ethiopian Renaissance Dam (GERD) has unsettled efforts at regional stability, economic cooperation, and peace. The lack of transparency with the project and political intransigency of the Ethiopian and Egyptian governments continue to heighten tensions in the region as the stakeholders bicker over current and evolving Nile basin treaties, compacts, and agreements.

The ambitious Renaissance Dam is Ethiopia's largest-ever engineering project. The expected energy output of 6,000 megawatts is triple the capacity of the country's second largest dam, Gilgel Gibe IV. Along with being viewed as vital to the country's economic development and future prosperity, the GERD has become a potent symbol of national pride and hope. Conversely, the dam has many critics—environmental and political—with the most vocal found in Egypt. As the last of the downriver countries, Egypt fears a serious disruption to the quantity and quality of water from the river. The reservoir will contain an amount equal to 1.3 times the annual flow of the Blue Nile. A diplomatic row has ensued with factions in Egypt going so far as to threaten to sabotage the project. For their part, Ethiopians respond by accusing their neighbors of living in the past and being unwilling to accept Ethiopia's economic and political development. Outside observers and critics, such as the nongovernmental organization International Rivers, note that from its inception the project has been veiled in secrecy. The government of Ethiopia and the Ethiopian Electric Power Corporation have been accused—not just by its harshest critic, Egypt—of suppressing internal discussion and criticism. It doesn't help Ethiopia's position that it has yet to produce an environmental or social impact assessment and has failed to consult downstream nations prior to construction. Indeed, very little is known about the ecological or hydrological impacts.

An independent field report commissioned by International Rivers has helped to shed light on some of the most important social and environmental issues (International Rivers 2012). The report, undertaken without consent of the Ethiopian government, confirmed many of the concerns of the downriver countries and international observers. According to the report, between 5,000 and 20,000 people will be forced to resettle as villages in the flooded area will be completely inundated. This directly contradicts the government's estimate of only 800 people. As the report notes, none of the affected people were consulted about the dam and their fate. Indeed, the political climate in the country continues to make it risky for civil society and conservationists to question the government's plans, and there are legitimate fears of government persecution. For instance, in June 2011, Ethiopian journalist Reeyot Alemu was imprisoned after she raised questions about the project. During their efforts to assess the impacts of the dam, International Rivers staff members received anonymous death threats (Pottinger 2013). Matters were made more tense when, at a 2011 conference of the International Hydropower Association in Addis Ababa, the then–prime minister Meles Zenawi called critics of the project "hydropower extremists bordering on the criminal" (Bosshard 2011).

Along with the social impacts and continued lack of transparency and discussion with the project, there is a long list of environmental impacts that the Ethiopian government continues to deny or dismiss. Geomorphologically, the Ethiopian highlands have been heavily deforested and subsequently are highly sensitive and prone to mass erosion. Sedimentation, a common problem for all dams and reservoirs, is a major risk for the GERD. There have been no studies to quantify the sedimentation risk or the impact it would have on the dam's power output, maintenance, and lifespan. Further, a soil conservation plan for the watershed is nonexistent. The Benishangul-Gumuz region, where the GERD and its reservoir are being constructed, is one of the few places in the country that is still forested. The community, which has long lived and depended on the resources of this forest system, will see, seemingly by government fiat, 90 percent of this forest flooded. Scientific studies have documented more than 150 endemic freshwater fish species in Ethiopia's portion of the Blue Nile (Pottinger 2013). The fish are a staple for many who live along the banks of the river. Yet again, there have been no significant efforts to document the effects that the dam is expected to have on fish habitats—a lake is a very different place to live for fish who have evolved in a riparian environment. Doubtless, those locals directly affected by these impacts, without adequate compensation, will have a hard time sitting back idly and watching their livelihoods destroyed despite appeals to their sense of national pride.

Though President al-Bashir of Sudan in 2012 said he supported the building of the dam, Egypt continues to be unsettled by the project and demands to be allowed full inspection and consultation on the design and implementation. Ethiopia vehemently denounces Egyptian involvement unless Egypt relinquishes its veto on water allocation as stipulated in the 1959 Nile Waters Agreement—an agreement, administered by former colonial power Britain, that clearly favors Egypt and Sudan by granting them exclusive rights to the waters of the Nile. Tensions have escalated as Egypt holds fast to the parameters of the 1959 treaty whereas Ethiopia and other upper basin countries proceed under the 2010 Nile Basin Initiative and the Cooperative Framework Agreement.

In an effort to diffuse tensions, Egypt, Ethiopia, and Sudan established an international panel of experts to review and assess the project. However, diplomatic efforts were dealt a serious blow when Egyptian president Mohammed Morsi, in a June 3, 2013, unknowingly televised discussion with the panel, suggested methods to destroy the dam, including support for antigovernment rebel factions in Ethiopia. The Morsi government subsequently apologized for the comment, though the gaffe clearly destabilized relations and reinforced Ethiopia's mistrust of its northern neighbor. Indeed, the Morsi government confirmed in 2013 that "Egypt's water security cannot be violated at all" or "endangered" and that "all options are open" (BBC 2013).

Despite the criticism and concern, Ethiopia continues to move forward with the GERD. With a scheduled opening date sometime in 2017, observers can only hope that the harsh political rhetoric will abate and a workable framework can be crafted where all members of the Nile basin feel part of the process. The future demands it, for Ethiopia has plans to build a host of dams and reservoirs along its portion of the Blue Nile, and doubtless Egypt will not willingly let them continue to act so arbitrarily.

Mark Troy Burnett

Perspective 2: The Renaissance Dam—A Platform for Peace

The Grand Ethiopian Renaissance Dam represents a new dynamic and new discourse in the regional history of the Nile basin. Examination of the amount of news events related to the basin in the years since construction began shows significant discussion about water rights, historic claims, economic development and cooperation, national interest, and even the greater good. Though dramatic headlines about the dam, such as the March 20, 2013 article entitled, "Egypt, Ethiopia Headed for War Over Water" in *Al-Safir Al-Monitor*, Lebanon's leading newspaper, suggest that Egypt and Ethiopia are at risk for water-based conflict, the reality is that the governments of Egypt, Ethiopia, and Sudan are engaged in ongoing, civil dialogue about dam feasibility and regional institutions of cooperation. The Renaissance Dam decision, although taken unilaterally by Ethiopia in 2011, is inspiring an unprecedented level of diplomacy between the three Blue Nile basin countries and offering a resources-based platform for dialogue that has the potential to lead to regional peace and stability. This essay considers the aspects of the Renaissance Dam that lend themselves to promotion of peace and cooperation between Nile basin countries.

Doubtless, this new era of cooperative dialogue in North and East Africa still suffers from many of the same, historical antagonisms—political instability, weak economies, cultural conflict, and natural disasters. For instance, the regional economies do not have strong international trade agreements with neighboring countries. Nine of the 11 basin countries are in the United Nations' List of Least Developed Countries—a list that maintains information on about 50 of the world's lowest economies. Populations are expanding but reliant on the same amount of resources. Egypt's High Aswan Dam on the Nile generates a large percentage of that country's energy. Sudan's Merowe and Roseires Dams account for 75 percent of generated domestic electricity supplies. Regardless, there are several motivations for the three basin countries to put these familiar challenges aside and move ahead with diplomatic discussion over the dam. The fact is, the Renaissance Dam will impact downstream flows of water and sediment to Egypt and Sudan, will generate 6,000 MW of electricity for domestic and export use, represents a political power shift in the basin in relation to the Nile Treaty, and marks a new era of social development for upstream countries.

The Nile Treaty was originally drafted and signed in 1929 by Britain and revised in 1959 by Egypt and Sudan just before sweeping independence for African countries from colonial powers. This treaty was developed to allocate Nile River waters to both Egypt and Sudan only. The original treaty is a product of the colonial era, with Britain signing on behalf of the basin countries. Today it is not considered fair or legitimate by many of the upstream countries. In fact, colonial era treaties in general in Africa are fraught with cultural feelings of illegitimate rule from colonial interests and powers to exploit African nations. The Renaissance Dam opens the conversation beyond theoretical ideas of national development because it violates Egypt and Sudan's exclusive water rights. Egypt and Sudan are in a situation where they are being asked to engage with their upstream neighbors outside the terms of the original treaty and, positively, they are complying. These two countries are extremely arid and have few water resources alternative to the Nile River. While open conflict is an option to promote their interests, they clearly stand to benefit more from engaging diplomatically with each other. This cooperation could then set a precedent for future cooperation and collaboration. Further, each of the basin countries has Nile development plans of their own that will require a cooperative mood to come to fruition.

Arguably, the Renaissance Dam offers direct benefits to the downstream countries. If the Blue Nile River is managed cooperatively, there is an opportunity to enhance these benefits. Egypt's Lake Nasser, which has incredibly high evaporation rates, may not be the best place to store water, but the reservoir behind Renaissance may offer storage with lower evaporation rates, enhanced flood and sediment control systems, and expansion of available hydroelectric energy. The region also currently experiences frequent energy shortages. Sudan, recognizing the potential, has already signed a contract with the Ethiopian Electric Power Corporation to import electricity. Available and reliable energy is a precursor to sustained economic growth. Further, available and reliable freshwater resources are also vital to growth and political stability. Most assuredly, the Renaissance Dam will increase energy and water availability in the region.

From a social perspective, the Renaissance Dam provides a salient example of development divorced from colonial influences, especially in the context of the Nile Basin treaty. The shift has provided upstream Nile basin countries a voice in discussions. The discussions began with the Nile Basin Initiative. A working group of basin country representatives drafted new language and amendments to the original, colonial-devised treaty. Included in the new teaty is the establishment of an international riparian commission to help steer development and cooperation in the basin. The initiative further includes a component known as the tripartite talks—a forum for ongoing diplomatic discussions between Egypt, Ethiopia, and Sudan.

Given that the Grand Ethiopian Renaissance Dam has already resulted in more than one year of structured and peaceful negotiations and diplomatic relations among the Blue Nile basin countries, the dam arguably has the capacity to serve as a platform for peace. As highlighted, there are many potential benefits that this project brings: politically, in empowering upstream countries previously denied via British imperialism; economically, in sharing benefits of energy generation and helping to elevate the Ethiopian population out of poverty; environmentally, in allowing for scientific investigation and cooperative mitigation of impacts; and socioculturally, by igniting cross-cultural dialogue between basin countries who for too long have not had or made opportunities to engage in joint development efforts.

Seizing upon the experience with the Renaissance Dam, one can only hope that this era of cooperation between the Nile Basin countries will spill over into other social and political arenas.

Jennifer C. Veilleux

REFERENCES AND FURTHER READING

Abdelhady, Dalia et al. "The Nile and the Grand Ethiopian Renaissance Dam: Is There a Meeting Point between Nationalism and Hydrosolidarity?" *Journal of Contemporary Water Research & Education* 155, no. 1 (2015): 73–82.

BBC News. June 10, 2013. "Egyptian Warning over Ethiopia Nile Dam." http://www.bbc.com/news/world-africa-22850124.

Bosshard, Peter. "Sustainable Hydropower—Ethiopian Style." 2011. International Rivers. http://www.internationalrivers.org/blogs/227/sustainable-hydropower-%E2%80%93-ethiopian-style.

Gebreluel, Goitom. "Ethiopia's Grand Renaissance Dam: Ending Africa's Oldest Geopolitical Rivalry?" *Washington Quarterly* 37, no. 2 (2014): 25–37.

International Rivers. "Field Visit Report, GERD Project." 2012. http://www.internationalrivers.org/files/attached-files/grandren_ethiopia_2013.pdf.

Pottinger, Lori. "Field Visit Report on the Grand Ethiopian Renaissance Dam." 2013. International Rivers. http://www.internationalrivers.org/resources/field-visit-report-on-the-grand-ethiopian-renaissance-dam-7815.

Tvedt, Terje, and Ebrary Academic Complete (Canada) Subscription Collection. *The River Nile in the Post-colonial Age: Conflict and Cooperation among the Nile Basin Countries*. London: I. B. Tauris, 2009.

Veilleux, Jennifer C. "The Human Security Dimensions of Dam Development: The Grand Ethiopian Renaissance Dam." *Global Dialogue* (Online) 15, no. 2 (2013): 42.

10 WATER AND CULTURE

OVERVIEW

This chapter shows how water is intertwined with life and its cultural forms. The case studies describe the role of water in creation myths, cleansing rituals, and the rites and religious beliefs of major religions, while the perspectives section discusses the pros and cons of cultural beliefs and cultural exceptions for whaling among indigenous groups.

We see from this chapter that water is a vital part of humans' life cycle from cradle to grave, and it is an integral part of ancient myths, legends, creation, and flood stories—even in distinct and geographically separated cultures around the world. Indigenous peoples' belief systems in particular have strong connections to water and kinship with the life force of oceans, rivers, springs, or wells.

Within this life cycle, the early development stage of a human begins within a mother's womb, where amniotic fluid envelops the fetus in a liquid protective blanket. As life continues, water flows through numerous religious and prayer rituals—baptism, coming of age, name-giving, adult initiation, and regular prayer rituals, all the way up to end-of-life anointing and burial washing rites.

Water and Creation Narratives

Some epic and mythical narratives, which explain creation, have passed down orally from tribe to tribe, group to group, family to family over the millennia. Ever since the first humans gathered around fires to keep themselves warm and protect themselves from wild animals, people have been telling stories about their creation and their cultural or spiritual beliefs—and passing them down to the next generation.

In many myths, the creation of the world begins with rains or life-giving fluids; and the demise of the world occurs through floods. Many of these events also signify

Modern architectural design interpretations of ancient waterwheels grace the center of Salzburg, Austria. (Camille Gaskin-Reyes)

spiritual cleansing, God's wrath or forgiveness, and humankind's redemption from its own folly. One particularly familiar Bible story, a classic tale of destruction and salvation, is of Noah's ark in the Old Testament, where God causes flooding to occur and instructs Noah to build an ark to save and regenerate the human race.

Water and the Arts

Water is a vital element in the arts and architecture—from the ancient hanging gardens of Babylon to fountains in the palaces of the Mughals, to water works in the inner sanctuaries of imperial palaces, and water features within different architectural and building styles. The oceans have inspired numerous masterpieces of art, countless novels, and volumes of poetry worldwide. Dutch painters such as Willaerts, Weissenbruch, and Toorop have portrayed scenes of shipwrecks, canals, and marine environments.

Countless novels, including *The Martian Chronicles* by Ray Bradbury, contain water themes, and poems written by the English poet Henry Wadsworth Longfellow include references to the cycle of rain from birth to death and the many moods of water. Throughout the ages, architectural design has incorporated water into religious structures, mosques, museums, urban fountains, rain gardens, artistic cisterns, house design (e.g., Frank Lloyd Wright's Fallingwater), and statues and memorials to honor both the dead and the living.

Water and Human Survival

In the past, nomadic humans were acutely aware of the importance of water sources for their survival and as locations to find and hunt animals. As hunter-gatherers became more sedentary, they started domesticating wild plants and associating rain and the seasons with bountiful harvests and the eruption of new life in plants, seeds, and grasses. Early civilizations might have delighted in life-regenerating rains for fields, but they would have been fearful of destructive floodwaters and devastating droughts. Many legends, myths, and rituals arose from the desire to explain seasonal growth cycles and the role of life-giving rains.

In arid climates, water was so important for survival that its presence or absence meant feast or famine. Water mismanagement, floods, and droughts were factors that contributed to the demise of many groups. It has been suggested that the old Khmer civilization of Angkor Wat, Cambodia, might have collapsed in the eighth century due to water mismanagement and the lack of canal maintenance. Catastrophic drought and excessive water erosion also played a role in the destruction of the Central American Mayan civilization in the ninth century. However, such events were often interpreted by early civilizations as god-induced occurrences in response to human misbehavior.

Water and Native Beliefs

Early societies held water in high esteem. There were numerous water gods, and some cultures hailed water as a god itself. Literally thousands of religious practices, legends, creation stories, and rain rituals, e.g., rain dances, revolved around water or lack thereof. The origins of primal, i.e., native, religions and practices (before the principal religions of Islam, Judaism, and Christianity) go back so far that it is difficult to put exact time frames on their emergence worldwide.

However, we know that when humans could not explain natural events, they usually assigned responsibility for them to deities or God-related factors. Over time, there was even some coexisting or blending of native beliefs with one another and with the later mainstream religions. Examples from the Americas come to mind, i.e., the blending of beliefs and water rituals of Portuguese and African deities in Brazil, and the integration of indigenous Aztec, Inca, Mayan, and other native practices into Catholic religious rituals in Mexico, Peru, Guatemala, and some islands of the Caribbean.

Sumerian Rain God Rituals

Around 3200 BCE, Sumerians settled around the Euphrates and Tigris River valleys, domesticating plants and trading surplus food. Without further need to hunt and gather food afar, some people could dedicate themselves to nonagricultural tasks and sacred pursuits, which led to the emergence of priest cults and high religious castes. These elites had time to contemplate matters beyond daily routines, i.e., the origin and purpose of creation, life, death, the afterlife, the presence of God or gods, and interpretations of natural occurrences through divine interventions. Priests either functioned as intermediaries with deities or defined themselves as gods. With the expansion of priest elites and the increase in rituals, more workers from the masses were pressed into service for temple construction and offerings to gods.

In time, religious practices and interpretations of priests and rulers flowed into myths and legends that were handed down to later generations. Most narratives contained the importance of water for life and the fertility of the land. The periodic emergence of natural disasters such as floods and droughts was often attributed to the anger or displeasure of the gods, who needed to be appeased or pleased. In locations prone to drought, societies therefore developed complex solicitations to rain gods such as rain dances and violent or nonviolent rituals to provoke deliverance of rains or termination of floods.

There were very specific rain gods. In Mesopotamia, according to legend, storm gods ranked as the highest of all gods; they produced rains when people appeased them and responded with drought, floods, and infertility when people angered them. The rain god Iskur was responsible for droughts and floods; he wielded a lightning bolt and rode a bull through a storm in the sky. The gods' practice of regulating human good and evil through rain seems to foreshadow the later Christian immersion or baptismal rituals, which symbolize the washing away of sins and the new start of life.

Native Myths

In the Aztec culture, priests sacrificed children to the rain god Tlaloc and smaller rain gods to bring rains to parched lands. In Australia, the Aborigines bled members of their group they considered rainmakers, then drizzled their blood over other men in the clan to induce rain. The stories of Hindu gods also depict a native rain god, called Indra. When enraged, she sent down angry rains to earth to punish people. In the pantheon of gods, the beautiful, peaceful goddess Ganga was important: she was the goddess of the most sacred river in India, the Ganges, or Ganga, River. She represented the celebration of river bathing rituals in the Ganges and held a water lily in her right hand to symbolize rain and fertility. The water pot in her left hand immortalized the importance of water.

Although indigenous or native peoples differed in their cultures and according to local, regional, or geographical contexts across the world, they shared similar ideas of harmony and affinity for nature and natural environments. Communities who lived

Hindu pilgrims purify themselves through ritual bathing in the Ganges River. India's Ganga Mahotsava festival honors the Ganges for providing for the millions of farmers who grow rice on its neighboring plains. Hindu worshippers dip themselves into the sacred waters of the Ganges during this annual festival. (Corel)

close to the sea or survived mainly through fishing regarded water not as a mere source of livelihood but as a life-giving spirit. The stories, chants, and legends of the Kumulipo people of Hawaii reflect strong spiritual connections and kinship bonds with the ocean, which they considered a place of living gods. To this group, the ocean was the common amniotic fluid and giver of life for all people and countries sharing its coasts.

Water as a Gift

The Maori people in New Zealand regarded the sea as a gift from Mother Earth, which humans needed to protect and use in a sustainable manner. Micronesians and many other indigenous peoples did not have a concept of private ownership of seas, rivers, or common waters in the contemporary sense of ownership; rather they perceived that water was a collective domain for all creatures and an extension of the spirits of land, forests, plains, animals, and plants. Therefore, all people had a common responsibility to protect water to ensure survival of the species and the spirits.

Creation and Flood Legends

Creation myths and legends were widespread across the world's early cultures and civilizations. Stories attempted to explain how earth and people emerged, which gods were responsible for what tasks, and what connections existed between gods and people. In the Hebrew Bible, water plays a prominent role in the relationship between God and humans. In a well-known story, God tells Abraham to stop worshipping other gods and move to the Promised Land. In return, God promises to take care of him and lead his people out of arid Mesopotamia into the rain-fed valley of Canaan. The abundance of water would be the reward for obedience.

The Old Testament also records that some of Abraham's descendants went from Canaan to Egypt, where they became slaves of the Egyptians. Moses later led them out of bondage back towards Canaan, with God intervening to part the Red Sea and allow them to flee from the pursuing Egyptians. In Canaan, God promised the Jews rain, if they obeyed his commandments and worshipped him alone. Along with this promise, however, came God's warning to rescind the pledge of rain if the Jews took other gods. This story also solidifies the image of a rewarding or punishing God using the instruments of rain or drought.

The story of the Great Flood is a classic example of the importance of water. This narrative is shared by Judaism, Christianity, and Islam, and is contained in the Old Testament (Genesis 1:6) and in the Koran with slight variations (Surah 21:30). In this story, God becomes so upset with his people's wickedness—10 generations after he created Adam and Eve—that he resolves to destroy them by sending rain for 40 days and 40 nights to flood the land. But first God instructed Noah, the only righteous one in God's eyes, to build an ark for his family and take a pair of each living creature on board. According to the narrative, when the flood subsided, Noah and the ark came to rest on Mount Ararat and were saved. This story shows once more how an angry God punished and redeemed humans using raging waters.

Gothic statue of Noah holding the ark, Cologne Cathedral, Germany. (Vladimir Wrangel/Dreamstime.com)

The story also recalls that afterwards a relenting God promised never ever again to destroy the earth by flood.

The flood story mirrors a narrative from the Epic of Gilgamesh, recorded by the Sumerians in written form on cuneiform tablets. This Sumerian story recalls that the god Enlil, having decided to destroy humans by flood, warns the priest Ziusudra of the coming flood. He instructs Ziusudra to build a great ship to carry all beasts and birds on earth to escape the impending catastrophe. The details of this story are very similar to the later story of Noah, including the final settling of the boat on top of a mountain and the promise of God never to destroy humanity by flood again.

Flood stories appear to have passed from generation to generation and among cultures and geographical areas, but they might also have developed independently in different continents: flood legends reappear in many different cultures, such as the Hittite, Greek, Roman, Hindu, and Scandinavian belief systems, among others. In Greek mythology, the most powerful god was Zeus, lord of the sky and the rain. The myth indicates that Zeus was so upset with humans that he caused a huge flood to wipe them out; only Deucalion and his wife, Pyrrha, escaped in an ark. The same tale circulated among the Romans, whose god of the rain, Jupiter, unleashed rain and caused the sea god Triton to whip up waves and drown humanity. Only the pious Deucalion and Pyrrha, who lived on the summit of Parnassus, were able to escape. Plato, the Greek philosopher, also described the story of a lost continent called Atlantis, claiming it was buried under the sea by a great flood.

Scandinavian legends and flood myths narrate how icy waters from the wounds of the slain giant Ymir drowned most of the world; however, the giant Bergelmir escaped with his wife and children in a boat made from a tree trunk and lived to repopulate the earth. Celtic legends also mention heaven and earth as great giants, with heaven laid over earth, causing people to be crowded and unhappy in the darkness. This caused the boldest people to incite a rebellion to slay the giants (heaven and earth) and spill their blood. Their blood caused a great flood, killing all humans except for one couple, who escaped with a boat.

A Lithuanian flood legend also recounts how the supreme god Pramzimas was angered by wars and the injustice of humans to each other. He sent two giants, Wandu (water) and Wejas (wind) to destroy the earth. While Pramzimas was eating nuts, he looked down upon the earth to check on the destruction process, and one of the nutshells fell onto the earth and landed on one of the highest peaks. People scrambled into the nutshell and were saved to repopulate the earth, once again rescued by divine intervention.

From Babylonia a flood myth recalls that the god Enki instructed all the other gods to destroy humans with a flood. Enki instructed a man called Atrahasis to build an ark to escape. He took cattle, wild animals, birds, and his family into the ark, and they survived to repopulate the earth. In Chaldean myths, the god Chronos warned Xisuthrus, the 10th king of Babylon, of a coming flood. He ordered him to build and provision a vessel for himself, his friends, family, and all types of animals. After the flood Xisuthrus landed his ship in the Corcyraean Mountains in Armenia, and future human survival was guaranteed.

Further north in Siberia, a flood story speaks of three groups, the Samoyed in north Siberia, the Yenisey-Ostyak in north central Siberia, and the Kamchadale in northeast Siberia. God punished them all by sending floods to wipe out humanity. However, as the story goes, the good people were once again saved to repopulate the earth.

Sculpture of the Hindu god Brahma, who is said to have created life from the cosmic ocean. Brahma is the creator god of Hindu mythology, and, along with Vishnu and Shiva, one of the three aspects of the divine. (Hector Joseph)

In Hindu flood myths, the god Brahma changed himself into a fish to warn his son Manu (the first man) of a worldwide flood. Manu built a large boat and gathered different types of seeds to save the species. Manu alone survived the flood and made offerings of butter, milk, and curds to the Gods. A woman appeared with whom Manu mated to repopulate the human race. The Zoroastrian legends recall the importance of floods as well. The God, Ahura Mazda, warned a person called Yima that destruction in the form of a flood was threatening the sinful world. He gave him instructions to build a boat, in which humans and pairs of small and large cattle, humans, dogs, birds, and plants and foods were to be saved.

Across the ocean in the Americas, flood myths of the Incas and

Aztecs were very prominent. In Inca mythology, God sent a flood to make all things perish, with the exception of a man and a woman who floated in a box. When the floods receded, the box was driven by the wind to a site called Tiahuanaco, where the creator told them to repopulate the earth. Aztec legends recount that a pious man named Tapi lived in the Valley of Mexico. The creator told Tapi to build a boat and take his family and a pair of every animal that existed on board. When Tapi finished construction, it began to rain, and the flooding of the valley submerged everyone except Tapi. Once again, God redeemed humans and rebuilt the world.

In Tanzania, Africa, as the flood story goes, when the rivers began to flood, God told two men to get into a ship and take seeds and animals with them. After the floods covered the mountains, the men sent out different birds to check whether there was dry land. When the floods were over, the men disembarked with the animals and the seeds, and life was revitalized. In Thailand and Laos, the legend recalls that the god Thens demanded a sacrifice of food from humans. When they refused, God punished them by flooding the earth. However, three men were forewarned, and they were able to build a raft and survive.

Water Monsters and Strange Marine Creatures

In other myths and legends, strange and terrifying creatures such as water monsters and mermaids were said to live in or under water. In ancient times, when sailors first ventured across oceans or into unknown waters, they feared these water creatures so much that superstitious beliefs ruled ships and the behavior of the crew. Other myths covered stories of lake and river creatures such as Nessie, the famous Loch Ness monster near Inverness, Scotland, or Leviathan, the water monster from the land of Canaan which was associated with Satan, the devil. In Germany, the Lorelei legend described a water monster located near the rocks on the Rhine River, which could trap and shipwreck ships. Greek mythology was also rife with mermaids and sea creatures of all shapes and sizes.

In other continents, there were legends, i.e., Panlong, a mythical Chinese sea creature in China; Rusalka, a Slavic sea monster; and the Selkie, a Scottish sea creature. Myths also described Tahoratakara, a Polynesian sea monster; Bunyip, an indigenous sea creature of the Australian Aborigines; and a wide array of countless beasts, monsters, and dragons that were said to have populated oceans, seas, and lakes across the world.

Sacred Rivers and Healing Waters

Water is associated with numerous sacred rivers and lakes, healing places, fountains, religious rites, and sites across the globe. The Nile, the Zambezi, the Congo, the Yangtze, the Mekong, the Ganges, the Rhine, the Aral Sea, Lake Fundudzi in South Africa, the Phiphidi Falls in South Africa, and the Okavango Delta in Africa are all locations where stories and legends about water spirits, sprites, and gods have emerged. In Germany, operas were composed about the Nibelungen, or water spirits, and myths centered around the Rhine River.

Narratives about miracle or healing waters have been passed down throughout the ages. There are literally thousands of sources of waters said to be healing waters around the world. To the Hindus, the Ganges (or Ganga) River is the principal water of life and is associated with the concept of *moksha*, or liberation. In general, Hindu temples are located near water sources, since adherents are under a strict obligation to wash or bathe before entering a temple. A morning ablution with water is an everyday duty for Hindus to carry out cleansing and spiritual purification rituals before chanting mantras. The Ganga Dussehra is a festival in Hardiwar, India, where Hindus approaching death can bathe in the river to ensure a good rebirth.

In Catholicism, the reported sighting of the Virgin Mary by Bernadette Soubirous in Lourdes, France, in 1858, created a holy site to which people with illnesses and incurable diseases still flock for a miracle cure after immersion in the holy waters. At the Tsubaki Grand Shrine in the Mie Prefecture in Japan, believers perform a Shinto ritual called *misogi shuho* to wash away their impurities and cure their ailments. At the waterfall of the Kiyomizu Temple in Japan, another purification ritual is called the *Misogiharae*, where people dowse themselves under the falls to wash away illnesses. And at the time of the Laotian New Year, offerings are made to the Mekong River as "the mother of waters" during the festival of Ban Pi Mai. This is a three-day festival during which people pray while awaiting the coming of the rainy season.

The Jordan River, a river in the Middle East; the Styx, a mythical Greek river; and the earth's galaxy, the Milky Way, were all thought at one time or another to be water pathways through which souls crossed from life to salvation in the afterlife. For example, ancient Greek mythology contained the belief that the souls of the dead were ferried to rest across the dark waters of the mythical River Styx. It was thought to be the river that separated the world of the living from the world of the dead.

Nighttime view of Kiyomizudera ("Pure Water Temple"), one of Japan's most celebrated Buddhist temples. Founded in 780 CE at the Otowa Waterfall east of Kyoto, it was named for the falling water. Kiyomizudera was designated a UNESCO World Heritage site in 1994. (Shutterstock)

The Sacred River Nile and the Gods

The Nile River is special. Out of ancient Egypt came a number of myths, legends, and gods directly related to this sacred river.

The Nile meant life, and the seasonal flooding of the Nile was essential for irrigation, agriculture, life, or death. To measure the seasonal height of the Nile ancient Egyptians developed a tool called a nilometer, or Nile measuring stick, also called an *ankh*. The ankh looked like a cross with a circle on top, and was later adopted as the symbol of the Coptic, or early Christian, cross. Water at a certain height signaled flooding, sedimentation, prosperous crops, and survival for the coming year. It also determined the amount of tax imposed by the Pharaoh on his subjects.

The Nile was not only sacred to ancient Egyptians, it was also a god in Egyptian beliefs. The river possessed godlike qualities, and all Egyptian pyramids and burial places were located on its west bank. Hence, the east bank signified life; the west bank meant death. The euphemism of someone going west was used in ancient Egypt to signify someone's death. Of the many Nile gods, Hapi was the creator god, who lived in a cave near the sacred cataract, Aswan, surrounded by crocodiles and the goddesses Anuki and Satis.

Animal sacrifices were made to Hapi to solicit the arrival of the Nile floods and the advent of fertile soil. Khmun was another creator god of the Nile, who fashioned all things, including the annual Nile inundation, while Osiris rose to prominence as the Nile god of flood and fertility. Regular Nile inundations meant that the chief god Maat—the final judge of Egyptians in the afterlife—approved of the natural order of things. Bountiful floods and abundant crops appeased Maat and all the Nile gods. Irregular flooding and harvest disruptions angered them.

Egyptians hunt with boomerangs and spear fish from the Nile River in this scene from the tomb of Nakht. The Nile River, from antiquity to the present day, has been the heart of civilization in Egypt, providing water and soil deposits for agriculture, energy, fish, and transportation. (Corel)

Cleansing Baths in Judaism

In Judaism, the mikvah was a special body of water used as a source of cleansing and rejuvenation (see further detail in case study). The best cleansing waters were considered to be the "living waters," the rivers and streams in the Promised Land, i.e., the Jordan River and the Dead Sea. The Red Sea had special significance for Jews since it was the place of the miraculous parting of the waters as described in the Hebrew Bible, where God helped them to escape from the pursuing Egyptians. This miracle was God's reward for the faith of Moses and his efforts to lead the chosen people out of slavery.

The Red Sea parting showed the power of God over a mighty sea and illustrated the use of water as an instrument of punishment for the Egyptians—and a blessing for the Jews. In Jewish thought, as expressed in Jeremiah 2:13, God himself was the fountain of living waters. This predates the words of Christ by hundreds of years, but the idea continued in the New Testament, where the spirit of God or eternal life is represented with such words as "living water" and the "water of life."

Water in Chinese, Babylonian, and Persian Myths

In ancient China and other cultures, water played important sacred roles as well. The water fountain at Pon Lai was believed to confer a thousand lives on those who drank it, according to Wang Chia, writing in the Chin Dynasty (265–420 BCE). Water sources were considered to be the source of life in the specific den of the dragon, which in China is still regarded as a sacred and special creature. Assyrian-Babylonian myths stated that gods and all beings arose from the mixture of salt and sweet water. The tears of the god Tiamat created the River Tigris and the River Euphrates, which were considered sacred and a source of miracles. In Persia, Anahita was an ancient water and fertility goddess who reigned over all rivers and lakes, symbols of birth, life, and miracles. People offered prayers and thanks to Anahita to gain fertility.

In other Mesopotamian legends, Abzu was the most prominent god of water. Civilization only became possible after Mesopotamian gods controlled water, and certain gods had specific roles and responsibilities. For example, Enki was the lord of water, and Ninurta was the god in charge of rain, floods, and fertility. Legends recall that Ninurta used stones to keep the Tigris River from flooding. Ninurta is also credited with introducing irrigation to Sumerian civilization. Other Sumerian stories expound that survival was only possible through the divine oversight of the fertility of humans, plants, animals, and agriculture. In the Sumerian pantheon of gods, other gods believed to be significant were Enbillulu, the god in charge of the Tigris and Euphrates Rivers; Nanna, responsible for the southern swamps in the area; Nanshe, for the sea; and Ishur, for the rain.

Water in Islamic Rituals

Muslim texts extol the sacredness of water and its association with the divine. In Islam, water is a symbol of God's life-giving spirit. The well of Zamzam is considered a sacred site in Islam. When Haggar and her son Ishmael (the son of Abraham) were wandering through the desert, it is said that Allah made water come out of the well to save them. According to the Hadith (the words spoken by the prophet Mohammed during his lifetime), when Mohammed was asked what was the most praiseworthy deed, he replied that it was to give a gift of water to another person. In countries practicing Islam, water is still considered a precious gift to bestow on visitors, a practice that is slowly changing in areas affected by severe drought and aridity.

Camille Gaskin-Reyes

REFERENCES AND FURTHER READING

Barnett, Cynthia. *Rain: A Natural and Cultural History*. New York: Crown Publishers, 2015.

Beversluis, Joel, ed. *Sourcebook of the World's Religions*. Novato, CA: New World Library, 2000.

Crim, Keith R., Roger A. Bullard, and Larry D. Shimm. *The Perennial Dictionary of World Religions*. San Francisco: Harper, 1990.

Levinson, David. *Religion: A Cross-Cultural Encyclopedia*. Oxford, UK: Oxford University Press, 1998.

Newton, David E. *Encyclopedia of Water*. Westport, CT: Greenwood Press, 2003.

Pearce, Fred. *With Speed and Violence: Why Scientists Fear Tipping Points in Climate Change*. Boston: Beacon Press, 2007.

Redford, Donald B. *Oxford Essential Guide to Egyptian Mythology*. New York: Oxford University Press, 2003.

Roaf, Michael. *The Cultural Atlas of Mesopotamia*. New York: Checkmark Books, 1990.

Smith, Huston. *The Illustrated World's Religions: A Guide to Our Wisdom Traditions*. San Francisco: Harper, 1995.

Willis, Roy. *World Mythology*. New York: Simon & Schuster, 1993.

CASE STUDIES

The case studies in this section develop the theme of water in creation myths in a vast array of cultures in different geographical parts of the world. Most of these myths have a watery element or watery void, whether in the sky or on land, from which the gods created the world or people to populate the world. The second and third case studies turn to the significance of water in Abrahamic and other world religions. In most of the religions explored in the case studies, water or rivers play a cleansing, purifying, or enlightening role throughout the life cycle of specific cultures and religions. In some of the rituals, water is combined with fire or baptism, whereby water is often used to signify new beginnings or awakenings as well as the final passage from life into death.

Case Study 1: Creation Myths in Different Cultures

A myth usually describes a story that has been handed down orally from generation to generation without any judgment of verification of its truthfulness. Creation stories that explained the origins of gods or interpreted natural phenomena eventually became part of the culture of a group, even if they were subsequently disproved by facts or scientific data. Most of them contain water or a watery element. As outlined below, cultures around the world have a wide range of creation stories; some are remarkably similar in spite of different geographical locations.

Babylonian Creation Story

The Babylonian saga of creation begins with two watery beings, Apsu (sweet water) and Tiamat (salt water) representing male and female. Their union brought forth sea monsters and gods. Tiamat's descendants opposed her and chose Marduk, the god of Babylon, to challenge her rule. Marduk, armed with hurricanes and tempests, killed Tiamat and her evil accomplice, Kingu, in battle. He then split Tiamat's body into two; with one half he created heaven for himself and other gods to reside in; with the other half he created earth.

African Creation Narratives

Africa overflows with water creation stories. The Boshongo group of the Congo claims that in the beginning there was only darkness and water, until the god Bumba developed a stomachache and vomited up the sun. The sun dried the water and left the land habitable for humans. In Nigeria, the Yoruba believe that in the beginning there was only the sky above and the water below. The chief gods were Olorun, who ruled the sky, and Olukun, who ruled the waters. Olukun created earth from a watery primeval liquid and breathed life into beings. In Zimbabwe, the creator of life was Modimo, who belonged to the water element. In southern Africa, the Zulus' creator was called Unkulunkulu, or "The Ancient One." He came from reeds in a watery swamp and brought forth people and creatures.

Another group that lived in Africa was the Dogon. Heaven was their god, known as Amma. In the beginning of time, Amma lived in the celestial region on the star Sirius and created everything. The stars represented the body parts of Amma and the constellation Orion, which was called the seat of heaven. Amma became split, creating Ogo, which was disorder, and the chaos of water. Ogo then descended in an ark along the Milky Way to the watery earth.

Egyptian Creation Narrative

The ancient Egyptians' creation story revolved around the Nile and its sacred waters. In the beginning there was watery chaos, called Nu; out of Nu, creation emerged. In this watery environment, Atum, the first god created a hill, from which he made other gods, humans, and living creatures. Another Egyptian myth recalls the sun god Ra, who transformed himself into the scarab Khepri, the creative force of the universe. Heaven and earth did not exist, so Ra raised everything out of Nu (creation). Ra made the watery abyss, from which he created everything else, including other gods. From the tears of Khepri's loneliness, men, women, and animals emerged.

Creation Myths in the Americas

In North America, the Cherokee are unique among indigenous peoples; they refer to themselves as the "water people." Their legend states that at the beginning there was only water. All creatures lived in the watery sky, which became overcrowded. A water beetle volunteered to go below the sky to find land or mud. When he found mud, he began spreading it everywhere, forming the earth. Earth became attached to the sky, but the animals could not see earth in the dark. Therefore, the sun was created to shine on the land of mud.

In South America, the creator of the Incas was the god Viracocha. Legend has it that he emerged from Lake Titicaca to bring light to the darkness and chaos by creating the sun, stars, and moon. His original people were disobedient, so he killed them through a flood. Remorseful, he allowed the water to recede, then tried again to re-create people, using clay. He then walked the earth as a beggar teaching his creations how to live better.

The Guatemalan Quiche-Maya in their sacred text, the Popol Vuh, reported that the gods attempted creation three times. On the final attempt, the creator god shaped four men, called the old ones. They used corn and clay from the water world to make other beings. While the new corn and clay formations slept, the old ones made women companions. From the first three creation attempts, the Mayans continue to honor the names of water, fire, and thunder.

Greek and Roman Water Creation Myths

The myths and philosophy of the Greeks and the Romans are well known. The Greeks believed in a pantheon of gods and developed creation myths associated with these deities. From these gods, Zeus emerged as the chief deity who brought the rain. The Greeks thought there was a void out of which Gaia, the Earth, arose. Prometheus, another god, angered Zeus when he smuggled fire into the world, so Zeus sent a flood to put out the fire, from which two humans escaped. When the flood subsided, the Oracle at Delphi told the two humans to throw the bones of their ancestors into the watery void; from the bones, more humans were created. In spite of their belief in the gods, the Greeks also had independent thinkers such as Socrates, who thought that the clouds, not Zeus, sent the rain. Socrates interpreted rain as water in constant motion.

Scandinavian Creation Myths

Scandinavian peoples had Norse creation myths, which reflected the cold climate of Northern Europe. In Norse mythology, there was nothingness in the beginning; gradually the space filled with water, which froze, then melted. This chaos was called Ginnungagap, and it had boundaries of fire and ice on either side. Drops from the melting ice revealed a giant called Ymir and a cow called Audhumbla. From their armpits they created people, who killed the aged Ymir. From Ymir's flesh they made earth; from his skull, the heavens; from his blood, the sea; and from his bones, the mountains and trees.

Japanese Creation Stories

In ancient Japan, myths emphasized the creation of the first emperor, not the first human. According to these myths, the earth was a floating amorphous mass of water. The eighth pair of gods, Izanagi and Izanami, created the eight islands of Japan. The first emperor, Jimmu-Tenno, emerged as a descendant from the great-grandson of the sun goddess. Three treasures symbolized his rule: a necklace, a mirror, and a sword—imperial symbols still represented in the sanctuary of Shinto shrines.

Australian and Pacific Ocean Creation Myths

Dreamtime was the cosmos of the native people of Australia. It encompassed the vast Pacific Ocean that included the Polynesian, Melanesian, and Micronesian islands. Given their oceanic location, their creation myths all appear to involve water, whereby Tangaloa, a creator god, sent a bird messenger over the endless sea and threw a rock to allow the bird to rest. Out of the many rocks he threw, islands emerged in the Pacific Ocean.

Hindu Creation Myths

Hinduism has many creation myths. Each village had its own god and each god its own creation story. Above all the gods in all the villages stood the triumvirate of Brahma, Vishnu, and Shiva, and above them was the god of all gods, Brahma. The creation myth involves the god Brahma, who was alone in a watery void, and by the force of thought created water. In the water, he deposited eggs, which grew into one golden egg. He then allowed himself to be born in the egg. Then, by force of thought once more, he split the egg, the two halves becoming heaven and earth.

Camille Gaskin-Reyes

REFERENCES AND FURTHER READING

Beversluis, Joel, ed. *Sourcebook of the World's Religions*. Novato, CA: New World Library, 2000.

Buck, William. *Mahabharata*. Berkeley: University of California Press, 1973.

Dalley, Stephanie. *Myths from Mesopotamia*. Oxford, UK: Oxford University Press, 1989.

Hart, Merriam C. *The Dawn of the World*. Lincoln: University of Nebraska Press, 1993.

LaHaye, Tim, and John Morris, *The Ark on Ararat*. Nashville: Thomas Nelson, 1976.

Norman, Howard. *Northern Tales, Traditional Stories of Eskimo and Indian Peoples*. New York; Pantheon Books, 1990.

Sproul, Barbara C. *Primal Myths*. New York: HarperCollins Publishers, 1979.

Waters, Frank. *Book of the Hopi*. New York: Penguin Books, 1963.

Willis, Roy. *World Mythology*. New York: Simon & Schuster, 1993.

Case Study 2: Water in Abrahamic Religions

Water has been a part of religious rituals for as long as we can trace human activities that pertain to the realm of the sacred. Most often, in offering a cleansing of the body, the use of water symbolizes the cleansing of the soul. An early instance of this is found in the Persian world of antiquity, perhaps as far back as the sixth century BCE, where Zoroastrianism developed as an important faith. Among its most significant features was the use of fire and water for ritual purposes: to signify the pure intentions of those who worshipped Ahura Mazda, the all-creative supreme being, and to accompany their larger behavioral patterns in pursuit of *druj* (straight/good), as opposed to the impure intentions of those following Ahriman, the adversary of Ahura Mazda, and *ash* (crooked/evil).

At least some Judeans exiled to Mesopotamia by the Babylonians in 586 BCE—and redeemed from exile through the conquest of Babylonia by the Achaemenid Persians in 538 BCE—may have been influenced by Zoroastrianism. Ceremonies in the temple had already included the act of sprinkling water as part of god-addressing rituals before the exile. In the centuries after the return, however, the priesthood in the temple came to be viewed by some Judeans as morally impure. A group (or possibly, groups) of them withdrew from the Judean mainstream and lived in monastic communities in the wilderness. The most renowned such community was at Qumran, along the northwest corner of the Dead Sea.

There, they awaited divine intervention, which, they believed, would bring to an end the spiritual impurity and moral corruption of nearly all of humankind—both enemies of the Judeans and mainstream Judeans alike. When the Qumran community, instead of experiencing victory over its many opponents, was destroyed or dispersed (probably during the time of the Judean revolt against the Romans, in 65–70 CE), it left behind many parchment scrolls, sealed in clay jars and hidden in caves near the monastic complex. These are called the Dead Sea Scrolls.

Dead Sea Scrolls

It is from the Dead Sea Scrolls that we learn of these ascetics' vision of a final battle between the forces of good, led by the Good Teacher (God's servant) and the forces of evil, led by the Wicked Priest (serving the adversary of God). The idea of such a final battle distinctly echoes Zoroastrian traditions. The key difference is that in the latter narrative the forces of ash, even the defeated Ahriman himself, turn back in the end to druj, whereas in the Qumran text, the enemies of the Lord are altogether destroyed.

Among the scrolls is one that describes the process of joining the sect and the ceremonies essential to membership. One ritual of particular importance is that of immersing one's body in water—the Greek verb for this is *baptizein*—to purify the body and to signify one's spiritual purity as a member of the group that will champion and be championed by God and God's angelic hosts at the end of time with its Final Battle.

One of the intriguing questions that emerge out of the Judean world as Judaism and Christianity slowly spread is: what happened to the Qumran community after the apparent destruction of its monastic complex? There is, of course, the possibility that they all simply disappeared. Perhaps, however, they or at least key elements of their beliefs and teachings helped shape Judaism or Christianity or both. John "the

Forerunner" (as he is known particularly in the Eastern churches), who came out of the wilderness proclaiming the imminent end of the current world and the articulation of a new reality, may even have been a member of the sect associated with Qumran. In the West he is best known as John the Baptist, due to the key act associated with his preaching: immersing those in water who were becoming new members of his group.

Baptism Rites

What eventually became Christianity placed and continues to place emphasis on baptism as an instrument for being reborn into a new, positive reality that looks beyond all the negatives inherent in our everyday material world. It would be defined, among other things, by a specific canon of books considered sacred, approved as such by a council of bishops that met at Hippo, North Africa, from 393 to 397.

Christian baptism reinforces the symbolism of the Israelites crossing the Red Sea to a new beginning. The concept of baptism was further fortified by the full immersion of Jesus by John the Baptist in the River Jordan. Upon his resurrection, Jesus commanded his disciples to continue performing baptism in the name of the Father, the Son, and the Holy Ghost (Matthew 28:19–20).

Today, baptism in the Catholic Church is rarely done by full bodily immersion, baptism or full body immersion may still be prevalent in Baptist, Orthodox, or Jehovah's Witness rituals. A priest performs regular Christian baptism through the pouring or sprinkling of water over the head. It symbolizes the cleansing power of water or liberation from the oppression of sin that separates people from God. The Catholic Church regards baptism as a life-changing sacrament, where the stain of original sin is removed from the individual.

Holy water is water blessed by a church leader for certain rites such as baptism, or for holidays such as Easter. This practice goes back to the fourth century and was adopted in Western countries in the fifth century. The custom of sprinkling water at mass began around the ninth century. At that time, churches arranged basins of water called stoups for people to sprinkle themselves upon entering the church. Christian rites still use water for priests to wash their fingers and communion vessels after the communion rite. While in the past, water was an integral part of an exorcism rite to cast out devils, today it is mostly used for blessings, dedications, and burial rituals.

Water in the New Testament is connected with the gift of eternal life, as stated in John 4:14, "But whosoever drinketh of the water that I shall give him shall be in him a well of water springing up into everlasting life." In Revelation 21:6, it says, "I am the Alpha and the Omega, the beginning and the end. I will give unto him that is athirst of the fountain of the water of life freely." John 4:1–42 tells an important story about the relevance of water; it recounts the story of Christ meeting a Samaritan woman at Jacob's well. He offers her living water so that she will never thirst again.

Jewish and Muslim Ritual Purity

Judaism's great emphasis on ritual purity stems from the Torah. In Judism, water is a symbol of God's blessing and spiritual refreshment (Isaiah 35:6-7). Ritual cleansing is important from womb to tomb. Rituals include washing hands and feet and immersion in "living waters," whether in a sea, river, or a *mikveh* (bath)—the specific dimensions, depth, and flow of which are carefully prescribed in the rabbinic literature. In early Judaism, when there was a central temple, ablutions were done by the priestly caste as part of initiation rites, and priests washed at their consecration (Exodus 29:4). On the Day of Atonement (Leviticus 16:4, 24, 26), all are required to wash with water to remove pollution (Leviticus 11:40, 15:15 and Deuteronomy 23:11). Most typically, men and women alike (in gender-separate groups) immerse themselves in the ritual bath prior to the Sabbath in order to be properly cleansed to receive the glorious day of peace—the body ideally reflecting the state of the soul. So, too, women immerse themselves after menses to be cleansed—ritually and physically—of the blood associated with life and death. In Jewish burials, the body is washed in clean water, dressed in a shroud, and placed in a plain wooden coffin.

Even more frequent than these total immersions is the obligatory washing of the hands before a meal—accompanied, always, by a blessing. In the course of Jewish history as a diasporic phenomenon, since the destruction of the temple in Jerusalem in 70 CE, the synagogue has been treated as a temporary temple until such time as the Temple will be rebuilt in the messianic era. The home, a place where, for example, one might pray formally three times a day, becomes a secondary sacred space. The dining room table assumes a sacred identity. Before eating and after eating particular blessings are recited, and before approaching the table to eat one washes while reciting a blessing. This ritual turns the act into a sacred as much as a hygienic one.

In Islam, Muslims must be ritually pure before approaching God or Allah. Some mosques have a courtyard with water, but most have places of ablutions outside mosques. In Islam, ritual purity is mandatory before all religious duties, especially prayer by the devout five times daily. There are three forms of ablutions. The first is *ghusi*: the major ritual for washing the body in pure water before weekly Friday prayer meetings, two main feasts, touching the Koran, and burying the dead. The second is *wudu*: a minor ablution to remove daily impurities before each of the five daily prayers: washing the face, rubbing the head, and washing the arms up to the elbows and the feet up to the ankles (Koran: Surah 5:7 and 8). The third type is performed where water is unavailable, using clean sand.

Camille Gaskin-Reyes and Ori Z. Soltes

REFERENCES AND FURTHER READING

Bible, The. King James Version. New York: Thomas Nelson, 1982.
Bowker, John. *World Religions*. New York: DK Publishing, 1997.
Gaster, Theodor H. *Myth, Legend, and Custom in the Old Testament*. New York: Harper & Row, 1969.
Panati, Charles. *Sacred Origins of Profound Things*. New York: Penguin Books, 1996.
Pollock, Robert. *World Religions*. New York: Falls River Press, 2008.
Quran, The. New Delhi: Good Word Books, 2009.
Wilkinson, Philip. *Religions*. New York: Sterling Publishing, 2008.

Case Study 3: Water in Other World Religions

Water and Hinduism

Hinduism is sometimes called a water religion since water has deep spiritual and cleansing powers. Hinduism is associated with five elements: earth, water, fire, air, and ether. Seven sacred rivers stand out: the Ganges, Yamuna, Godavari, Sarasvati, Narmada, Sindhu, and Kaveri. Hindu myths explain that the origin of the Ganges, the most sacred river, is the matted hair of the god Shiva, the Destroyer. In India, the Kum Mela is held every three years at four different locations on four different rivers. It is the largest religious gathering in the world. At the final meeting, a Maha (great) Kum Mela is held at Allahabad, which is located at the confluence of three holy rivers. At that time, millions of people converge upon the area to bathe in the Ganges.

The Ganges is understood by Hindus to purify those who immerse themselves in it according to prescribed rituals. Drinking from it and the bathing of images of deities within it reflect the importance of the river as a gift of God. Individuals even pray to it; for some the river is itself a goddess, the daughter of the mountain god, Himalaya. The Ganges' water connects the devotee to God. It is bottled by pilgrims and given to the ill and the dying, and also to wedding couples to cleanse them of sins and bring them good luck in this life or the next.

In the life cycle of a Hindu, water use in worship begins immediately after childbirth. Prescribed rituals in birth, coming of age, marriage, and death all involve water. A relative of the newborn pours drops of water on the baby's body, and a sprinkling of water is performed on the third day after birth. A girl's attainment of puberty is celebrated with a ceremonial bath followed by a feast. In Hindu temples, gods, or *murtis*, are also given periodic baths.

Morning ablutions are obligatory in Hinduism. A worshipper draws water through the nostrils and cups water in his or her hands while reciting a mantra and pouring water back into the river. In daily life, Hindus preface almost every action by offering food and water to a god. Devotees must wash before entering a temple. The deceased are cremated close to rivers. On the third day after cremation, the ashes are collected; on the seventh day water rituals are performed; and on the tenth day the ashes are cast into a holy river.

Water and the Zoroastrians

Zoroastrians believe that pollution was evil and pure water was sacred. Its narrative claims that Zoroaster, the founder, received his revelations from Ahura Mazda, the god on the river. Zoroastrians pray five times a day and perform ritual washings before prayers. Minor ablutions involve washing and prayers. The most serious form of pollution is considered contact with a corpse, requiring a nine-day cleansing ceremony involving prayers and constant washing. Believers observe the High Holy Day, called *Haurvatat*, by keeping water unpolluted and praying at a natural source of water.

Water and Taoism

In Taoism everything in the world is comprised of five elements: water, fire, wood, metal, and ether. These five elements correspond to the five major organs in the body.

From Taoism, the modern use of feng shui has emerged, which involves placing a bowl of water in homes to create harmony. In Buddhism, however, water does not play as significant a role as in other religions. External rituals are understated in favor of internal enlightenment. The main use of water in Buddhism is in the funeral ceremony: water is poured into a bowl placed before the monks and the dead body. As the water fills the bowl, the monks chant sacred chants and prayers for the departed.

Water Rituals and the Cherokees

Similar to Judaism, North American indigenous peoples have a long history of recognizing the importance of water to connect humans to the forces that created and maintain our world. The Great New Moon Ceremony of the Cherokee, for instance—the first of seven central rituals within that tradition—is performed at the time of the autumn new moon and includes a purification ceremony consisting of immersing oneself seven times in water (this is called "going to water"). A very different significance of water is its centrality in the more widespread sweat ceremony, during which prayers of thanks are offered in four cycles. Each time, water is poured onto hot "grandfather" stones, yielding billows of hot steam that cleanses and purifies one's body, reducing the toxins within it and expressing a parallel cleansing of one's spirit.

Exception to the Use of Water for Purification Rites

Nearly every tradition across humanity with ready access to water absorbs it into ideas and ceremonies that pertain to purification. One of the more interesting exceptions to this was the medieval Albigensian movement. Located mainly in the south of France between the 11th and 14th centuries, the Albigensians also called themselves Cathars, from the Greek word *katharos*, meaning "pure." Their sense of purity included the idea that any material substance is, by definition, impure. Thus, water is as impure as anything else in our world, and baptism is a false sacrament to use water. The mainstream church, not surprisingly, saw the Cathars as heretics, and they did not survive beyond the 14th century.

Camille Gaskin-Reyes and Ori Z. Soltes

REFERENCES AND FURTHER READING

Bowker, John. *World Religions*. New York: DK Publishing, 1997.

Eliade, Mircea. *Myth and Reality*. New York: Harper Torch Books, 1968.

Emoto, Masaru. *The Hidden Messages of Water*. New York: Atria Books, 2004.

Evans, Bergen. *Dictionary of Mythology*. New York: Dell, 1991.

Hall, Manley P. *The Secret Teachings of All Ages*. New York: Penguin, 2003.

Heidel, Alexander. *The Gilgamesh Epic and Old Testament Parallels*. Chicago: University of Chicago Press, 1949.

Hooper, Richard. *Jesus, Buddha, Krishna, & Lao Tzu: The Parallel Sayings*. New York: Bristol Park Books, 2013.

Kramer, Samuel Noah, ed. *Mythologies of the Ancient World*. Garden City, NY: Anchor Books, 1961.
Panati, Charles. *Sacred Origins of Profound Things*. New York: Penguin Books, 1996.
Pollock, Robert. *World Religions*. New York: Falls River Press, 2008.
Swami Prabhavananda. *The Upanishads*. New York: Signet Publishing, 2002.
Wilkinson, Philip. *Religions*. New York: Sterling Publishing, 2008.
Wilkinson, Philip, and Neil Philip. *Mythology*. New York: Sterling Publishing, 2007.

ANNOTATED DOCUMENT

River Ganga at a Glance: Identification of Issues and Priority Actions for Restoration (2010)

Background

The document on restoration of the Ganges, or Ganga, River basin, excerpts of which are outlined below in this section, illustrates the irony of the situation of the Ganges River. Hindus believe the river is holy in nature; it has had religious and mythological significance in India's culture for millennia. However, pollution, abuse, and overexploitation of the Ganges has led to severe contamination of these holy waters. In fact, of the 14 major river basins in India, the Ganges has the highest level of pollution.

This mismatch between legend and reality prompted India's Central Pollution Control Board (CPCB) to formulate a Ganga Action Plan in the 1980s to address Ganges pollution and set in motion key measures to address the problem. However, in spite of financial support and involvement of many NGOs and communities, the program had only limited public impact.

In 2009, the government established the National Ganga River Basin Authority (NGRBA) to oversee state and federal pollution abatement and implement an environment management plan in the river basin. The management plan was to include measures to accommodate water and energy needs of the growing Indian population and economy, while protecting the sanctity of the river system. The vision of the Ganga River Basin Management Plan is that the river:

- Must continuously flow
- Must have healthy connectivity with the environment surrounding and feeding into the river
- Must not be viewed as a carrier of waste
- Must have adequate space for its various functions
- Must function as a habitat

The management plan is an early step in the government's National Mission for Clean Ganga (NMCG), which aims to eliminate the dumping of sewage into the Ganges River by 2020. Restoring the Ganges to minimum acceptable levels of water quality for bathers and river users is a necessary goal, given the cultural importance of the Ganga for mass bathing rituals, cremations, devotional rites to gods, goddesses, and the river itself, as well as the health hazards posed by high levels of contamination.

Summary

The NGRBA produced several documents to provide information on the Ganga River Basin Management Plan, including *River Ganga at a Glance.* This document outlines the many issues facing the Ganges, including pollution from sand and stone mining and other industrial discharge, riverbed farming, netting of fish, wallowing animals, open defecation, clothes washing, and religious rituals such as cremation on the river. Challenges to the river's level and flow, such as diversion to hydroelectric projects and agricultural irrigation, as well as seasonal flooding, are also addressed.

The excerpt below describes the cultural, environmental, economic, and political significance of the Ganges River and identifies the objectives of river system restoration, which include the preservation of aquatic flora and fauna and the improvement of the aesthetic quality, health, and sanitation levels of river communities through reduction of industrial and other wastes at the source. Proposed measures include the development of zero-waste pilgrimage destinations, tourist restrictions, the conservation of river dolphins, and the reduction of sewage dumped into the river.

Camille Gaskin-Reyes

Selected Excerpts from *River Ganga at a Glance: Identification of Issues and Priority Actions for Restoration* (2010)

River systems have been the birthplace of civilizations all over the world. They are woven into the social and economic fabric of society and penetrate deep into the psyche of the people living around them. Nowhere is this more evident than in India where the Ganga, Indus, Narmada, and other rivers possess the cultural identity transmitted down the ages through sacred literature, the Puranas and the Vedas, as well as through popular myths and legends.

The River Ganga (commonly called as Bhagirathi in the stretch Gangotri to Devprayag and Hubli in the stretch Farakka to Ganga Sagar) occupies a unique position in the ethos of people of India. Emotional attachment to the river and the centers of pilgrimage on its banks runs deep and long in the Indian History.

The Ganga originates from the ice caves at Gaumukh (N 30°55′, E 79°7′) at an elevation of 4100 m. Alaknanda, its main tributary in the mountainous stretch, rises beyond Manna Pass, 8 km from Badrinath (N 30°44′, E 79°41′) at an altitude of 3123 m, and meets at Devprayag. The Ganga traverses a distance of ≈2510 km from its source to its mouth (Ganga Sagar), draining eleven states of India. In her course she is joined by many tributaries, important being Bhilangana, Alaknanda, Ramganga, Kali, Yamuna, Gomti, Ghagra, Gandak, Kosi and Sone.

The entire stretch of river Ganga (main stem) can be viewed into three segments:

A. Upper Ganga ≈ 294 km Gaumukh to Haridwar
B. Middle Ganga ≈ 1082 km Haridwar to Varanasi
C. Lower Ganga ≈ 1134 km Varanasi to Ganga Sagar

These three segments not only differ in their geomorphology, ecology, and rheology but are different in terms of issues that need to be addressed:

A. The river in the upper segment flows on steep and narrow bed, mostly rocks and boulders, carries cold water, is subjected to much less anthropogenic pollution, has highly sensitive and fragile ecosystem and biodiversity, and most importantly considered to have potential for harnessing hydropower.

B. The river in the middle segment enters and flows in plains, meandering mostly on bed of fine sand, has wide river bed and flood plain, and most importantly modified through human interventions in terms of huge quantities of water diversion/abstraction and subjected to high degree of pollutant loads from domestic, industrial, and agricultural activities.

C. The river in the third segment has experienced considerable changes in the sediment transport and deposition, causes wide spread flooding, undergoes frequent changes in her channel path, and most importantly is subjected to international disputes on flows and interventions made and/or are being carried out/planned.

Upper Ganga Segment: Gaumukh to Haridwar

. . .

Suggestions and Recommendations

- Gangotri Valley to be viewed as a place of pilgrim tourism and spiritual activities. All commercial activities in the vicinity of Gangotri (say within 500 m) to be transformed into eco friendly activities. Plan for environment protection and preservation of natural and pristine conditions (e.g. hotel culture to be changed to hut culture; severe restrictions on overnight stay of tourists, promotion of pilgrim tourism than commercial tourism, facilitating “Pad Yatra”, encouraging use of locally available materials, provision for segregated collection of entire solid waste of all kinds, entirely eliminating disposal of any kind of waste from anthropogenic sources in the valley, complete recycle/reuse and conversion into acceptable products of wastes generated; sanitation and bathing facilities with no direct/indirect discharge into river valley, control of noise and artificial lighting, etc.)

- Harshil to be developed as nature friendly, zero waste terminal pilgrim tourist spot with facilities of ashrams, guest houses, parking, internet, etc.
- All existing hydro electric projects may be redesigned and operated based on requirement of E flows.
- Community toilets and wash rooms with zero discharge of solid and liquid wastes to be developed at number of places as per approved plan for the entire UGS.
- Detailed studies and documentation on changes due to implementation of hydroelectric projects in the region. . .

Middle Ganga Segment: Haridwar to Varanasi

. . .

Suggestions and Recommendations

- River Bank and River Water Quality Management Plan to be prepared as described . . .
- Industries directly or indirectly discharging their solid/liquid wastes into Ramganga, Kali, and Ganga must be directed to follow best available practices for managing solid/liquid wastes and attain complete recycling of water and proper disposal of solid wastes/sludges as per norms enforced by the regulating agencies.
- Ganges Dolphin Conservation Zone Garhmukteshwar—Narora Barrage to be declared as "NO GO AREA" for which detailed studies have been done by WWF-India.

Lower Ganga Segment: Varanasi to Ganga Sagar

. . .

Suggestions and Recommendations

- Hydrological and geomorphological studies for managing sediment transport and water resources. . . .
- River Bank and River Water Quality Management Plan to be prepared as described . . .
- Industries directly or indirectly discharging their solid/liquid wastes into Ganga must be directed to follow best available practices for managing solid/liquid wastes and attain complete recycling of water and proper disposal of solid wastes/sludges as per norms enforced by the regulating agencies.

Source: Ministry of Water Resources, River Development & Ganga Rejuvenation. *River Ganga at a Glance: Identification of Issues and Priority Actions for Restoration* (2010). Available online at https://nmcg.nic.in/writereaddata/fileupload/33_43_001_GEN_DAT_01.pdf.

PERSPECTIVES

Should Cultural Exceptions Be Made in Whaling Laws?

Overview

Whales are some of the largest creatures on earth. Their high levels of intelligence and sociability have made them Hollywood stars and poster children for the environmental movement. These marine mammals can be found in nearly every ocean, and societies around the world have found use for them. Due to their heft—a blue whale can reach up to 100 feet—and versatility, cetaceans have long been hunted as a source of food and resources. The earliest evidence for dolphin hunting dates back to 6300–5300 BP, on Santa Cruz Island off California. Archaeologists have found similarly ancient signs of whaling along the North Pacific Rim and nearby Arctic regions in places like Japan and Chukotka, Russia, areas where people have continued to hunt whales into the modern era for commercial and subsistence purposes.

Today, whaling can be placed into three main categories. First, there is commercial, for-profit whaling, begun by the Basques in the 11th century. Living in present-day Spain and France, these people hunted right whales and bowhead whales as far west as Newfoundland, Canada, and as far north as the Svalbard archipelago, in the Arctic Ocean. Whaling became modernized in the 19th century, when harpoon gun-fired grenades allowed boats to swiftly kill large numbers of whales for their oil, blubber, and baleen, an activity which peaked in the 1960s.

Today, only Norway and Iceland officially carry out commercial whaling, objecting to the moratorium placed on the activity by the International Whaling Commission (IWC) in 1986. While these two Nordic countries are members of the IWC, their reservation against the moratorium contradicts what has become an increasingly accepted international norm. The adherence of most of the world's countries to the moratorium has massively reduced whaling, helping all whale species recover from the risk of extinction. Whereas 66,000 whales were killed in 1961, only 326 were killed in 1989—yet that number grew to 2,000 by 2008 (Hurd 2012).

Part of the recent increase has taken place under the second of two categories: scientific whaling and indigenous whaling. So-called scientific whaling is permitted by the IWC, which allows countries to "kill, take, and treat whales for purposes of scientific research" (IWC 1946). However, Japan's scientific whaling in the Southern Ocean off Antarctica was ruled illegal in March 2014 by the International Court of Justice (ICJ) in a case brought by Australia, which argued that the Asian nation was violating its obligations under international law (Strausz 2014). No such ban, however, has followed for the country's scientific whaling in the North Pacific Ocean, though many environmentalists hope that one will follow suit.

Harder to justify on ethical grounds, yet arguably necessary in order to maintain the overall health of whale stocks, would be the banning of the third category: aboriginal subsistence whaling. The IWC explains that this practice "does not seek to maximize catches or profit," and the organization "ensures that hunts do not seriously increase the risk of extinction" while also allowing "native people to hunt whales at levels appropriate to cultural and nutritional requirements (known as 'need') in the long-term" (IWC 2014). The multilateral organization manages indigenous whale hunts by the Inupiat and Yupiit in Alaska, the Chukchi and Yupiit in Russia, the Inuit in Greenland, and the indigenous peoples of Bequia in Saint Vincent and the Grenadines. The IWC also oversees whaling by the Makah in Washington State, but lawsuits brought by environmental organizations have halted this hunt. Indigenous whaling also takes place outside the auspices of the IWC in Indonesia, and by communities in Canada, where the constitution guarantees the Inuit people's right to whale. Neither of these countries are members of the IWC.

As the first essay explores, those in favor of aboriginal whaling highlight the practice's sustainability. Indigenous peoples use the whole body of a whale from its flesh to its rib bones, which have even been used as structural supports for dwellings in Chukotka (Gusev et al. 1999). It was not until the advent of industrialized whaling that species began to precipitously decline. Further, if indigenous peoples cannot hunt whales, it might be difficult and costly to fly in supplies to make up for the amount of sustenance and materials that would otherwise come from a whale. Aside from simple nutrition, whales have important cultural and spiritual meaning to indigenous peoples such as the Inupiat, who call themselves the "People of the Whales." To deprive these peoples of their ability to hunt whales would be to deprive them of their way of life.

On the contrary, those opposed to indigenous whaling observe that a dead whale is a dead whale regardless of the motives behind its killing. The second essay assesses the point of view of those who contest this activity, condemning it as brutal and uncivilized. Contrary to popular belief, indigenous whaling has become modernized in many cases, with hunters using motorized boats and harpoon guns. In places like Greenland, whale meat has been found being sold to tourists both on the island and as far away as Copenhagen, violating IWC regulations. Thus, although indigenous whaling is smaller in scale than industrialized modernized whaling, environmentalists still protest it, particularly its creeping commercialization. They argue that, just as indigenous peoples have adapted to the modern era by wearing high-tech parkas and driving snowmobiles, they can also adapt by discontinuing the archaic practice of whaling.

The two essays reveal that while aboriginal subsistence whaling enjoys somewhat more sympathy from the global community than commercial whaling, it is still a highly contentious issue, especially since the distinction between subsistence and commercial whaling is not always clear. Still, unlike in the waters off Antarctica, where Sea Shepherd boats have rammed into Japanese whaling boats in heated conflicts, environmental activists have so far refrained from entering into open skirmishes with indigenous subsistence hunters. Yet the issue remains a bitterly divided one, entangled with fights over indigenous rights and animal rights. Both sides have vocal proponents, ensuring that the debate over aboriginal subsistence whaling will rage into the 21st century.

Mia Bennett

Perspective 1: In Defense of Cultural Exceptions for Whaling

For thousands of years, people in a wide range of environments from the frigid Arctic Ocean to the warm waters off Indonesia have hunted whales for subsistence. While it is mistaken to believe that indigenous peoples inhabit an ahistorical world untouched by modern civilization, in many cases their practices still reflect ancient traditions closely tied to sustainable use of the earth's resources. With numerous rights already taken away from indigenous peoples, they should not have to cease their remaining activities, such as whaling, due to the flagrant overhunting of whales by industrialized societies. As such, cultural exceptions should be made in whaling laws when the practice is carried out for aboriginal subsistence purposes. Yet close monitoring of indigenous hunts is required to ensure transparency and guarantee that they are being conducted for their stated purposes.

Whaling is integral to several indigenous societies around the world, as the practice allows them to provide for the needs of their communities. In Alaska's North Slope, the Inupiat way of life revolves around the hunt of the bowhead whale. In the town of Barrow (population 5,200), a single bowhead whale carcass can feed a community for at least a year. Whale meat is a welcome traditional and local addition to a diet that otherwise must rely upon expensive foodstuffs like milk flown in from hundreds of miles away (DeMarban 2013). In Indonesia, the Lamalera use every part of the approximately six sperm whales they catch annually for everything from food to currency (BBC 2011). Without the ability to hunt whales, several cultures would lose their traditional ways of life along with a sustainable, local means of feeding themselves and providing for their needs.

Indigenous subsistence whaling thus differs starkly from commercial whaling. This industrialized practice has killed thousands of whales per year since the 17th century, expanding enormously in the 20th century to meet growing global demand for products like whale oil. In two Antarctic seasons alone (1959–1960 and 1960–1961), two Soviet ships killed 25,000 humpback whales (Clapham and Ivashchenko 2009). In light of the decimation of whale stocks, the International Whaling Commission (IWC), established in 1946, placed a moratorium on whaling in 1986. The organization currently allows aboriginal subsistence whaling in Greenland, Russia's Chukotka Autonomous Okrug, Saint Vincent and the Grenadines, and Alaska and Washington State (IWC 2014).

Today, only Iceland and Norway, both of which object to the moratorium, whale for commercial purposes despite IWC resolutions asking them to cease. Japan, after ceasing its commercial hunt in 1988, has since carried out what it claims to be a scientific whaling program. After the cetaceans have been ostensibly analyzed, the whale meat is distributed in supermarkets and restaurants across Japan even though national consumption of the product has declined (Blok 2008). While the argument could be made that Japanese whaling should be granted a cultural exception, in actuality, an "elite-driven countermovement, encompassing powerful actors and organizations from the bureaucratic, political, industrial, and cultural spheres" (Blok 2008, 60) actually drives the whaling industry, which is no longer a tradition integral to contemporary Japanese culture. Japanese whaling in the Southern Ocean is an industrialized activity that kills up to 950 minke, fin, and humpback whales yearly while reaping a profit of $50 million (Gales et al. 2005), making commercial interests seem to be the real motivation—not cultural ones.

Reflecting the codification of antiwhaling sentiment into international law in March 2014, the International Court of Justice (ICJ) ruled against Japan's scientific whaling activities in the Southern Ocean Whale Sanctuary off Antarctica, finding that "scientific output to date appears limited" (ICJ 2014). So, whereas Japan hunts whales for the benefit of a small group of politically powerful stakeholders, indigenous peoples hunt whales to fill dietary and cultural needs otherwise hard to meet.

Indigenous whaling is not without its issues. First, while cultural exceptions should be made for indigenous whaling, the claims justifying the right to whale and the activities themselves must be subject to close scrutiny and monitoring. The IWC, for instance, allows a total catch of 140 Eastern North Pacific gray whales by Chukotka's Yupiit and Chukchi indigenous peoples and Washington State's Makah people (IWC 2014). Yet it was discovered that the gray whales caught in Chukotka were being fed to foxes (Reeves 2002). It is unsurprising that the meat went to the dogs, so to speak, since Chukotka's indigenous peoples view gray whale hunting as less prestigious than bowhead whale hunting. This misuse of whale meat is worrying given that bowhead whaling, reintroduced after the collapse of the Soviet Union, has allowed the Yupiit to allegedly "strengthen their cultural identity, self-sufficiency, and hence self-respect at this time of extreme social, physical, and emotional distress" (Freeman 1998, 82). The IWC and the indigenous peoples claiming a right to whale must therefore closely communicate to ensure mutually satisfactory policies that enable the continuation of sustainable aboriginal sustenance hunts while ensuring that no whales are killed unnecessarily.

A second issue with permitting cultural exceptions is that while some IWC limits pertain to the number of strikes a community can carry out—that is, how many whales can be legally hit with a harpoon or rifle—others pertain to the total number that can be caught. Setting strike limits maintains whale populations more effectively than total catch limits, as struck, injured whales can still later die, depleting the population the same way a successfully hunted one would. The IWC should therefore turn all quotas into strike limits as a best practice. Ultimately, effective management of whale stocks affects more than just whales. The global decimation of these cetaceans has negatively affected other marine mammal populations, such as seals, sea lions, and otters, highlighting the need to maintain healthy whale stocks for the integrity of the entire marine ecosystem.

Allowing cultural exceptions for aboriginal substance whaling fosters the continued existence of traditional ecological knowledge, which has largely enabled a sustainable way of living with the earth, unlike industrial practices. After all, whale populations did not begin to decline dramatically until the introduction of commercialized whaling. With indigenous peoples sustainably hunting whales for thousands of years prior, according them the right to continue their practices while adhering to a robust and scientifically informed management system does a small amount of justice to peoples whose ways of life have already been trampled by the relentless onslaught of capitalism, industrialization, and globalization.

Mia Bennett

Perspective 2: No! A Dead Whale Is Still a Whale

From Pinocchio's Monstro to Captain Ahab's Moby Dick to the biblical leviathan that consumed Jonah, it is not that long ago that whales were viewed as monsters—sinister forces that lurked in the deep, dark oceans. The imagery shifted when it was discovered that whales had utilitarian value. Viewed as an exploitable resource, beginning in the 17th century the creatures were hunted en masse.[1] As technology and skills improved, by 1940 more than 50,000 whales were being killed annually. As our perception and understanding evolved, it was realized that whales were not monsters, but intelligent and long-lived beings—in fact, these sea mammals are more akin to humans than fish. Global efforts to protect the species soon took off as evidence of their looming extinction mounted. Unafraid of death, conservationists from Greenpeace and Sea Shepherd courageously placed their boats and themselves between a whaler's harpoon and its target. "Save the Whales" emerged as the slogan of a massive awareness-raising campaign that eventually helped inspire the worldwide ban on commercial whaling in 1986 as well as the formation of the International Whaling Commission (IWC) to oversee conservation efforts. Unfortunately, there are countries that have chosen to ignore the ban (Japan, Norway, and Iceland) as well as countries that continue to hunt by invoking "cultural exemptions" (Denmark, Russia, and Saint Vincent and the Grenadines).

The current IWC regime governing the hunting of whales via "cultural exceptions" is flawed, corrupt, ineffective, and in serious need of reform. Frankly, so-called cultural exceptions, which include anything from science to religious ceremony, are thinly disguised attempts to continue the practice of commercial hunting. The time has come to ban all forms of, and reasons for, whale hunting. "Culture" and "tradition" are not reason enough to slaughter some of Earth's most majestic and threatened creatures. The excuse of "culture" has long been used to defend unsavory practices from slavery to racism to misogyny.

Further, the "subsistence" and "for food" arguments in support of whaling are weak and anachronistic. Due to the biomagnification of toxins up the food chain and into the flesh of whales and other marine mammals, the consumption of whale meat in aboriginal communities, such as the Inuit of Greenland, is at historic lows. Further, by exposing the cruel practice, awareness-raising efforts have been successful in dramatically decreasing the demand for whale meat in Japanese markets—for too long a popular delicacy in sushi restaurants. The Norwegian and Icelandic publics as well, upon learning that the purpose of their tradition of whale hunting and IWC-invoked "cultural exception" was to supply Japanese markets, have come out in droves to protest their countries' behavior and demand an end to the practice.

Since they resumed whaling in 1993, Norwegian commercial hunters have killed more than 10,000 whales. Fueling the demand to end the hunt is the fact that, despite considerable investment and research by the Norwegian government and whalers themselves, there is admittedly no humane way to "harvest" a whale. Indeed, the method for killing a minke whale, with a penthrite harpoon, has changed little in over 150 years. Visibility, swells, and movement make it difficult for even the most experienced whaler to kill instantly. According to Norwegian government statistics, 20 percent of the harpooned whales are hit in nonvital areas and end up suffering long and agonizing deaths.

In the case of Japan, proponents of whaling rationalize and defend the practice by invoking cultural tradition and national heritage. The Institute of Cetacean Research in Tokyo is the country's premier institution defending and monitoring the whaling industry. The institute claims it supports whaling for "scientific" reasons, when the fact is that particular clause of the IWC regime is a smokescreen for the continued sale of whale meat to Japanese consumers. Moreover, ardent defenders believe that efforts to end Japan's whaling are nothing more than Western arrogance. As Hoshikawa observes, "The Japanese whaling industry has cunningly used the term 'culture' as a get-out-of-jail-free card—by framing this as an issue of culture or sovereignty, it aims to make any anti-whaling group look like they are colonialist and discriminatory" (quoted in Hunter 2009). He and a small but vocal faction in Japan believe that "the hunt is senseless slaughter in service of fake science, a dead industry, and nationalist posturing. . .the whales should not bear the punishment for our foolishness" (Ibid.). Hoshikawa and others are convinced that whaling is not even commercially viable but persists because it is framed as an issue of national sovereignty: "countries like Japan that still run whaling hunts now see it as a political defeat to cave in to international pressure" (Ibid.).

The monetary value of whales—as much as $100,000 for a single minke whale—along with the stubborn invocation of national sovereignty and cultural tradition seemingly guarantees that the practice will continue. Though begun with good intentions, the IWC has proven to be a weak and ineffective body, its members prone to bribery and cajoling by Japan to support whaling or at least not make much of a diplomatic row when it flaunts the rules. The IWC defends its approach by arguing that if it eliminates the "cultural exceptions" clause the whole agreement and general ban would unravel, paving the way for a full-scale resumption of commercial whaling. While this may, in a sense, be possible, though unlikely, the real issue is not the survival of a failed bureaucracy but the survival and flourishing of whales. Industrial scale hunting when combined with habitat degradation and climate change has pushed many Cetacean species to the brink of extinction. Whales as mammals are slow to reproduce yet can live between 70 and 150 years. This makes them vulnerable to population collapses and extinction. The biggest animal to have ever existed, the blue whale, despite being protected has yet to recover to a stable population size. The West Pacific gray whale was hunted so effectively that it is presently the most endangered in the world with less than 100 individuals, though its close relative, the East Pacific gray whale, has recovered. If given enough time and space, many of the species may recover. To ensure this recovery all forms of whaling should be banned and no country should be allowed to invoke cultural exemptions. Further, the world's nonwhaling countries, such as the United States, should more vocally press for a complete and absolute ban. Doubtless, just as our present generation views slavery with malice, so too will our descendants feel about whaling.

Mark Troy Burnett

NOTE

1 Whaling can be conceptualized as the act of hunting whales for their meat, bones, and blubber, or for sport. The parts of a whale are used to make numerous products and chemicals such as kerosene, transmission fluid, margarine, candles, tools, and jewelry. Scholars suggest that small-scale whaling may have begun between 5,000 and 9,000 years ago.

REFERENCES AND FURTHER READING

BBC. "Sustainable Fishing: Lamalera Whale-Hunters in Indonesia." BBC video, 5:11. First broadcast March 16, 2011. http://www.bbc.co.uk/learningzone/clips/sustainable-fishing-lamalera-whale-hunters-in-indonesia/11954.html.

Blok, A. "Contesting Global Norms: Politics of Identity in Japanese Pro-whaling Countermobilization." *Global Environmental Politics* 8, no. 2 (2008): 39–66.

Chiropolos, M. L. "Inupiat Subsistence and the Bowhead Whale: Can Indigenous Hunting Cultures Coexist with Endangered Animal Species?" *Colorado Journal of Environmental Law and Policy* 5 (1994): 213.

Clapham, P., and Ivashchenko, Y. "A Whale of a Deception." *Marine Fisheries Review* 71, no. 1 (2009): 44–52.

DeMarban, Alex. "Fermented 'Stink Whale' Landed in Barrow, and It's a Blessing during Tough Season." *Alaska Dispatch*, July 18, 2013. http://www.adn.com/rural-alaska/article/fermented-stink-whale-landed-barrow-and-its-blessing-during-tough-season/2013/07/19/.

Freeman, Milton M. R. *Inuit, Whaling, and Sustainability*. Vol. 1. Walnut Creek, CA: Altamira Press, 1998a.

Freeman, M. M. R., L. Bogoslovskaya, R. A. Caulfield, I. Egede, I. I. Krupnik, and M. G. Stevenson. *Inuit, Whaling, and Sustainability*. Walnut Creek, CA: Altamira Press, 1998b.

Gales, N. J., T. Kasuya, P. J. Clapham, and R. L. Brownell. "Japan's Whaling Plan under Scrutiny." *Nature* 435, no. 7044 (2005): 883–884.

Gusev, S. V., A. V. Zagoroulko, and A. V. Porotov. "Sea Mammal Hunters of Chukotka, Bering Strait: Recent Archaeological Results and Problems." *World Archaeology* 30, no. 3 (1999): 354–369. doi:10.1080/00438243.1999.9980417.

Hunter, Emily. "Whaling the Latest Culture War." THIS. May 5, 2009. http://this.org/magazine/2009/05/05/whaling-culture-war/.

Hurd, I. "Almost Saving Whales: The Ambiguity of Success at the International Whaling Commission." *Ethics and International Affairs* 26, no. 1 (2012): 103–112.

International Court of Justice (ICJ). Judgment. Whaling in the Antarctic (*Australia v. Japan: New Zealand Intervening*). March 31, 2014. http://www.icj-cij.org/docket/files/148/18136.pdf.

International Whaling Commission (IWC). *International Convention for the Regulation of Whaling*. December 2, 1946. The Avalon Project, Yale Law School. http://avalon.law.yale.edu/20th_century/whaling.asp.

International Whaling Commission (IWC). "Aboriginal Subsistence Whaling." 2014. http://iwc.int/aboriginal.

Kalland, Arne. *Unveiling the Whale: Discourses on Whales and Whaling*. Vol. 12. 2009.

Krupnik, I. I. "The Bowhead vs. the Gray Whale in Chukotkan Aboriginal Whaling." *Arctic* 40, no. 1 (1987): 16–32.

Pfister, B. "Sequential Megafaunal Collapse in the North Pacific Ocean: An Ongoing Legacy of Industrial Whaling?" *Proceedings of the National Academy of Sciences* 100, no. 21 (2003): 12223–28.

Reeves, R. "The Origins and Character of 'Aboriginal Subsistence' Whaling: A Global Review." *Mammal Review* 32, no. 2 (2002): 71–106. http://onlinelibrary.wiley.com/doi/10.1046/j.1365-2907.2002.00100.x/full.

Savelle, J. M., and N. Kishigami. "Anthropological Research on Whaling: Prehistoric, Historic and Current Contexts." *Senri Ethnological Studies* 84, no. 1 (2013): 1–48.

Strausz, Michael. "Executives, Legislatures, and Whales: The Birth of Japan's Scientific Whaling Regime." *International Relations of the Asia-Pacific* 14 (2014): 455–478. doi:10.1093/irap/lcu007.

Tripp, E. "Whale Meat Being Sold Illegally in Greenland and Denmark." *Marine Science Today*. December 11, 2012. http://marinesciencetoday.com/2012/12/11/whale-meat-being-sold-illegally-in-greenland-and-denmark/.

11 WATER AND GENDER

OVERVIEW

It has been said that women hold up half the sky. In many villages and cities of developing countries women also hold up other things: they hold up water containers and clay jars on their heads, carrying them from distant water sources to their homes, refugee camps, villages, and squatter settlements several hours a day. In some countries, girls and women spend more time fetching and managing water than going to school, studying, or setting up businesses.

This chapter reviews key issues of water and gender, looking at the differentiated roles of women and men in the water sector. It looks back at historical patterns of water use and change—based on the Celts in England—but also includes contemporary examples of gendered roles in water and social and economic activities. The case studies discuss water rights, or lack thereof, of vulnerable groups, the role of women in aquaculture or fish farming, and the experiences of women explorers and pacesetters in ocean travel. The section on perspectives compares examples of gender issues in development projects.

Traditional Emphasis: Engineering and Economic Aspects of Water

In the past, the typical approach to water policy, planning, and practice was to emphasize two main aspects: a) physical engineering works, i.e., construction of dams, reservoirs, pumping stations, pipes, and wastewater treatment plants; and b) economic variables, i.e., calculation of construction costs, water demand and supply, pricing, tariffs and water consumption levels. Social, community, and gender aspects of water ownership and the equity of water rights and allocations received scant attention until the 1980s. Since then, development institutions, policymakers, researchers, communities, and NGOs have become more aware of gender implications in design and execution of water projects.

New Gender Perspective

In the last two decades it has become more common to view the water sector more comprehensively, i.e., linked to other sectors and gender and social complexities. Gender is often equated with women, but a gender perspective in the water sector does not look only at women's roles in water. Rather, it means understanding the different perspectives, roles, and needs of women and men, their interrelationships, their decision-making processes, and any gender differences in ownership and control of water access points. Gender analysis also provides a lens through which we can look at ascribed water roles and expectations given to boys and girls from birth onward, and how such expectations could impact their opportunities, future lives, and livelihoods.

This chapter analyzes how gender roles in water are shaped by historical, cultural, social, political, and economic factors, traditions, or beliefs, which change as societies evolve. Ownership of land and water rights is a key issue. It is difficult to fully understand women's roles in the demand, supply, and access to water without addressing female/male ownership of water and irrigation systems and equitable distribution of water-related assets. This is especially critical in dry regions where water equals survival. Additional gender complexities are also linked to caste, religion, or ethnicity issues, as case studies in this chapter show. Cultural, social, and economic factors also matter. Evolution of society and changes (slow or rapid) in the fabric of life shape gender roles and norms as well.

Water for All

The United Nations (UN), in its 2005–2015 International Decade for Action: "Water for Life," highlighted the importance of clean drinking water and sanitation for all people in the world. In 2000, the UN Millennium Development Goals (MDGs) also set a target to cut in half the proportion of people without access to safe drinking water and sanitation by 2015.

However, due to inadequate resources and poor follow-up on this pledge by many nations, this goal was not fully met. Worldwide, 2.1 billion people gained access to improved sanitation, but out of 193 UN member countries only 95 countries met the sanitation target. Of those 193, 147 countries met the safe drinking water goal and 77 countries met both targets (UNDP 2015; Sachs 2015, 492).

While the MDGs acknowledged the differentiated roles of men and women in development and their unequal access to water and sanitation services, the MDGs' objectives were broad and they did not specifically address structural barriers to women's participation and decision-making frameworks in the water sector. They also lacked problem diagnostics and tools to help poor women access clean water, increase female leadership in the water sector, and mitigate the gender impacts of water privatization.

The MDGs did not disaggregate gender outcomes or make recommendations on how to dismantle gender inequities. There was insufficient attention to gathering comparative, disaggregated data on male and female access to water, cost, and time taken to fetch water, and the gender equity of public and private water-provision methods. These factors impeded reliable measurement of specific gender-related advances towards MDG goals.

It was left to individual countries to develop their own national MDG plans over the 15-year period and secure financing for specific policies and activities to address gender imbalances in water and sanitation services. The United Nations International Children's Emergency Fund (UNICEF) and other aid agencies also developed specific programs to help countries respond to women and girls' water and sanitation needs, but in general funding was inadequate to cope with the severity of the problem in the poorest countries.

Gender Roles in the Past

Looking back in time, we see that gender-specific patterns of water use and management emerged in all societies. For example, hunting and gathering groups developed division of labor and complementary roles between men and women in household and community tasks. Women fetched water, gathered and prepared food, and reared children, while men took care of hunting, clan protection, and warfare duties. Over time, these gendered roles adjusted with the shift from hunting/gathering communities to more permanent settlements and the corresponding changes in socioeconomic systems and technologies.

Water was a vital element of human survival from day one. The nomadic groups that exercised control over strategic sources of water possessed greater advantage in the dominion of wide swaths of land for hunting and foraging. In roving tribes, women's knowledge of the location of water wells and streams was a crucial font of wisdom and a key factor for group survival. Women, aided by girls, were in charge of fetching or storing water and safeguarding the quality of the source. This role passed from generation to generation, society to society, and was evident in ancient Rome, Greece, Egypt, Troy, Babylon, and other cultures.

In the past (and in African villages today), wells were important places of social interaction for women. Images on Grecian or Roman urns, paintings, and mosaics depict women carrying water containers and socializing at watering places. Women also had predominant roles as mermaids and water goddesses in legends, myths, and religious practices across the world, although some high-level water gods were male, e.g., Neptune, the Roman god of freshwater and the seas (see chapter 10).

As early hunting and gathering groups turned to more sedentary life, patterns of human movement changed—and with this change, water supply arrangements shifted as well. Early European towns developed around central water sites, wells, rivers, and watermills, as these sources of water took center stage in community life. Women's roles in water gathering and water management became less critical, and eventually they lost their accumulated historical knowledge of distant sources of water. They also lost their female, social spaces at former water holes.

Ultimately, centralized water supply points in villages, towns, and emerging cities completely assumed the function of water provision and distribution in an increasing number of settlements. A significant gender shift in decision-making occurred with the construction of central wells, water mills, and increasingly complex systems of pipes and water connections. Water technology became the domain of men; women lost ground.

Celtic Water Culture and Gender Roles

Historical accounts of Celtic culture and its process of modification in Britain illustrate this process. In 5000 BCE, both female and male deities played important roles in water worship and religious rites of Celtic tribes, called the Durotriges, or water dwellers. Water was a key element in the configuration of Celtic structures and ceremonies. It is thought that the River Seine in Paris was named after the revered Celtic water goddess Sequana.

The Celts celebrated religious rites at holy wells and trees. Archaeological evidence suggests that these two symbols stood for relatively equitable gender roles: wells, stone circles, and water represented the feminine; trees and henges (standing stones) stood for the masculine. However, changes in religious practices of local Celtic deities, mainly due to the influx of other Celts from France as well as Roman occupation, would lead to shifts in ancient water rituals and undermine the role of women in water gathering and management.

The Celtic culture of honoring both male and female deities changed drastically when Celts from France moved into Dorset, England. They introduced agriculture and animal herding and slowly changed communal land ownership to private property. Celtic women lost power in this shift because French Celtic males positioned themselves to accumulate land and animals, and gradually exercise dominion over water sources.

Increasing Male Domination under Christianity

This gender switch in land and water control reduced women's status and socioeconomic standing, compared to previous patterns of greater gender balance in Celtic England. The situation worsened when the Romans invaded Britain in 55 BCE. They superimposed a militaristic structure over French-Celtic arrangements and introduced new hierarchical patterns based on male dominance and male access to weaponry. Women further lost decision-making powers in this transition.

After invading England, the Romans continued to worship water goddesses, but those were based on their Roman gods. For example, the water goddess Minerva replaced Sulis, the Celtic goddess of healing. Gender roles were to change yet another time: the new Christian Church of the Romans was male-dominated, and water and water worship became a male domain.

Pope Gregory decreed the conversion of Celtic holy water wells and temples into Christian places, renaming the water spaces of female goddesses after places linked to male saints. Male Christian symbols such as the fish and the staff became predominant.

Abbeys and different orders of monks built chapels and water channels around former female-managed water sources, changing them into holy sites for Christian pilgrimages. Early Celtic female importance as water goddesses and water managers fully declined during the rise of Roman Christianity.

The continuous reorganization of land and water in Dorset, England—first by French Celts, then Romans, and later by Saxons in their 1066 conquest of England—completed the full dismantling of women's original water roles. Agriculture intensified and enclosures (consolidations of small landholdings under private ownership) brought more land under individual large-scale cultivation.

Male Saxon settlers reshaped the landscape with more sophisticated farming and water technologies, further accentuating gender divisions. Under Saxon rule, individual land and water rights passed down through male hereditary lineage. Collective (female and male) water and land ownership soon became a thing of the Celtic past.

Male appropriation of land and water rights excluded women from water rights and male power structures. Male landowners modernized wells and streams through constructing water pumps and more complex water systems under their control. Male-run abbeys continued to accumulate vast amounts of land and control access to water and water distribution to their tenants. In the Dorset area, it was male landowners who started the practice of building watermills to grind corn and grain (Strang 2005, 26).

By the 1600s, male farmers had controlled every stretch of river and source of water in Dorset through the construction of reservoirs, sluices, pumps, and pipes. This new era of agricultural and technical advancement and consolidation of land and property rights excluded women outright. Traditional water roles of women in traditional Celtic life were forgotten in the dustbin of history. Women were no longer water carriers, goddesses, or distributors. As they lost their active roles in water management, women became more passive recipients and consumers of water. It is interesting to note that some artists in 19th-century Britain portrayed women as nude bathers in silent, passive poses.

Women's Water Roles in Developing Countries

In the developing world, particularly in Africa and Asia, women continue to play a major role in the collection and household management of water. In Africa, less than 50 percent of households have access to safe sources of water, and only 4 percent of households have piped water. In South Asia, only 12 percent have domestic connections (UNDESA).

In most rural areas and low-income settlements in Africa, women are the main collectors, providers, and managers of water. The care of ill family members—especially those with HIV/AIDS and other pandemic illnesses—has added an extra burden on women to collect additional water for patient care. Women, too, fall ill or die from AIDS, often leaving young daughters in charge of household water needs.

In some African villages, women and girls have to walk long distances, sometimes for several miles or three hours a day to fetch water and firewood for domestic tasks. It is estimated that worldwide women and girls spend more than 200 million hours daily collecting and hauling water (Kaye 2012). Time allocated to these activities is

Amhara women carry heavy water jugs down a dirt road in an arid zone, Ethiopia. (Dr. Gilbert H. Grosvenor/National Geographic/Getty Images.)

time lost for girls to be in school and women to be in productive, educational, or leisure activities. Due to deforestation, the time needed to find firewood to boil water for drinking and cooking has increased significantly. With few prospects for education or careers, poor girls are caught in a vicious trap of poverty, illiteracy, little education, and early marriage.

In sub-Saharan Africa, refugee crises continue to evolve due to conflicts, severe drought, and pervasive shortages of water. In refugee camps in Africa (also in the Middle East), severe threats to the personal security of women and girls have been reported. They are continually exposed to the threat of rapes, kidnapping, and violent attacks when they leave the camps to gather water and firewood. Even though UN agencies have implemented guidelines to address violence against women refugees, e.g., safety lighting and patrols, the reality is that most refugee camps lack the financial and human resources to effectively guarantee security.

Incorporating Women's Roles in Development Planning

In the past, national and international water development plans and projects in developing countries have tended to emphasize women's roles as water collectors and consumers but underestimate their roles as working women in agriculture and small businesses, and as household managers in the health, water, and sanitation sectors. More recently, water and gender issues have moved into the human development and global aid arena. Attention in UN documents and national policies is more focused on water and sanitation projects that not only emphasize engineering aspects, but also incorporate gender analysis and consideration to gender roles and impacts of water sector programs.

Nonetheless, it is difficult to translate documents and plans into practice. Many countries continue to pay lip service to UN declarations but cannot muster the funds to implement programs. In fact, some developing countries have turned to privatization of public water utilities as a means of divesting control over water services and dealing with financial problems. However, this process has been a double-edged sword.

Increasing privatization of public water utilities in some countries has made water more costly—almost prohibitively so—for poor women, and especially for female-headed households. In many places, this situation has led to outright conflict. Women, who are the main protagonists in household water management, have usually been in the vanguard of protests against increase of water tariffs due to privatization. The conflicts around this issue are discussed in chapter 6, but experience teaches us that there are differentiated gender impacts of water shortages and private operation of water utilities.

Addressing Stakeholder and Community Concerns

In many low-income countries, political and ethnic conflicts add additional dimensions to the discourse around water, power, community relationships, and gender equity. Women are generally underrepresented in decision-making and power structures in the water sector. One of the stumbling blocks to addressing the problem of gender representation is the lack of methodologies and specific data to incorporate gender indicators into analyses that inform policy formulation, project design, and execution.

In recent years, methodologies such as stakeholder analysis and gender toolkits for the water sector have become available to analyze and uncover structural aspects of water use before programs are designed. Better information provides a more robust foundation for informed policy action and resolution of conflicts.

These instruments for robust stakeholder and gender analyses pose key questions at the household or individual level. A cross-section of such queries might be the following: who fetches and manages water for households; who owns water sources; who uses water to care for the sick and for religious practices; who irrigates crops; who owns agricultural land or can inherit land; who owns pumps and water trucks; who can afford to pay for water and who cannot; who benefits from or controls water technology; and who is impacted by higher tariffs or private control of water. Stakeholder analysis also looks at potential impacts of water regimes, drought or seasonal rainfall on gender roles, female-headed households, household poverty levels, and different ethnicities and social groups.

Using the toolkits, planners can gain more insight into water conflicts, the situation of vulnerable or marginalized groups, participation of women and men in village water committees and decision-making structures, and gender differentials in land ownership, technology, and irrigation use. However, at the operational or project level, the real challenge and litmus test for planners is how to actually use the results of stakeholder and gender analyses to maximize benefits for men and women, enhance good stewardship of water, and expand women's leadership and decision-making roles, where lacking.

New Technologies to Reduce Women's Water Burdens

In addition to fostering women's leadership, it is important to address structural, societal constraints that relegate women's roles to heavy water-lifting tasks. While it is difficult to rapidly change ingrained structures, some efforts are underway to reduce women's physical burden of fetching water and also the time taken on this task. It is still a common sight to see women in African or Indian villages trudging long distances with heavy water containers balanced on their heads, but many organizations, foundations, and NGOs are working on new water carrying technologies to relieve the pressures on women and girls.

Innovative designs are attempting to replace women's head-held traditional water containers with larger plastic drums and water rollers that can be filled with water and rolled or pushed by villagers to households or settlements. These containers have many names such as Q Drums, Hippo Water Rollers, and all-terrain Water Wheels, and can carry more water, thus reducing the number of trips women and girls take per day. It is also hoped that these new technologies might help reduce the stigma of water collection as "women's head-carrying water duties," and that more males would be inclined to fetch water.

Unfortunately, many of these containers still remain unaffordable for very poor households, since their costs range from $20 to $100 dollars (Rahman 2011). With further development of technology and mass production, it is anticipated that the costs could be dropped, and these new methods could become reasonable options for poor families. Some corporations and NGOs have set up foundations to help provide funds for low-income women and families to acquire these new water transportation devices and to gain better access to clean water and improved sanitation.

Gender Roles in Mongolia

While women still remain the primary water collectors at the household level in Africa and India, this pattern may vary in different parts of the world. Research on gender and water roles in Mongolia indicates that men are the main agents of water collection in urban settings (Hawkins and Seager 2010). In Ulaanbaatar, the capital of Mongolia, men carry out the water fetching from outside sources, except that they regularly use small trucks or animal carts. In urban areas, women are still mainly responsible for managing water domestically.

However, in rural Mongolian households, the patterns are more traditional: in more than half of all villages, women handle both external collection and domestic management of water. As in most developing societies, Mongolian men are more predominant in decision-making in the water sector. Mongolian women, in spite of relatively high levels of education and literacy, are less involved than men in formal water management and decision-making in both urban and rural areas.

Gender and Caste Roles in India

In India, there is an intersection of water, gender, and caste roles: the position of women and men in the water sector is related to and constrained by the Hindu caste system. The caste system prescribes that water can be polluted by the touch of those considered an impure caste such as the Sudras, and its subgroup, the Dalits (formerly called the "Untouchables").

For women who are Dalits, this situation signifies double jeopardy. In the hierarchical caste system, Brahmins occupy the ranks of the purest castes with prescribed roles in scholastic, ritual, and religious activities. Male and female Dalits, considered to be the lowest caste, are assigned occupational roles such as cremating the dead, handling sewage and dead animals, and washing and cleaning waste products of higher castes. Female Dalits are considered to be the lowest on the social ladder.

Women draw well water in India. Though there is an abundance of freshwater in the world, it is not always available in the regions most affected by intense population growth. Women often bear the burden of carrying and providing water for their families and villages. (Corel)

Impurity resulting from caste lineage is considered irreversible in India. Changing a person's ascribed caste is not possible, although Dalits can and do achieve some social mobility in society. It is no longer legal to punish intercaste marriage in India, but prevailing customs and social norms pose obstacles to social mixing of upper and lower castes. Discrimination continues to persist along caste and gender lines.

A gendered analysis of water in the Indian context cannot ignore the origin and evolution of religious belief and cultural systems governing water as well as the ongoing social and cultural modernization of the country. In early Vedic texts of more than 3,000 years ago, water is regarded as a purifying and sacred element. This concept of purification is primarily spiritual rather than physical: contact with sacred water leads to attainment of pure consciousness.

Traditional beliefs in India laid the foundation for the later concepts of untouchable castes and purity or impurity of water related to layers of caste and social stratification. Social and economic groups within this rigid caste system also became linked to inheritance-based rights and privileges over land and water, passed down from generation to generation. Any caste considered impure was not allowed to touch and defile pure water and was doomed to handle impure or soiled water. Women in the lowest castes had the least rights to water or land resources and were among the most disadvantaged in society.

There is a school of thought that gender bias in Hindu society does not truly reflect the earliest Vedic texts, which, it is argued, showed respect for women's rights, allowed women to plead their cases in court, and fostered equitable religious and secular roles of both men and women. They point to archaeological findings, which indicate that hunter-gathering groups of central India in the Mesolithic period (about 10,000 BCE) were more gender-balanced than in the later caste system. This view also cites the high powers associated with Hindu goddesses such as Durga, Kali, and Lakshmi.

In many Hindu festivals today, these goddesses continue to be revered and celebrated. The name of the city of Calcutta (now Kolkata) stems from the goddess Kali; the word "Shakti" denotes female power, and Jal Devi is a water goddess who is believed to possess power to keep water flowing in traditional wells. In Hindu myths, snakes appeared in water sources to indicate the goddess's wrath, causing water to dry up.

Along the same lines, it has been argued that the once equitable roles of men and women in ancient India changed to the caste system as a result of male misinterpretation of Vedic texts. Proponents of this view believe that land and water ownership benefitted men with the passage of time, mainly due to the impact of Islamic culture and male-dominated British colonial administration. This view of Indian society interprets that the rigid caste, gender, and class stratification structures and gendered inheritance of assets was not inherent in ancient Vedic society but imposed by male-dominated political structures (Laungani 2015).

Over time, wealthy, upper-caste landowners drew upon the labor of lower castes for agriculture and delegated tasks considered impure to those at the bottom of the pecking order. To enforce the rules, upper castes prohibited mobility of persons across caste boundaries and maintained barriers through strict rules of separate living areas, restrictions on intercaste marriage, and caste separation of water use and eating practices. Lower castes faced severe penalties if they extracted water from upper caste wells.

To maintain the purity of water, upper castes excluded Sudras physically and socially from village boundaries and segregated them in their own caste areas. Within the caste stratification process, women's tasks slowly changed from the unrestricted, outdoor gathering of food in former nomadic societies to more homebound domestic and reproductive roles. Society became more patriarchal: women resided in the family homes of their husbands after marriage and became subordinate to the rules of the new home.

Upper-caste men were considered the purest members of society, followed by upper-caste women. Moral codes prohibited physical association of upper caste women with lower caste men, especially Sudras: violations were subject to severe punishment for both parties. Since lower caste women could not touch food or water served to upper castes, upper-caste woman gradually became more homebound and in charge of overseeing food and beverage preparation and serving.

Under traditional norms, upper-caste members, if touched by a Sudra, had to take purifying baths; if they looked at a Sudra, they had to cleanse themselves by looking at the sun, moon, and stars and rinsing their mouths with water. Since Sudras were in charge of washing and handling corpses and wastewater, no higher caste could touch them. Water sources were considered contaminated by a Sudra's touch; hence strict physical boundaries for the use of water sources avoided the problem of upper castes having to constantly undergo elaborate cleansing rituals.

In modern India, poor women and girls, including widows, carry the brunt of water collection in villages. This role starts from childhood as girls help their mothers, and continues until old age. One of the first rituals of a bride upon entering her husband's home is to fetch water for the extended family. Marriage reinforces the water burden of poor girls: they still have to fetch water for washing, cooking, and tending to small domestic gardens. If men are observed carrying water for households, it is usually for payment or as assistance to older or disabled female relatives.

Winds of Change

Notwithstanding the historical patterns and cultural norms described above, strong winds of change are blowing in contemporary India. India's globalization and modernization process, including the rapid growth of high-tech industries and call centers, has been rapidly changing women's roles and caste immobility. Women of all castes are graduating from universities in unprecedented numbers and entering the modern labor force, especially in the high-tech sectors. Legislation has helped to reduce discrimination against women in general and Dalits in particular.

Some advances in Indian society have occurred in the water sector as well. The public sector often provides water to local Dalit communities in rural villages, and schools and educational institutions promote educational opportunities for Dalit girls. However, there is still a long way to go before long-standing, structural caste and gender discrimination practices in the water and other sectors change. It still remains to be seen what other shifts are needed to fundamentally transform the excluded status of Dalits beyond water.

Water Roles in Sudan

In Sudan, women's roles in water are not only linked to cultural and religious traditions, as in India, but also to patterns of drought in this arid country. There are differentiated gender relationships within various ethnic groups. A study of the Gedaref area, located in east-central Sudan close to the Ethiopian Highlands, looks at the cross-section of all these factors. Three distinct population groups have lived in Gedaref for centuries: indigenous Arabs, groups from western Sudan, and West Africans from outside the region (Coles 2005, 75).

Men in all these three groups collect water since wells are deep and require greater physical effort. However, the extent to which men take complete charge of water collection is related to culture and gender roles. Some wells are located far away due to drought and water depletion, and require long distances across dry terrain to access these sources.

In all three groups, women are the poorest and the most domestically restricted, but Arab households seem to have the most patriarchal structures. In established Arab villages, male heads of households assume full responsibility for taking care of women and children and fetching water for the households; the cultural expectation is that women should remain in the domestic sphere and avoid the possibility of meeting strange men at distant water sources.

Due to drought and groundwater depletion, and ever-longer distances to travel, the role of women as water carriers in the group of Muslim villagers is declining. Men and boys fetch water from outside sources, and their responsibility for these tasks keeps female family members at home. Animals are important for water transport: for distances greater than three miles, males use donkeys and camels. Whereas women can individually lift about four gallons of water, donkeys carry up to 70 gallons in a leather pouch.

However, women's responsibilities in domestic management of water in all three groups have not changed. They are still in charge of conserving and using water for domestic tasks such as cooking, bathing, cleaning, laundry, and small gardens. While water availability and consumption of the household is related to economic status of male breadwinners and the size of households, women are the ones who prioritize water use. They decide on which standards of hygiene could or should be relaxed to preserve water, and they control gifts of water to visitors—a traditional act of religious charity in arid areas.

In general terms, drought and the reduced water supply in the Sudan is disrupting village life and sparking ethnic and other conflicts. Entire communities are vying for limited water; some are encroaching upon the water rights of others. Herders are pitted against farmers. The chronic lack of water forces families to migrate or purchase water from water trucks, a practice which often leads to crippling household debt.

Camille Gaskin-Reyes

REFERENCES AND FURTHER READING

Coles, Anne. "Geology and Gender: Water Supplies, Ethnicity and Livelihoods in Central Sudan." In *Gender, Water and Development,* edited by Anne Coles and Tina Wallace, 75–93. Oxford, UK: Berg Press, 2005.

Geoforum. "Gender and Water: Good Rhetoric, But It Doesn't 'Count.'" Geoforum Editorial, *Geoforum* 41 (2010): 1–3.

Green, Andrew. "Women in South Sudan: 'They Attack Us at Toilets or Where We Collect Water,'" *Guardian*, September 11, 2014. http://www.theguardian.com/global-development/2014/sep/11/women-south-sudan-sexual-violence-camps.

Hawkins, Roberta, and Joni Seager. "Gender and Water in Mongolia." *Professional Geographer* 62, no.1 (2010): 16–31.

Hutchings, Anne, and Gina Buijs. "Water and Aids: Problems Associated with the Home-Based Care of AIDS Patients in a Rural Area of Northern Kwazulu-Natal, South Africa." In *Gender, Water, and Development*, edited by Anne Coles and Tina Wallace, 173–188. Oxford, UK: Berg Press, 2005.

Joshi, Deepa, and Ben Fawcett. "The Role of Water in an Unequal Social Order in India." In *Gender, Water, and Development*, edited by Anne Coles and Tina Wallace, 39–56. Oxford, UK: Berg Press, 2005.

Kaye, Leon. "Improving Women's Access to Water: Why Business Must take a Role." *Guardian*, September 24, 2012. http://www.theguardian.com/sustainable-business/improving-womens-access-water-business-role.

Kristof, Nicholas, and Sheryl Wudunn. *Half the Sky: Turning Oppression into Opportunity for Women Worldwide*. New York: Knopf, 2009.

Laungani, Nirmal. "Culture: Women's Status in Ancient India." *Hinduism Today* web edition. January/February/March 2015. http://www.hinduismtoday.com/modules/smartsection/item.php?itemid=5566.

Rahman, Maseeh. "Can India's Women Cast Off the Burden of Water Carrying?" *Guardian*, September 2, 2011. http://www.theguardian.com/world/2011/sep/02/india-women-burden-water-carrying.

Sachs, Jeffrey D. *The Age of Sustainable Development*. New York: Columbia University Press, 2015.

Strang, Veronica. "Taking the Waters: Cosmology, Gender and Material Culture in the Appropriation of Water Resources." In *Gender, Water, and Development*, edited by Anne Coles and Tina Wallace, 21–38. Oxford, UK: Berg Press, 2005.

UNICEF. "Water, Sanitation and Hygiene: WASH and Women." Updated April 30, 2003. www.unicef.org/wes/index_womenandgirls.html.

United Nations Department of Economic and Social Affairs (UNDESA). "International Decade for Action 'Water for Life' 2005-2015." www.un.org/waterforlifedecade/background.shtml.

United Nations Development Program (UNDP). "The Millennium Development Goals Report 2015." July 6, 2015. http://www.undp.org/content/undp/en/home/librarypage/mdg/the-millennium-development-goals-report-2015.html.

Water Resources Systems Division. "Hydrology in Ancient India." http://www.nih.ernet.in/rbis/vedic.htm.

CASE STUDIES

Two of the three case studies in this section portray roles of women at work in developing countries, whether in Madagascar through their farming, water-carrying and animal-herding roles, or in Asia through their involvement in shrimp and fish farming. In both cases women are very vulnerable to changes in water availability and quality, droughts, floods, and shifting climate regimes, or to their precarious social and economic status in aquaculture. The third case study looks at groundbreaking efforts of women from developed countries to set the pace and break records in ocean rowing, swimming, sailing, and polar exploration.

Case Study 1: Water and Gender in Madagascar

While the availability of water in developed countries is a fairly normal occurrence, access to safe drinking water is not a given for many people in developing countries or in the most remote or arid places in the world. In many areas, women and girls fetch and transport water for cooking, bathing, care for animals, and cultivation of home gardens.

Over the past 20 years, a growing number of international agreements have highlighted the need for better water and sanitation for men and women. In 1992, the Rio Declaration on Environment and Development called for enhancement of the role of women in natural resource and water management. In 1995, the Beijing Declaration and Platform for Action proposed a new international commitment to the goals of equality, peace, and development for all women. While this declaration was laudable, the commitments were nonbinding and measures to attain progress towards the goals were unclear.

In 2000, the Millennium Development Goals (MDGs) set targets for clean water and sanitation for all. Since then there have been some advances in improving women's access to clean drinking water and sanitation services. But there are major challenges ahead. In 2015, the decade of the MDGs came to an end, and a new decade of Sustainable Development Goals (SDGs) began, continuing with some of the unfinished social and economic development agenda items of the MDGs.

Goal #6 of the SDGs underscored the importance of clean water and sanitation. It set forth goals to promote access to equitable sanitation and hygiene services by 2030, with particular attention to gender equity and the needs of women, girls, and communities in vulnerable situations. It highlighted the need to reduce pollution and improve water quality and efficiency and conservation. In addition, this goal stressed policies and measures to address water scarcity and the depletion of freshwater reserves such as desalination, wastewater treatment, and recycling. Improved transboundary cooperation was seen as an important activity to protect shared water resources, expand community and international collaboration, and improve capacity of developing countries to carry out water- and sanitation-related activities.

The analysis in earlier parts of this chapter has indicated that women and men have different roles in managing water. In many developing countries, women and children are mostly responsible for domestic water use and community collection of water resources, but are not well represented in leadership positions. Women are also very vulnerable to changes in water availability and quality, droughts, floods, and shifting climate regimes.

Madagascar is a case in point. It is the fourth largest island in the world, located off the coast of Africa. Of the island's 19.7 million people, 60 percent do not have access to safe drinking water. Madagascar possesses some of the world's most unique ecosystems, including a total of eight plant families, five bird families, and five primate families that live nowhere else on the planet. Its tropical forests and marine environments are home to endemic species of flora and fauna; however, 15 species are now extinct due to deforestation and loss of biodiversity.

While Madagascar is rich in freshwater resources, many people still lack clean water. It is also one of the poorest nations in the world. Most people are forced to survive on less than two dollars per day. Like many other vulnerable communities around the world, villagers in Madagascar depend on the availability of healthy freshwater ecosystems for food, potable water, agricultural production, and protection from damaging floods and living organisms. Women and children are particularly affected by these occurrences.

A United Nations Children's Fund (UNICEF) study in 2002 estimated that more than 3.5 million school days were lost per year on the island due to diarrhea and other preventable waterborne diseases. This is due to the fact that only 18 percent of the urban population uses latrines and 10 percent in rural areas. More than 30 percent of the population defecates in the open air, according to a 2010 World Health Organization (WHO) report. This situation has led to ecosystem degradation and human health and sanitation problems.

Conservation International (CI), an international nongovernmental organization, has been working in Madagascar since 1987. Its goal is to improve biodiversity and contribute to improved welfare of the population. CI is a private, nonprofit environmental organization with programs in more than 40 countries. Its mission is to empower people to responsibly and sustainably preserve ecosystems, and to foster human welfare by building on a strong foundation of science, partnership, and field demonstration. For CI, there are strong links between ongoing and emerging conflicts over scarce resources and impacts on human well-being and environmental protection.

CI seeks to demonstrate that just as deteriorating environmental conditions can bring about or heighten conflict, abundant natural resources and healthy ecosystems can serve as the scaffolding for healthy, prosperous, and peaceful societies. There is growing awareness and recognition by global leaders, national governments, and institutions that collaboration and the constant search for workable solutions will be important factors in addressing the challenges of resource conservation and sustainable water management.

CI's work in the forest corridor Natural Resource Reserve in Southern Madagascar (called COFAV) has focused on generating benefits for biodiversity and the population. Deforestation rates in the corridor decreased from 1.5 percent in 2000 to 0.40 percent in 2010. Since 2005, CI has helped to establish protected area management systems and build capacity among community-based organizations and associations to conserve the island's valuable biodiversity. CI's goal is to help protect the sources and flows of freshwater by promoting practices that integrate land- and water-resource management and sustain natural watersheds to ensure the long-term availability and quality of freshwater (Conservation International).

In the remote communities of the COFAV, community members have to walk for more than a day to get to health services or a market. There are very few safe water sources, and communities need to gather water for everyday uses. Recognizing these needs, CI Madagascar, with support from the U.S. Agency for International Development, partnered with two local organizations to strengthen community capacity to build water point sources or wells (USAID). At the same time, the project promoted improved family planning in addition to water and environmental sanitation activities. In order to increase capacity for effective conservation, CI and partners empowered communities to manage their resources in an environmentally sustainable manner.

In each of the five COFAV villages in the project, women participated as full members of the local water management committees. Water management committees were the mandated structures within the national Madagascar water policy framework. These committees were instrumental in making decisions about maintenance of water wells and actions to protect water as a valuable resource.

As extension agents, women also took leadership in the conduct of informal education seminars with children and mothers to increase awareness of safe water management and healthy behaviors. Women and men together were responsible for ensuring regular monitoring of the water quality from the point sources and reporting to local authorities. They also developed awareness programs and messages about healthy families, healthy ecosystems, and ways to protect the environment.

By the end of the project, CI and its partners had strengthened community capacity to better manage the island's health and natural resources. More than 21,000 people listened to an educational integrated message or attended a presentation on healthy behaviors. Community members helped to install eight boreholes/water sources serving more than 595 households (about 2,836 people), and built more than 20 latrines. The members of the eight local water point management committees also increased the participation of women in decision-making. Women were fully involved in all aspects of the project. Since the construction of the new latrines and water sources, community members have been observing a reduced incidence of diarrhea and waterborne diseases (Conservation International).

Integrated conservation projects that address the water sector needs of countries and integrate women and girls in countries like Madagascar are critical to sustainable development. The recommendations for future projects that emanate from the Madagascar experience include the following: consideration of women's needs upfront in the project design phase and during implementation; use of monitoring and evaluation (M&E) frameworks and development indicators disaggregated by gender; and integration of gender considerations in all programs to attain gender-equitable results and address the different needs of men and women.

Another lesson learned was that projects of this nature should pay special attention to how anticipated project outcomes are likely to affect men and women differently. Systematic gathering of information is needed to identify differences and inequalities that men and women face, and to understand gender nuances such as the division of labor, differential access to resources and power; and different opportunities or constraints that men and women face in terms of water rights, education, employment and social welfare.

Patricia Biermayr-Jenzano and Janet Edmond

REFERENCES AND FURTHER READING

Conservation International. "Climate Adaptation for Biodiversity, Ecosystem Services and Livelihoods in Rural Madagascar." Action Pledge, 2011. http://www.conservation.org

/publications/Documents/CI_Madagascar_NWP_Submission_Climate_Adaptation_for_Biodiversity_Ecosystem_Services_and_Livelihoods.pdf.
Conservation International. "Madagascar Communities Profit from Forest Conservation." 2016. http://www.conservation.org/projects/Pages/madagascar-communities-profit-from-forest-conservation-ambositra-vondrozo-corridor.aspx.
UNICEF. "Invest in Children—Advance Sustainable Development." http://www.unicef.org/events/wssd/.
United Nations Development Program (UNDP). "Goal 6: Clean Water and Sanitation." http://www.undp.org/content/undp/en/home/mdgoverview/post-2015-development-agenda/goal-6.html.
United Nations. "Sustainable Development Knowledge Platform." https://sustainabledevelopment.un.org/.
USAID. "Impacts of Climate Change on Rural Livelihoods in Madagascar and the Potential for Adaptation." http://pdf.usaid.gov/pdf_docs/PNADP632.pdf.

Case Study 2: Women's Participation in Shrimp Farming in Asia

The rise of aquaculture around the world, particularly in Asia and some parts of Latin America, has led to the increase of women working as laborers in large-scale shrimp and fish farming, but also to some negative impacts of such enterprises on low-income rural women.

In Asia, particularly in Indonesia, China, India, and Thailand, small-scale aquaculture was practiced for decades. Traditionally, the cultivation of shrimp was carried out by rural villagers, women together with men working in rice fields and fishponds. The small-scale and sustainable cultivation of tilapia, post-larvae shrimp, and milkfish provided important sources of protein for local communities and for sale in village markets without need for extra feed, chemical compounds, or antibiotics in aquaponds. In Indonesia, traditional shrimp farming in freshwater ponds was recorded more than 2,000 years ago on East Java (Siregar 2004)

However, these traditional, low-impact shrimp and fish farming systems changed drastically in recent decades due to increase in market demand and the influx of large companies and international corporations into aquaculture on an industrial scale in China, Thailand, the Philippines, and Malaysia. Large transnational corporations, encouraged by governments and funding from international agencies, such as the World Bank and the Asian Development Bank, shifted to more intensive shrimp monoculture. This shift led to the destruction of mangrove forests and use of these spaces to impound coastal water areas for aquaculture. Small-scale fish and shrimp farming was affected.

High-Impact Shrimp Farming

The shift to large-scale enterprises also had gender implications. For centuries women had been independently harvesting shrimp in Indonesia from local mangroves, which were natural nurseries for baby shrimp and crustaceans. Mangroves yielded many products, including fish, small game, honey, medicinal plants, and wood for fuel.

In Papua New Guinea, indigenous women were largely in charge of food production and harvesting products from mangroves using time-tested traditional methods. The shift from local production and consumption into export-oriented aquaculture caused the loss of traditional livelihoods, mangrove destruction, depletion of freshwater, pollution of coastal waters, and the outright blocking of local women's access to coastal resources and coastline areas filled with extensive impoundments.

In Indonesia, many women were forced to sell or leave their lands following the takeover or expansion of large-scale shrimp or fish farms. Most enterprises operating large ponds came from abroad or outside of the community; local villagers, including the women who had formerly made their living independently through shrimp harvesting in natural mangroves, were forced to work as unskilled and low-paid laborers in the aquaculture industry. Economic shifts led to gender shifts.

Vulnerability of Women and Children

Women and children were the most vulnerable groups in the changing economics of aquaculture. Only a few corporations or industrial groups controlled large hectares of fish ponds and farms—as opposed to the traditional system of small farmers managing one hectare. Most small female farmers eventually became wage laborers or workers in shrimp-processing factories. In Indonesia, such large-scale aquaculture operations were supported and subsidized by a government policy financing transmigration programs of people from Java to outer islands such as Kalimantan.

In Indonesia, large companies own shrimp farms that are as large as 30,000 hectares, and the trend is towards even larger sizes. Up to 80 percent of all Indonesian shrimp exports are currently in the hands of a few large corporations. In the world of industrial-type aquaculture, men are also low-paid workers in the shrimp farms, but women are more vulnerable. Women, who still eke out a living as small shrimp farmers, are heavily indebted and mostly end up losing their small areas of land to creditors and larger farms when they cannot pay off their debts.

This precarious situation for local men and women has generated social tensions and conflicts between local communities and outside investors. The business model of the large shrimp farms continues to be based on the conversion of large tracts of mangroves or wetland ecosystems to shrimp farms. Most of these large operations set up agreements with local farmers to work in the ponds for a number of years or until such time as they are able to pay back their debt to the company for inputs and other supplies.

Gender Roles in Bangladesh Aquaculture

In Bangladesh, another country with extensive aquaculture areas, large companies are also engaged in expanding territory through destruction of mangroves and displacement of small farmers. In the area around Khulna in Bangladesh, studies have documented negative impacts on women caused by the spread of shrimp aquaculture ponds (Swedish Society for Nature Conservation 2011). Due to destruction of natural mangrove forests, women are forced to walk further to gather firewood, water, and food.

Since large aquaculture farms often employ more men from outside local communities than local villagers, the influx of a labor force from other areas has created a local demographic imbalance (more men than women) and brought unease and conflicts to traditional village societies. Allegations of harassment and rape of local women are commonplace, and are part of a broad pattern of abuse and gender inequity across the shrimp sector in Bangladesh (Swedish Society for Nature Conservation 2011).

Proponents of the shrimp industry in Bangladesh, however, argue that the shrimp industry has a positive impact on women's lives since it employs some women in fry catching, pond clearing, and factory processing of shrimp. Critics argue that the pay is extremely low and the shrimp farming industry uses child labor, robbing children of the opportunity to attend school and combat poverty. Women fry collectors in Bangladesh often work for no income since they fall into bonded labor relationships with fry traders, who become their creditors for life.

Heavy Work Burden

Even if women are not employed in the shrimp industry in Bangladesh, their work burden is still increased because the expansion of industrial farms in former mangrove areas forces them to walk further to look for fuel, nonsalinized water, and alternative sources of food. While some shrimp-processing plants employ women as factory workers, they are usually hired under casual or temporary labor contracts, which prohibit them from joining labor unions or fighting for better working conditions.

In the absence of financial security and stable employment, many women fall victim to sexual harassment of male factory supervisors who misuse their power. These negative conditions have led to greater marginalization of women who have few alternatives to obtain productive work.

The impact on women working in aquaculture has been particularly acute due to their position at the lowest paid rung of the ladder. They are also more vulnerable to physical and sexual assault and intimidation of factory bosses and managers. In processing plants, women are deprived of fair wages and other labor rights and benefits, since they are mostly on short-term payrolls. It is estimated that women in the shrimp-farming processing plants earn around $46 a month, 60 percent of the wages of male workers (Shrimp News International 2015).

To assist women to improve their livelihoods and empower themselves, the United Nations Industrial Development Organization (UNIDO) has started gender-focused training courses for women in shrimp farming areas in Bangladesh to improve their technical capacity and incomes by enabling them to become small independent shrimp farmers. Courses include training in shrimp farming, shrimp farm design, water quality management, shrimp feeding, harvesting and post-harvesting techniques, record keeping, and accounting methods.

In coastal Bangladesh, development aid agencies and government institutions continue to view export-oriented shrimp farming as a mechanism to propel development in remote rural regions and improve local livelihoods. The shrimp industry remains the domain of large enterprises with large sums of investment capital and provides some employment for communities, but directly and indirectly it is eroding the foundation of village life for very small farmers and shrimp laborers, forcing them into low-paid worker positions or lengthy debt servitude.

Camille Gaskin-Reyes

REFERENCES AND FURTHER READING

Shrimp News International, Bangladesh. "Women in the Shrimp and Prawn Farming Industry Treated Unfairly." April 19, 2015. https://www.shrimpnews.com/.

Siregar, P. Raja. "Large Scale Shrimp Farming and Impacts on Women." World Rainforest Movement: Mangroves and Shrimp Farming, 2004. http://wrm.org.uy/oldsite/deforestation/shrimp.html.

Swedish Society for Nature Conservation. "Murky Waters: The Environmental and Social Impacts of Shrimp Farming in Bangladesh and Ecuador, 2011." http://www.naturskyddsforeningen.se/sites/default/files/dokument-media/murky_waters.pdf.

World Rainforest Movement. "Honduras: Shrimp Farm Expansion Within a Ramsar Site and Protected Area." WRM Bulletin, no. 153, April 2010. http://wrm.org.uy/oldsite/bulletin/153/Honduras.html.

World Rainforest Movement. "Mangrove Restoration Is Necessary, Mangrove Monoculture Plantation Is Not." WRM Bulletin, no. 151, February 2010. http://wrm.org.uy/oldsite/bulletin/151/Mangrove.html.

Case Study 3: Women Explorers and Pacesetters

When we think about past ocean explorers who carried out expeditions around the world, the people who might immediately come to mind are Magellan, Cook, Pizarro, Cortes, Vasco da Gama, Francis Drake, Heyerdahl, Shackleton, and others. We are less likely to have at our fingertips the names of women pacesetters, who dared the seas, oceans, and poles as explorers and travelers.

However, in recent years, a number of intrepid and athletic women have emerged as polar explorers, ocean rowers, sailors, and long-distance swimmers. With determination and grit, women are rewriting the history of exploration through their achievements and the setting of new records in their fields. This case study outlines some of these feats.

Rowing Feats: One Woman: Three Oceans

One example is in international rowing. In December 2010, Katie Spotz set an ocean record by becoming at the age of 22 the youngest person ever to row solo across an ocean. She took 70 days and six hours to row across the Atlantic from Dakar, Senegal, in Africa, to Georgetown, Guyana, in South America. Katie Spotz's 2,817-mile ocean journey had the objective of raising funds for the Blue Planet Run Foundation, which funds drinking water projects around the world.

Spotz prepared for this challenge by running marathons, completing endurance cycling trips, and rowing on Lake Erie. She estimated she would take about

100 to 110 days to cross the Atlantic, but took less time, riding strong trade winds without major storms. Her 19-foot wooden rowboat was outfitted with GPS tracking and communications systems, solar panels, and saltwater desalination equipment to provide freshwater along the way. The boat was designed to withstand hurricane force winds and waves of up to 50 feet, an improvement over past ocean rowers. She raised more than $70,000 for the Blue Planet Run Foundation for drinking water projects around the world (Maag 2010).

Another impressive rower in transoceanic travel was Roz Savage, the first woman to row solo across three oceans: the Atlantic, the Pacific, and the Indian Oceans. *National Geographic* named her Adventurer of the Year in 2010. Roz Savage covered over 150,000 miles through ocean rowing and spent over 500 days at sea in a 23-foot rowboat, becoming the first woman in 2005 to complete the Atlantic Rowing Race alone.

In this first Atlantic row, she started at the Canary Islands and ended up in Antigua after 103 days after coping with many challenges, including damage to her oars and the loss of her major support equipment (camping stove, stereo, cockpit navigation instruments, and satellite phone) to huge storms.

In 2008, Savage—after an unsuccessful attempt in 2007—left San Francisco on her way across the Pacific Ocean to Hawaii. She arrived at the island of Oahu after 99 days at sea, and entered the record books as the first woman ever to row from California to Hawaii.

In 2009, Savage continued her Pacific passage from Hawaii to the island of Tuvalu in the South Pacific. This was the second stage of her Pacific journey, but this stage posed a number of risks including the malfunctioning of her water-making equipment and extremely strong winds. These problems forced her to land on the island of Tarawa in Kiribati instead of Tuvalu.

In 2010, she started the third stage of her Pacific journey, leaving Kiribati on April 19, 2010, for the final solo journey across the rest of the Pacific. After 48 days at sea with an average daily progress of 49 miles, she arrived at Papua New Guinea, first passing through the Solomon Sea. With this achievement she set a record as being the first woman to row solo across the Pacific.

In 2010, Savage continued with her rowing feats and conquered the Indian Ocean, her third, under her steady hand. On this trip her equipment worked, allowing her to listen to 270 audiobooks.

Women on Polar Expeditions

Another female pacesetter, who has captivated world attention, is the British explorer, Rosie Stancer. Over the past 10 years or so, she has participated in major expeditions to the Arctic and Antarctic. Her achievements are impressive: in 1997, she belonged to a team of 20 women participating in an all-woman expedition to the North Pole. It consisted of five teams of four women each working in a relay format that reached the North Pole in 73 days. This expedition raised money for the St. John Ambulance organization. In 1999, Stancer and four other women completed a woman-only expedition to the South Pole without a professional guide and with only one resupply drop. This expedition raised funds for the British Special Olympics.

In the 2003–2004 Antarctic season, Rosie Stancer completed a solo, unsupported expedition in Antarctica from the Hercules Inlet to the South Pole in approximately 44 days, beating the previous record by over seven days. In 2004, another British female explorer, Fiona Thornewill, completed a solo expedition to the South Pole in a shorter time, completing the route in 41 days and eight hours.

In the spring of 2007, Stancer set out from Canada to reach the North Pole solo with the objective of becoming the first female to complete the journey. However, after covering most of the route, due to extreme challenges (including self-amputation of two frostbitten toes), she was forced to abandon the expedition with only about 100 miles to go. Stancer is expected to make another attempt in 2016 (International Polar Foundation).

Women in Swimming

In addition to women rowers across the oceans, and Arctic and Antarctic explorers across the ice, there are women who have braved the seas through solo swimming or sailing. Diana Nyad exemplifies the endurance and persistence of the long distance swimmer. At the age of 64, in 2013, on her fifth attempt, Nyad was the first person to complete the swim between Cuba and the United States without a shark cage. She braved powerful currents, sharks, and jellyfish on her 110-mile odyssey, which she completed in 52 hours, 54 minutes and 18 seconds.

Nyad's swim crowned a four-year period of intensive training and fundraising in preparation for her feat. Swimming from Cuba to the U.S. mainland without a shark cage had been considered one of the greatest swimming achievements possible due to the prevalence of sharks in the crossing. Nyad had already unsuccessfully attempted this swim 35 years before. Until 2013, her subsequent attempts had also proven elusive due to storms, shoulder pain, and respiratory distress caused by jellyfish stings.

Nyad's final conquest of the Cuba–Florida route at the age of 64 capitalized on plotting her swimming course to follow the Gulf Stream currents, using a specially designed bathing suit and face mask to protect against the jellyfish stings that had derailed the previous attempts. After Nyad's record-breaking swim, she raised over $110,000 for victims of Hurricane Sandy in the United States and also received a special award from the president of the United States.

Women in Sailing

While Diana Nyad proved that there are practically no upper age limits on solo and endurance swimming, a young 14-year-old Dutch girl also showed there is no lower limit of extreme endurance and sailing capability. In 2009 Laura Dekker sailed around the world on a two-year-long, 27,000-mile trip. The trip was controversial because many people around the world, including the Dutch government, were concerned about the potential perils of this solo voyage for an adolescent.

Even though the Dutch government legally intervened and attempted to block Decker when she was 13 years old, at 14 she prevailed and successfully completed the journey alone. She carried a camera to record footage on the trip, which has been converted into a documentary called *Maidentrip*.

Camille Gaskin-Reyes

REFERENCES AND FURTHER READING

International Polar Foundation. www.polarfoundation.org/.

Maag, Christopher. "Woman is the Youngest to Cross an Ocean Alone." *New York Times*, March 14, 2010. http://www.nytimes.com/2010/03/15/sports/15row.html.

National Geographic. "Adventurers of the Year." http://adventure.nationalgeographic.com/adventure/adventurers-of-thc-ycar.

NPR. "Katie Spotz's Solo Ocean Trip Sets World Record" March 23, 2010. http://www.npr.org/programs/talk-of-the-nation/2010/03/23/125065175/.

Reingold, Jennifer. "This Woman Rowed Across Three Oceans by Herself." *Fortune*, October 16, 2014. http://fortune.com/2014/10/16/this-woman-rowed-across-three-oceans-by-herself/.

Rothman, Lily. "A 14-Year-Old Girl Sailed Around the World—And She Brought a Camera." *Time*, January 16, 2014. http://time.com/894/maidentrip/.

Savage, Roz. Blog. http://www.rozsavage.com/.

Westchester, Simon. *Atlantic Great Sea Battles, Heroic Discoveries, Titanic Storms, and a Vast Ocean of a Million Stories*. New York: William Morrow, 2010.

ANNOTATED DOCUMENT

Mainstreaming Gender in Water Management, UNDP Resource Guide (2006)

Background

The 2006 Resource Guide was developed by the United Nations Development Program (UNDP) with a number of NGO partners to facilitate the analysis of gender perspectives in water resource projects and their incorporation into policies, programs, and projects. The Resource Guide also presented tools and techniques to address gender inequalities and strengthen water management and human development capacity for men and women.

Summary

One of the main concepts underlying the UNDP manual is called "Integrated Water Resources Management," or IWRM, which, according to the Resource Guide, is a systematic process for the sustainable development, allocation, and monitoring of water resources. Through IWRM, programs and projects can be developed using a multisectoral, comprehensive approach to water management to address competing demands for finite freshwater supplies.

Since both women and men are stakeholders in development, a gender perspective ensures that priority is given to the full participation of men and women and recognition of their ownership in resolving community water and sanitation problems.

This guide emphasizes the linkages between water and other areas of social and economic life. It advocates a coordinated and sustainable approach to management of water, land, and other natural resources without compromising their availability for future generations.

It provides a valuable toolkit for policy makers, analysts, international organizations, and governments to facilitate agreement on principles, and the setting of water policies and priorities and the implementation of specific initiatives that value the contribution of both men and women.

The Resource Guide calls for improved water governance and increased coordination and collaboration among various sectors, i.e., drinking water, sanitation, industry, irrigation, and ecosystem preservation. The guide promotes the development of technical know-how to minimize competition among different sectors that use water and reduce conflicts among individual, community, and government stakeholders.

Within this context, the guide emphasizes that sustainable use of water resources is a critical way to combat poverty and reduce gender and social disparities in access to and control over water and other resources. It therefore advocates for gender mainstreaming in decision-making structures.

Gender mainstreaming is interpreted as assessment of the potential consequences of planned actions for women and men. These may include: governance, technology, legislation, policies, projects, finance, and all levels of development programming. The Resource Guide provides a strategy for making women's and men's concerns an integral factor in political, cultural, social, economic, and societal spheres of life, so that inequality is not perpetuated, particularly in developing countries.

Water Subsectors

The guide builds on analyses of gender issues, knowledge of manuals, reference materials, methodologies and lessons learned from actual field experiences in the water sector and case studies. It sets forth guidelines for 13 water subsectors or specific areas of water use. The subsectors, in which gender and water management approaches should be integrated and mainstreamed, are listed as follows:

- Gender, Governance, and Water Resources Management
- Gender, Water, and Poverty
- Gender, Sanitation, and Hygiene
- Gender, Domestic Water Supply, and Hygiene
- Gender and Water Privatization
- Gender, Water, and Agriculture
- Gender, Water, and Environment
- Gender and Fisheries
- Gender and Coastal Zone Management
- Gender and Water-Related Disasters
- Gender, Water, and Capacity Building
- Gender Planning and Tools in Water Sectors
- Gender Responsive Budgeting in the Water Sectors

The following section contains some excerpts from the Resource Guide.

Camille Gaskin-Reyes

Excerpts of Key Concepts of the Resource Guide

Water should be treated as an economic, social, and environmental good.

Water policies should focus on the management of water as a whole and not just on the provision of water.

Governments should facilitate and enable the sustainable development of water resources by the provision of integrated water policies and regulatory frameworks.

Water resources should be managed at the lowest appropriate level.

Women should be recognized as central to the provision, management, and safeguarding of water.

Water is essential to human beings and all forms of life. But pollution and lack of access to clean water is proliferating the cycle of poverty, water-borne diseases, and gender inequities.

Water is an entry point for sustainable development, poverty eradication, human rights, reproductive and maternal health, combating HIV and AIDS, energy production, improved education for girls, and a reduction in morbidity and mortality.

And yet there are still 1.1 billion people without access to safe drinking water and 2.6 billion without access to adequate sanitation. This situation has an enormous negative impact on women and children.

There is deepening poverty worldwide, and the most vulnerable groups are women and children. Women experience poverty differently than men, as they are generally treated unequally.

It is estimated that, of the 1.3 billion people living in poverty around the world, 70% are women. Women work two-thirds of the world's working hours, produce half of the world's food, and yet earn only 10% of the world's income and own less than 1% of the world's property.

Why Gender, Water and Poverty?

Countries with the lowest gender-related development indices (Sierra Leone, Niger, Burkina Faso, and Mali) also had high poverty rates and little access to water, health, and education. Other countries with high poverty rates (Bolivia, Colombia, Guatemala, Honduras, Nicaragua, and Paraguay) also had high rates of social, gender, and ethnic inequality.

Access to water of sufficient quality and quantity will reduce the incidence of water-washed and water-borne diseases, improve health and productivity for women, and attendance in schools for children.

When there is competition for water resources, women and the vulnerable often lose their entitlements. Women's development priorities for water resources may be for sources nearer homes so that they are able to balance their productive and reproductive roles. If they are not consulted, then these priorities will not be considered.

Improved livelihoods and food security for women and the disadvantaged are also dependent on access to sufficient water resources. Participation in water management can also improve the dignity of women through giving them a voice and choice. It also improves targeting and efficiency.

Women are more vulnerable than men to chronic poverty due to gender inequalities in various social, economic, and political institutions. Such inequalities can be found in the uneven distribution of income, control over property or income, and access to productive inputs (such as credit), decision-making resources and water resources, rights and entitlements that often favor men in opposition to women. Women are also subject to bias and social exclusion in labor markets.

The HIV and AIDS pandemic, which is both a cause and a consequence of the vulnerability that is characteristic of poverty, has driven some countries to adopt home-based care approaches as health institutions fail to cope with the demand for services.

The home-based care approach implies that there should be water of sufficient quality and quantity to avoid secondary infections as well as to reduce the burdens of caregivers, who, in most cases, are women and girls. Women and children carry an unequal burden of deepening poverty.

Definitional Misconceptions

Poverty is multi-dimensional, location specific, and varies by age, culture, gender, and other socio-economic aspects. Perceptions of poverty also differ from women to men: for example, in Ghana men defined poverty as the inability to generate income, while women viewed it as food insecurity.

Poverty is not only about material deprivation; it also includes a lack of voice or power, vulnerability to crises and other adverse situations, and limited capacity to cope with such vulnerabilities. If water resources are located far away from homes, women and girls have to walk further to collect water, thus reducing the time available for productive work.

Effective water management offers social networks for women through management committees, but very often women end up doing unskilled and unpaid work related to water management. Continuing to link poverty to material well being masks other dimensions of poverty, such as powerlessness and exclusion from decision-making.

Gender, Poverty, and the Environment: A Three-Way Interaction

While separate Millennium Development Goals have been set for poverty, gender, and the environment (encompassing water and sanitation), they are interrelated and there is a three-way interaction among them. Water is essential for the well being of human beings, vital for economic development, and a basic requirement for the health of ecosystems.

Clean water for domestic purposes is essential for human health and survival and, combined with improved sanitation and hygiene, it will reduce morbidity and mortality, especially among children. Water is also vital for other facets of sustainable development such as environmental protection, food security, empowerment of women, education of girls, and reduction in productivity loss due to illnesses.

Water is a catalytic entry point for developing countries in the fight against poverty and hunger, and for safeguarding human health, reducing child mortality, and promoting gender equality and protection of natural resources.

Some Policy Implications

In integrated water resource management, water is viewed as both an economic, environmental and a social good, and thus in some cases it can be considered a commodity responding to the principles of supply and demand. It thus has a market value determined for certain uses.

The water sector is often divided into productive and non-productive water uses. The non-productive uses of water (health, domestic chores, and sanitation) tend to be the responsibility of women and are not considered in economic assessments. These should be incorporated into the assessment of relative economic values of water resources to allow for the understanding and consideration of the interdependence between productive and domestic water.

Water as a commodity implies that the development of water resources should be based on demand. However, poor women are generally unable to express their demands for services, nor do they have the capacity to defend their rights, especially if there are recognizable and transferable property rights over water.

In addition, female-headed households have even lower capacity to express, demand, and defend their rights. In order to meet the water demands of poor women, governments must collect sex-disaggregated data and develop gender-sensitive indicators in all sectors, including water, sanitation, agriculture, and irrigation.

The use of participatory tools is also important for engaging the voiceless and less educated who may have difficulties understanding written text. Only this way can priorities of the poor women and men and boys and girls be heard and understood.

Source: United Nations Development Programme and Gender and Water Alliance. *Resource Guide: Mainstreaming Gender in Water Management.* 2006. Available online at http://www.undp.org/content/dam/aplaws/publication/en/publications/environment-energy/www-ee-library/water-governance/resource-guide-mainstreaming-gender-in-water-management/IWRMGenderResourceGuide-English-200610.pdf.

PERSPECTIVES

Which of the Following Development Aid Projects Have Been Most Effective in Incorporating Gender Considerations into Water and Sanitation Projects to Achieve Intended Outcomes?

Overview

As described above, the UNDP Resource Guide presents key principles of mainstreaming gender perspectives into water management policies and programs. In development practice there is often a gap between theory and application. There are countless manuals and resource guides, technical guidelines, and handbooks that aim to provide blueprints for the successful implementation of projects. However, many handbooks and resource guides remain just that—blueprints on a shelf, not incorporated into programs and actual practice due to lack of funds or political commitment. This section reviews and compares the experiences of three case studies, which are included in the Resource Guide as examples of international projects using the gender-integrated approaches outlined in the guide. These examples are from Zimbabwe in Africa, Nicaragua in Central America, and Jordan in the Middle East.

Similarities and Differences

The three projects exhibit many differences and similarities due to important variations in the geographical area, cultural factors, and ecological conditions of the projects. This brings home the adage that one size does not fit all in development planning. The differences range from the arid environment of Jordan to the more water-rich areas of the Zimbabwean and Nicaraguan communities. The similarities are evident: in all the case studies, women, assisted by girls, bear the brunt of the burden to fetch, manage, and save water for the entire household, and articulate the need for help with that burden. In all three projects, funds come from outside the community to jump-start the process. However, the outcomes are different.

The examples in Nicaragua and Zimbabwe show that the project design took great care to integrate and engage both men and women in the achievement of water and sanitation goals. The training and outreach mechanisms included in these projects helped provide long-lasting technical and leadership skills as well as attitudinal change.

In these two projects, women benefitted from their high participation; overall they acquired valuable organizational and technical skills that would help them in other areas of life. This was especially the case in Zimbabwe, where women became experts in the drilling of boreholes and latrine repair skills that were transferable to other activities in the construction sector.

In both Nicaragua and Zimbabwe, men showed some initial resistance at first, but were eventually won over to the project and became more engaged over the course of project execution. While women's participation was very strong in Zimbabwe—simply due to the predominance of female-headed households and fewer men physically present in the village—in Nicaragua women's participation was under 40 percent. In Nicaragua, women took on coordinating and decision-making roles in water management committees, which might have helped to offset the relatively low level of broad-based engagement.

The example from Jordan is quite different from the two other projects. It appears that while this was a project conceived by women (The Women's Society), and managed by women, full participation of the entire community (both men and women) was not evident. The design was targeted to women and did not foresee the integration or active involvement of men, most likely due to cultural factors.

Even though women benefitted from the training and provision of skills, it is not clear if these advances spilled over to male members of the community or provoked any changes in male attitudes on women's water-carrying roles. The project provided water cisterns, and it appeared that, after the project, women continued to shoulder the full responsibility of drawing water from the cisterns and managing water for the village.

A common thread was that all three projects received external funding from international agencies. In the case of Nicaragua and Zimbabwe, the funds were outright grants. The example of Jordan (and, to some extent, Zimbabwe's case) was more interesting in terms of financial sustainability, since there was a loan or grant repayment mechanism through a revolving fund or community water fund. This measure assured the flow of financial benefits to a wide range of recipients and kept the funds going for a longer period.

In all three projects, the Resource Guide's principle of integrating water with other sectors is evident through the linkages of the water project with other activities such as market gardens, health and training, beekeeping, and agricultural pursuits. However, looking at the big picture of these projects, only the projects in Nicaragua and Zimbabwe mainstreamed the Resource Guide's recommendations for gender-integrated water management.

The Jordan project appears to be the most ring-fenced one for women. Due to cultural norms and the very design of the project, the project did not seek an integrated view of women's roles with those of male community members in the water sector. Men did not actively participate, though they received the benefits of additional water through the women's project. While the Jordan project appears to be the most financially sustainable project of the three examples due to the continuation of the revolving fund, the other two projects seem to be more socially sustainable and more integrated into the community.

All three projects avoided a top-down, conventional, male-oriented, or male-dominated approach to the water supply problems. Nonetheless, it appears that a longer-term female empowerment factor was more present in the Nicaraguan and Zimbabwean cases. Of these two countries, the Nicaraguan experience would appear to have the most social staying power in the future, since the community had a considerable presence of male counterparts that could join the project. The women in the Zimbabwe project were mostly from female-headed households and could not fully count on the support of male community members (through their lack of physical presence). Hence, these women were expected to experience more hardships in carrying out and sustaining their strenuous, multiple roles.

Camille Gaskin-Reyes

Perspective 1: Gender Mainstreaming in a Water Supply and Sanitation Project in Manzvire Village, Chipinge District, Zimbabwe

About a decade after Zimbabwe's independence in 1980, the country carried out an initial water sector reorganization program in 1993, encouraging women's participation in activities in the water sector. Even though the newly independent country had installed some water supply systems, most women and girls in rural villages still had to walk long distances to fetch water. This situation negatively affected women's productivity and girls' school enrollment levels. High female dropout rates ensued, as girls reached puberty and couldn't cope with the lack of sanitary facilities for females at most schools.

In 1997, the Chipinge district in Zimbabwe adopted a community-based management approach to water in Manzvire, a village with a population of just over 5,500 people. This district was plagued with water and sanitation problems: water quality and sanitation issues, use of simple pit latrines, and lack of surface water sources in the village. Since the nearest river was 10 miles away, water was tapped from groundwater using boreholes and shallow wells. An additional problem complicated matters even further: the loss of men to the AIDS epidemic, as well as male out-migration to cities to find work, meant that at least 80 percent of households were female-headed. In some cases, underage children were running households alone due to the loss of both parents from AIDS.

The Project

Responding to the severity of this problem, the United Nations International Children's Emergency Fund (UNICEF) contributed US $4,000 to the Chipinge Rural District Council (RDC) to repair its water supply systems, primarily the boreholes that provided water to residents. The RDC adopted the community-based approach outlined in the Resource Guide. Funds were used to support community involvement and the training of local women to sink wells and build latrines. The project attempted to integrate both men and women in planning and decision-making regarding the new water sources and the repair of existing systems. However, given the limited presence of men in the area, the task mainly fell to women.

The principal factor of the project was UNICEF's ability to mobilize women. Without this factor, the project would have had little chance of success. Women were the main stakeholders in the village, given their preponderance and existing roles in water collection and management. In Manzvire, women actively selected the technology and site locations for wells and latrines. The project also contained a savings and credit component, which provided revolving funds for the purchase of spare parts and greasing materials for pumps and boreholes. As a result of their participation, women felt a sense of pride and empowerment. They formed a cooperative garden and a community water fund, and also requested monetary help from the few males in the village or those far away from the area.

Outcomes

According to the account of the project in the Resource Guide, the injection of grant funding and UNICEF's support and expertise empowered Manzvire's women to become more actively involved in decision-making on water, an important factor for their day-to-day activities and survival. Women contributed labor to the borehole maintenance work on a voluntary basis once they had completed the mandatory training. At the end of the project, the women had rehabilitated 15 boreholes and constructed 60 new ones.

The availability of water closer to the village freed up the women to pursue productive activities such as market gardening, which improved their nutritional base and income levels. Women used the interest from their community fund to set aside funds for maintenance of the boreholes and the purchase of spare parts. Girls were able to stay at school longer since they reduced the time spent collecting water. Overall, the community learned new ways to reduce water contamination and received training in the proper disposal of rubbish and sewage from local households.

Key Success Factors

A review of this case study against the criteria of the Resource Guide and the examples of best practices indicates some important factors: UNICEF was critical as a funding agent, a catalyst, and provider of technical and organizational skills to the community. Women were the clear beneficiaries of the project since they received the training and applied their construction and maintenance skills acquired in the water project. The women's empowerment prompted them to venture into new areas of productive work and integrate water into work in other sectors, following the principles of the Resource Guide.

In the design of the project, water was not viewed as an isolated resource but was combined with health education, skills building, market gardening, a community fund, and high community engagement in decision-making. UNICEF held technical workshops to bring home the benefits of equal training of men and women in the development of water management and construction skills. To carry out the physical work, women had to wear construction worker–type overalls during latrine and borehole construction. For some women, it was the first time they had shed their long traditional dresses for "men's" work overalls. In addition to women's empowerment, the village benefitted from a significant decrease in diarrhea and waterborne diseases.

Obstacles Along the Way

The experience with the project indicates that it was not easy to change attitudes at first. At the beginning, the male minority in the village felt threatened by UNICEF's backing of the women and initially resisted the women's participation in decision-making. Eventually, the men of the village overcame their resistance and took on household tasks while the women were attending community meetings and training. One difficulty in this project was that very few men were available in the village to take on their fair share of the project's construction burden and to share the myriad of household and child-raising tasks with the women.

Commentary

The project appears to be sustainable, which is in line with one of the recommendations of the Resource Guide. The fact that the women were trained in the maintenance and operation of the system assures that the pipes and latrines are not likely to suffer the degradation and disrepair of the past system. Had the women not taken matters into their own hands to build and fix the latrines and boreholes, they would have spent more time in the long run fetching water in an endless, vicious cycle. Also, the community would have continued to suffer from waterborne diseases, and the women would not have acquired leadership and other skills from project work to help break that cycle.

Women also achieved greater economic empowerment through the establishment of market gardens, since they used the new sources of water to irrigate gardens for the sale of surplus produce. The proximity of the water also freed up more time for girls' schooling, helping to ensure that the next generation of women is likely to have a future with some career or social mobility prospects.

Camille Gaskin-Reyes

Perspective 2: Gender Equality as a Condition for Access to Water and Sanitation in Nicaragua

In 1998, Hurricane Mitch ravaged Nicaragua and led to more than 4,000 deaths. It severely affected Leon and Chinandega, two communities involved in this project example. In 1999, in the aftermath of the storm, these communities faced drought and contamination of their water sources. Although these villages had some groundwater resources, demand had already outstripped supply. Contamination problems also reduced the availability of water for most residents. The situation in the two communities mirrored the general situation of the Nicaraguan population (only 50 percent of village residents have access to potable water and sanitation).

The Project

Through the cooperation of the international nongovernmental organization CARE, headquartered in Atlanta, Georgia, and the Swiss Development and Cooperation Agency (COSUDE), funds were provided to support a partnership with the municipality of Leon in Nicaragua to implement the project. The project's goal was to support the construction of latrines and new water systems to improve access to water for the 17,000 inhabitants in 45 communities of Leon and Chinandega.

In the project area, women and children bear the brunt of transporting and managing water. Prior to this project, few efforts had been made to build in gender considerations and gender equality in water sector interventions. The presence of CARE and COSUDE guaranteed considerations of gender equality as an integral element of project design and implementation. Other goals were the achievement of community participation and project sustainability.

The project sponsors applied the principles of the Resource Guide through organizing gender sensitization workshops to raise consciousness of both men and women about gender roles in water and the value of understanding these roles in the planning, organization, construction, and management of the new water systems.

The training programs helped break down men's early resistance to the active involvement of women in the project and led to a clearer understanding of residents' responsibilities in the water and sanitation sector. More than 85 percent of the 687 male participants grasped the importance of reducing the water-fetching burden of women and children, and understood the importance of hand-made wells and household connections as secure sources of drinking water and measures to relieve women of an enormous burden.

Outcomes

After completion of the gender workshops and other training sessions, men's participation in the project increased, as did women's engagement in decision-making. Women were elected to more than 70 percent of the committees' posts, all of which were previously occupied by men. These positions were in areas of trust and responsibility—coordinators, vice-coordinators, and financial managers. CARE also trained women in the operation and maintenance of 276 waterworks. However, in this area there was a 37 percent participation rate of women, mainly due to the fact that women still had to carry out household and child-rearing tasks and could not consistently attend classes.

In addition, the project built in components to facilitate community discussion and dialogue on topics such as gender roles, women's self-esteem, female and male identity, women's rights and commitments, and women's participation in project benefits. The project report indicated that these activities appear to have changed men's perceptions about men's and women's roles in water management. However, it is not evident how sustainable this change in attitude would be in the long term.

Nonetheless, the training component was critical, since it represented a big shift in project design, away from the emphasis on physical works, such as construction, and toward awareness of gender roles and positive attitudes, which heightened the success of the intervention. The project also promoted the concept in the Resource Guide that access to water is a human right, and that all men, women, and children should have equitable opportunities to access services.

Key Success Factors

One of the project's key elements was the prior assessment of community needs, using a gender approach. An important aspect of this needs assessment (part of the stakeholder analysis methodology) is a survey's finding that most members of the community (men and women) trusted the dedication of women to lead the sanitation and water committees of the villages.

However, there was a large disparity between men's and women's perceptions of the importance of water. In the survey, most men could list only two uses for water while women could record up to 11. The survey indicated that reducing the female burden of carrying water was not in the forefront of the minds of male community members. This led to the perception that it was important to sensitize males through training sessions on women's roles in water as recommended in the Resource Guide.

The project developed specific methodologies for gender mainstreaming, which was consistent with one of the principles of the Resource Guide. It stressed coordination of all funding agencies, and partnership and collaboration with local governments and all stakeholders engaged in the project. These elements are deemed in the Resource Guide to be key factors to enable equitable and participatory access of communities to water. In addition, the project used the help of project promoters to help with critical aspects of social outreach and communication about the project's merits and expected benefits.

The project enabled rural women to become more aware of their rights and leadership potential. In spite of the fact that women's overall participation in numbers was less than that of men, the women gained more awareness about their strengths and earned the respect of men in the community through developing leadership and organizational skills. There was a relatively high participation of all age groups of women, from the young to the middle-aged to the elderly, in all the different cycles of the project.

Camille Gaskin-Reyes

Perspective 3: Jordan: Rural Women Securing Household Water through Installation of Water Cisterns in Rakin Village

Jordan is located in one of the planet's most water-stressed regions, the Middle East. Its economic development and the well-being of its population depends on the optimal management of scarce water resources. In addition, the country's geopolitical location makes it extremely vulnerable, not only to droughts and climate change but also social, military, humanitarian, and political conflicts, including those related to the sharing of water resources. Jordan's natural water scarcity has been worsened by high population growth mainly due to the influx of refugees from other countries. Securing fresh, clean water is a constant challenge for Jordan's communities, especially the poor.

The Project

The project arose when a local women's group in Rakin Village, with funding from the Small Grants Programme of the Global Environment Facility (GEF), planned a community-based project. The Rakin Women's Society was the entity that requested the funding and designed and executed the project using a revolving loan system. The project entailed the construction of water cisterns and improvement of water-harvesting systems in Rakin.

The beneficiaries were women, who were identified as in need of secure and sustained water resources for household use. Rakin, with a population of 5,500, is a low-income area located in the Karak Governorate in the south part of Jordan. Its economic mainstay is employment in services in the state's public institutions, the military, and agriculture. The village suffered from water shortages, soil erosion, and runoff, but produced some fruit and nuts, olives, barley, wheat, spices, and forest products. Herding of sheep and goats and beekeeping complemented the other activities.

The project was spearheaded by the Rakin Women's Society, which was founded in 1991 as a charitable group with the objective of improving social, economic, cultural, and health conditions of the women in Rakin. Using the principles of the Resource Guide, the sponsors applied gender analysis tools by identifying women's and men's roles in water management.

The analysis indicated that, in the village, women bore the burden of managing household tasks and water collection and use. Availability of water was crucial for the cultivation of small land plots and maintenance of household food security. However, there was not sufficient water to satisfy all of the residents' needs, including demand for human consumption, household, livestock, and irrigation purposes.

Rakin exhibited many of the same problems of poor communities in arid areas. It received piped water once every two weeks for only a period of six hours, which was woefully inadequate for the community's needs. Inhabitants therefore had to purchase water at high prices from outside vendors. The project's main aim was to use the funds to build cisterns for water storage. GEF funds also provided technical support to mange the project. A project steering committee was formed to implement activities such as training, management of loans and repayments, and organization of the selection criteria and selection of participants (in this case only females) according to certain variables.

Outcomes

The project came with a revolving loan, which meant that the finances were structured along the lines of a revolving system, providing 100 percent repayment so that funding would continue long after the cisterns were constructed. The project was extremely important to provide a more reliable source of clean water for village households, and to enable irrigation of fields and food security. It also reduced the high financial burden of the community to buy expensive water from tankers.

The project had positive outcomes for the women participants. It provided additional resources for their families and households because of the greater availability of water. More water meant an increased supply of products from women's gardens, which led to higher incomes. This improved their status as decision-making partners in the household and community at large. Women, who were involved in organizing and managing the grants, also acquired important leadership, organizational, and accounting skills.

Another positive feature of the project was the 100 percent payback on the loans that the women incurred to pay for the cisterns. This is consistent with other experiences in small-sized projects around the world, where women are seen to be reliable managers of credit with low levels of default. The funds that flowed back from the repayments were redistributed to other female beneficiaries in the community, since they were continuously replenished. The women used the revolving fund to expand beekeeping activities and install solar cells, thus diversifying their incomes and linking economic activities across sectors. This was in line with a Resource Guide principle of synergy. The project also generated positive health impacts and benefits for the entire community, since water was more available for safe drinking, washing, bathing, cleaning, and religious activities.

Conclusion

Only in two of the three projects reviewed did the project design take sufficient care to integrate men in the planning and execution of activities. The Jordan project only involved women, although the increased supply of water was vital to the entire community. All three projects furnished training and leadership skills to the women involved in the project, and in all three cases women capitalized on the training afforded by the original project to expand their economic activities even further.

All three projects received external funding from international agencies. In the case of Nicaragua and Zimbabwe, the funds were outright grants. Jordan's example (and, to some extent, Zimbabwe's case) was interesting in terms of financial sustainability, since the loan or grant repayment through a revolving fund or water community fund assured the flow of financial benefits to a wide range of recipients for a longer period.

All three projects pursued the principle of the Resource Guide of integrating water with other sectors. The Jordan project appeared to be more financially sustainable due to the revolving loan setup, but this feature only partially offset the project's lack of integration into mainstream village life. The Nicaragua project seemed to have the best chance for evoking the most social, cultural, and attitudinal change, since men in the community were able to change their mindset and participate in the project's activities and benefits.

Camille Gaskin-Reyes

REFERENCES AND FURTHER READING

CARE International. http://www.careinternational.org.uk.

COSUDE: Nicaragua: Gender Equality as a Condition for Access to Water and Sanitation. http://www.sswm.info/sites/default/files/reference_attachments/LANUZA%202003%20Gender%20Equality%20as%20a%20Condition%20for%20Access%20to%20Water%20and%20Sanitation.pdf.

Office of the Special Advisor on Gender Issues and Advancement of Women, Gender, Water, and Sanitation of the UNDP. *Case Studies on Best Practices*. New York, United Nations, 2006.

Swiss Development and Cooperation Agency (COSUDE) (German). http://www.deza.admin.ch/index.php?userhash?34814011&navID=1&l=e.

UNICEF. "Zimbabwe." http://www.unicef.org/infobycountry/zimbabwe.html.

12 A TIME FOR CHANGE AND A CALL TO ACTION

OVERVIEW

The topics covered in this book go to the core of water's importance for the social, cultural, and economic well-being of humankind and the sustainability of the planet. Throughout the various chapters we have seen that water is not just a mere liquid that falls from the sky or comes out of a tap. It is a life-nurturing substance, the hallmark of human society, and the bedrock of our collective existence. It is also an expression of the cultural, religious, or spiritual life of billions of people in the world.

This book underscores the importance of ocean exploration and the use of seas and oceans for transportation and other important human activities. From time immemorial, human societies have relied on rivers, seas, lakes, and oceans as important waterways for discovery, trade, warfare, communications, and settlement across boundaries.

In addition, people have harnessed water for energy, food production, and a myriad of social and economic activities, enabling humankind to achieve unprecedented levels of industrial development, globalization, worldwide trade, and human welfare compared to past millennia. Clean water has facilitated advances in health, medical and scientific progress, social well-being, and the reduction of waterborne diseases. Water has also enabled people to engage in religious rituals and cultural pursuits over millennia.

Notwithstanding these achievements, this book shows how human activities and intensive use and overexploitation of freshwater and marine resources have also had negative impacts and placed significant pressures on the world's ecosystems.

Pressures

The early chapters of this book examine the relationship between human-induced climate change and water, as well as the impacts of droughts, storms, loss of

Barge traffic on the River Seine in Paris attests to the river's importance for the transportation of goods. (Camille Gaskin-Reyes)

permafrost, melting ice, and rising sea levels across the world. They raise the issue of climate change mitigation and adaptation measures, threats to low-lying, vulnerable islands and coastlines, and the quest of developing nations for climate justice and climate change compensation measures.

While some chapters illustrate the economic and social benefits of water for society, others address growing threats and competition for water resources. These trends are occurring within the context of population growth, rising demand for water, and endangered freshwater supplies. There is a worldwide problem of growing water insecurity in nations, regions, communities, and many economic sectors in both developed and developing countries, from California to Chad.

Currently, one-third of the world's people live in countries experiencing moderate to very high water stress situations. Over a billion people in developing countries and 50 million in higher-income countries lack access to safe water for drinking, personal hygiene, and domestic use.

The book shows that droughts, excessive aquifer drawdowns, the damming of rivers for hydropower plants, and water pollution affect freshwater supply, often lead to water-stressed areas and conflicts, and inhibit harmonious sharing of transboundary water resources.

In some places, humans are tapping groundwater aquifers faster than rainfall can recharge them, or polluting or exploiting ocean resources and ecosystems faster than they can replenish themselves. Increasing amounts of water are needed for agriculture, industry, mining, and energy to satisfy the demands of a globalized, trade-connected world and the industrialization of developing countries, intensifying water scarcity in both arid and nonarid areas.

The steep rise in aquifer use and inappropriate irrigation practices has led to water table salinity and crop losses. Climate change and higher temperatures are also exacerbating the evaporation of rivers, lakes, or seas, lowering reservoir levels, and creating new dust bowls. Bottlenecks in freshwater supply and pollution of surface and groundwater sources have stimulated the growth of lucrative bottled water industries worldwide, and facilitated expansion in private water utilities management and control of water.

Cubans line up for water, distributed by truck, in Trinidad, Cuba, on May 14, 2011. Like many other countries, Cuba has been experiencing a severe drought for the last few years. (Dreamstime)

In the marine environment, human pressures on coastal and marine ecosystems, ocean warming, acidification, and pollution are causing severe impacts. These include depletion of fish stocks; damage to marine organisms; changes in migration cycles due to warming; mangrove destruction and expansion of aquaculture (following higher demand for protein and fewer fish stocks); coral damage and bleaching through climate change; reduction of biodiversity; and trafficking and smuggling of endangered marine species and other illicit items on the high seas.

In addition to already intensive oil exploration in coastal and ocean areas, countries and the private sector are extending aquaculture enterprises, mining the seabed, and tapping into common areas such as the melting Arctic Sea and permafrost for oil and as transportation lanes. These activities pose further threats to ocean contamination and acceleration of climate change.

Water, once considered an infinite resource in most people's minds, no longer flows boundlessly, as many nations, regions, or communities draw down from global bank accounts. On land and on sea, water problems are manifested in the growing "tragedy of the commons" crises. These are especially evident in freshwater reserved or marine resources such as fisheries, as private users race to tap resources before they run out. The emphasis on short-term gains is contributing to a vicious spiral of long-term depletion.

The areas of ocean governance, water-sharing arrangements, and conflict management touched upon throughout the book illustrate the need for nations and communities to better manage water and marine resources, paying attention to the needs of current stakeholders and future generations.

As the book points out, there is a wealth of international, national, regional, and local organizations, conventions, laws, regulations, rules, and guidelines that deal with water use and management at all levels. But there is little consensus on approaches and concrete action. There are also few robust enforcement mechanisms regarding the rules and regulations in the water sector, particularly on the high seas.

Therefore, in addition to the problems of water itself, there is the problem of coordination of the water sector. Sections of the book touch upon the complexity of the entire water sector and the difficulties of all parties to resolve conflicts or achieve consensus from international to local levels. There is no central water governance framework to monitor and address problems of freshwater loss and marine biodiversity, which increases the coordination difficulties.

We Are Not the Best Stewards of the Water Planet

It bears repeating that humans cannot survive without water, yet history teaches us that they are not always the best managers of their water assets. This is not likely to portend well for the future, unless people agree on fundamental changes in water conservation and stewardship strategies.

It is ironic that—through history up to now—even though humans know firsthand the importance of water in all facets of daily life, they still find it difficult to share water resources and coordinate governance of such a critical public good.

In past millennia, less population pressure enabled humans to adapt to changing climates and times of water and natural resource vulnerability. However, since the industrial revolution, rapid demographic growth and economic changes have tapped water reserves and outpaced our ability to keep up with current water problems and plan for the sustainability of resources for future generations.

Since the 1950s, a globalized and consumer-oriented world has created bottlenecks and inequities in the supply, distribution, and allocation of water resources in many countries. The rise of agricultural exports has also signified the rise in water exports (called virtual water), even from arid countries. Large volumes of water are needed to produce fruits, vegetable, grains, and beef for exports.

As sections of the book indicate, water management is complicated by the fact that there are different gender roles and gender patterns of ownership, distribution, and allocation of water. There are also gender differences in the impacts of water programs and other development interventions.

The book points to the social, economic, and political impacts in countries plagued by such water stress situations, including the aftermath of climate change, water pollution, droughts, megastorms, floods, and rising water levels.

Developing countries affected by these problems are experiencing out-migration flows of desperate communities (called environmental or climate change refugees). This situation is leading to pressures on those countries or regions receiving the influx of refugees and can cause the outbreak of potential conflicts.

Pushing against Planetary Boundaries

Scientists believe that humans are overstepping some limits of the earth's ecosystems and pushing against planetary boundaries. The main areas are the following: human-induced climate change through GHG emissions; ozone depletion of the upper atmosphere through industrial chemicals (chlorofluorocarbons); soil, water, and crop damage caused by fertilizers, salts, and toxins; overexploitation of ground and surface freshwater; conversion of forests to farms, pastures, and cities, and increased GHG; human-led loss of biodiversity; air pollution and urban smog; lake, river, and ocean contamination; ocean acidification and loss of marine organisms and fish stocks.

Scientific information from official UN reports on climate change and other studies have confirmed most of the above processes. Examples from this book describe how climate change, higher temperatures and rising sea levels, water scarcity, and pollution are playing out in scenarios such as lower crop yields, desertification, rising sea levels, displacement of people from flooded or water-deprived areas, rising global divide between water haves and have-nots, and higher risks to human health and the welfare of poorest communities. These signs are wake-up calls for humanity.

We Are Water Codependent

We are water codependent. The same water is flowing around and around in a cycle of currents across the world, touching us from ocean to ocean, sea to sea, river to river, pole to pole, raindrop to raindrop. The interdependence of water and the interaction of human lives provide a huge opportunity for societies to jointly embrace preventive, corrective, and adaptive actions to tackle humanity's problems.

What first began thousands of years ago as ripple effects of human interventions on the planet has become a roaring tsunami of impacts that threaten our welfare and that of future generations. An important theme that flows through the book, especially in the area of water governance and water sharing, is the need for less talk and more collective action among stakeholders at all levels: international, national, regional, and local.

Humans are faced with the challenge to put aside short-term national or individual interests and to think long-term and in a more holistic manner. Achieving consensus is not likely to occur overnight, but there are some medium-term steps that could provide quick gains for improved water management (discussed in the section below).

As long as there is scarcity of clean freshwater, its commercial value will continue, and the current international debate about whether water is a human right or a commodity will continue to play out. It is essential to resolve this contradiction to help nations and communities address freshwater depletion and the growing divide between the water haves and have-nots.

Water Is Everybody's Business

Water is everybody's business. If different sectors, institutions, nations, groups, and communities cannot work together for the common public good, then water becomes nobody's business in particular. Uncoordinated action is like water cupped in the palm of a hand. It tends to trickle through cracks and lose its cohesiveness. Many examples in the book have illustrated the merits of joint action.

Addressing the problems of the oceans is a serious obligation. Coming up with solutions requires robust collective action of all stakeholders and enforcement of international regulations. We don't lack the international laws to manage the common oceans; the problem is mustering the political will and commitment to action, and implementing joint actions to enforce regulations and tackle global problems.

A Time for Change

The discussions in the various chapters, case studies, and perspectives sections in the book all point to the need for the global community to manage water resources more sustainably. To remain true to our name, the "water planet", we need a more thorough understanding of the scale and scope of how water resource problems affect human welfare at all levels. In the past, the rise of civilization has always followed the availability of water. Ironically, we are now left to wonder if the deterioration of human welfare could eventually follow the lack of water.

In spite of the challenges outlined above, there are windows of opportunity for action. Humans have little choice but to act now. Without sufficient drinking water there is no chance for survival; without water for crops there is no food; without clean water there is no chance for a healthy life; without global collaboration on water there is little chance for worldwide peace.

There are clear links between sustainable water management, socioeconomic development, and poverty reduction. Without adequate freshwater in the world, it would be hard to end extreme poverty for millions of people who lack this resource and are vulnerable to climate change. Devastating impacts of climate change could create worldwide migratory movements, more severe pressures on countries, and security problems.

Future water security of the water planet is dependent on coordination, changes in human attitudes, long-term thinking, and improved collaboration among national, group, sector, and individual interests. The following sections provide glimpses of organizations and initiatives around the world that help mobilize action, provide responses to current threats to food and water security, and bring consensus on water management and conflict mediation strategies.

Camille Gaskin-Reyes

REFERENCES AND FURTHER READING

Dicken, Peter. *Global Shift: Mapping the Changing Contours of the World Economy*. New York: Guilford Press, 2015.

Hayward, Tim. "Human Rights Versus Emissions Rights: Climate Justice and the equitable Distribution of Ecological Space." *Ethics and International Affairs* 21, no. 4 (2007): 431–450.

Humphreys, Stephen, ed. *Human Rights and Climate Change*. Cambridge, UK: Cambridge University Press, 2010.

Posner, Eric A., and David A. Weisbach. *Climate Change Justice*. Princeton, NJ: Princeton University Press, 2010.

WATER INITIATIVES AND INTERNATIONAL EFFORTS

There are thousands of public, private, and nongovernmental groups around the world actively using water, managing water, or working in the water sector. However, public and private sectors activities are poorly coordinated; NGOs often vie for the same pool of donor dollars and sponsors to fund programs. In addition, not all public organizations, researchers, foundations, private sector institutions, industries, or NGOs use performance indicators to measure or assess the outcomes of their water-related activities, making it harder to gauge their effectiveness in resolving problems.

Newer actors since the 1950s have been many international development banks and funding agencies, which play a key role in financing water programs. These institutions are paying more attention to social and gender inequities in water supply, distribution, and allocation mechanisms. The examples below illustrate the activities of many different actors in water policies and programs.

The Pacific Institute for Studies in Development, Environment, and Security

The Pacific Institute for Studies in Development, Environment, and Security, based in Oakland, California, is dedicated to safeguarding the environment and promoting sustainable development. Peter H. Gleick, an expert on international water resources and cofounder of the Pacific Institute, serves as the organization's president.

Founded in 1987, the Pacific Institute provides independent research and policy analysis. The organization addresses global freshwater issues, including such problems as water shortages and aquatic habitat destruction. The staff's work encompasses not only research but also the publication of reports. The institute strives to provide solutions by working with influential decision makers, various advocacy groups, and the general public in order to change public policy.

The Pacific Institute takes an interdisciplinary approach in its research and analysis, consulting experts and sources from numerous fields of study. This is done in an attempt to integrate knowledge and possibly make connections that could be missed using a more traditional approach. The institute also brings opposing groups together to thoroughly examine potentially effective real-world solutions to difficult and pressing problems.

Five main areas of research are addressed by the institute: water and sustainability, environment and security, community strategies for sustainability and justice, economic globalization and the environment, and global change. Work done in the water and sustainability area concentrates on numerous topics. For example, the global water crisis topic addresses efforts to ensure basic access to clean water. The water efficiency topic concentrates on conservation as the most efficient way to save water for future needs.

In the early 1980s, Gleick did groundbreaking research on the impacts of climate change on water resources. Despite improved models, computers, and climate analysis during the past two decades, his work is still considered relevant and important. He suggested that dramatic changes would become evident in snowfall, snowpack, and runoff, and this research still forms the basis for attempting to understand what potential risks climate change will hold in store for society.

Gleick was appointed to the UN-Sigma Xi Scientific Expert Group on Climate Change and Sustainable Development. This group analyzes approaches and policies for adapting to and offsetting impacts of climate change.

Grenetta Thomassey

Climate Action Network

The Climate Action Network (CAN) is an association of 365 nongovernmental organizations that are concerned about climate change and are interested in developing strategies to limit and counteract human contributions to climate change. CAN views global warming as the greatest threat to life on Earth and wants to make sure greenhouse gases stay at sustainable levels. By working together, members have become more effective in forcing cuts in greenhouse gas emissions. The oil and coal industries are powerful lobbies, and CAN hopes that by joining together, members will project as powerful a voice.

CAN was established in March 1989. Representatives of nongovernmental environmental organizations from all over the world had gathered in Germany to discuss climate change. It was decided that to deal with the issue on an international level, there needed to be an easy way to pass information from group to group and to coordinate their efforts. CAN was created as a network because each member has its own style and methods, which is considered valuable because different countries and situations demand different approaches. A network also allows flexibility.

CAN's stated goals are:

To coordinate information exchange on international, regional, and national climate policies and issues, both between CAN groups and other interested institutions;

To formulate policy options and position papers on climate-related issues;

To undertake further collaborative action to promote effective nongovernmental organization involvement in efforts to avert the threat of global warming.

Some of CAN's members are Greenpeace, the World Wide Fund for Nature, and Friends of the Earth. CAN has observer status in the Framework Convention on Climate Change. It has regional offices in Africa, South Asia, Southeast Asia, Europe, Latin America, the United Kingdom, Canada, and the United States.

ABC-CLIO

United Nations Framework Convention on Climate Change

The United Nations Framework Convention on Climate Change (UNFCCC), a nonbinding treaty with 192 members, was the most substantial result of the "Earth Summit" held in Rio de Janeiro in June 1992. Created with the goal of reining in greenhouse gas emissions in an effort to halt global warming, the UNFCCC laid a foundation for creating future updated protocols on international cooperation for reducing human-driven atmospheric pollution. As a result of the UNFCCC's trailblazing, more than 180 countries have ratified the Kyoto Protocol, an amendment to the UNFCCC, since 1997, a dramatic attempt to reduce greenhouse gas emissions by 2012.

The UNFCCC had several key elements meant to initiate and maintain active participation by signatories in the pursuit of eliminating dangerous greenhouse gas emissions. One of the most important elements of the agreement was the creation of an international inventory of greenhouse gas emissions. Member nations are required to submit their countries' data on both emissions and clean up efforts; in turn, nongovernmental regulators give them feedback on their progress as they attempt to meet the UNFCCC's goals for emissions reduction.

The UNFCCC also created a physical location (currently offices in Bonn, Germany) where signatories may gather and share experiences, strategies, and ideas concerning greenhouse gas emissions reduction as well as pollution clean up that has already occurred. This is one of the most significant elements of the UNFCCC because it illustrates that member nations recognize the scientific fact that climate change and atmospheric pollution are issues that transcend national and regional interests. This is crucial because the UNFCCC is one of the first examples of nearly every nation in the world formally agreeing that countries must work together to slow down atmospheric pollution despite any political, cultural, or economic differences.

Nevertheless, the UNFCCC was divided into three categories, called Annexes, according to economic power. Annex 1 nations are industrialized nations that agree to reduce their emissions to target levels or buy emissions credits to offset their pollution. Annex 2 nations are developed countries that agree to assist Annex 3 countries—the developing nations—in the process of development according to rigid greenhouse gas emissions restrictions. In turn, Annex 3 nations agree not to sell offset credits to industrialized nations at the expense of their own pollution levels. In this way, the UNFCCC also provides incentives for developing nations to strive to be "greener" from the beginning of their industrialization, rather than become dumping grounds for the waste of the industrialized nations of the world.

Nancy L. Stockdale

WaterAid

WaterAid is the world's largest international nonprofit organization dedicated to providing safe water, sanitation, and hygiene in the world's poorest communities. Established in 1981, it is active in 37 countries worldwide, and its goal is to bring change to millions of lives every year through improvement of safe water and the development of sanitation projects. WaterAid's main goal is to improve the access of the poorest and most marginalized people to drinking water, sanitation, and hygiene. In 2014, WaterAid assisted 2 million people to obtain safe water and 3 million people to access sanitation projects as part of its mandate to lift people out of poverty and exclusion.

WaterAid works jointly with many partners to influence decision-makers at the policy level. It has several corporate sponsors that support the organization with donations and cosponsorships of projects throughout the world. WaterAid's hallmark approach is to partner with local communities to gain a deep cultural and social understanding of water issues; a key focus is to provide project leaders and communities with the skills and support to plan, design, and manage sustainable projects that meet their day-to-day needs.

WaterAid believes that water and sanitation are basic human rights that are essential to provide a solid foundation for health and education services, improvements in people's livelihoods, and the fight against poverty. The organization advocates locally and internationally to bring about structural changes in water policies and practices and ensure that the importance of water, hygiene, and sanitation in poverty reduction initiatives is recognized by decision-makers, donors, sponsors, and international agencies.

Camille Gaskin-Reyes

Charity Water

Charity Water is an international NGO based in the United States. The organization has a novel approach: it uses 100 percent of all donations to directly fund clean water projects and depends on other private funds and sponsors to cover staff salaries, office systems, rent, supplies, and other operating expenses. This is in contrast with many other NGOs, which use donation funds for their overhead. Charity Water works on thousands of projects in 24 countries in Africa, Asia, and Latin America. It collaborates with local partners to select the locations for water projects based on the assessment of needs.

Charity Water places a premium on building strong relationships with local communities and avoids duplicating the work of other organizations. For communities to participate and benefit from Charity Water's programs, they have to show commitment and willingness to participate in specific water projects through self-help mechanisms and direct involvement in the repair and maintenance of project works and water facilities. Charity Water selects the most appropriate technology possible in unison with local partners, taking care to incorporate physical limitations such as variations in terrain and cultural factors such as the community's comfort level with the technology used.

Charity Water carries out a meticulous selection process to ensure that local partners fully collaborate and take ownership of the provision of safe drinking water. To ensure funds are used properly and not mismanaged, Charity Water uses performance and sustainability indicators, which are important features to assess effectiveness of its efforts. Charity Water has a rigorous process in place to assess the operational capacity and technical skills of its local partners. It works with water experts to apply best practices and lessons learned from other projects to avoid repeating errors of the past. Charity Water believes a water project is not just a pipe with running water, but also the development of a community and the provision of leadership, ownership, and technical skills for during and after the project. Charity Water believes that these factors are as important as the hardware of the project and important for its sustainability.

Camille Gaskin-Reyes

Women for Water Partnership (WfWP)

The Women for Water Partnership (WfWP) is an international NGO headquartered in The Hague in Holland. It targets the improvement of water services for women and the incorporation of gender issues in the water and sanitation sector. WfWP is a global alliance of women's organizations and networks engaged in initiatives to resolve water and sanitation and foster sustainable development.

The organization's goal is to promote women's leadership across sectors and groups to achieve maximum impact of their development work. Recognizing the important role that women play in water collection and management, WfWP includes 26 women's networks in 100 countries in Africa, Asia, Latin America, and the Caribbean, as well as in Western Europe. These networks reach over 1 million women.

In line with the 1992 Dublin Principles, WfWP recognizes that women are active change agents who play critical roles in the provision, management, and safeguarding of water resources. The organization believes that women should be included in all aspects of water programs, especially in policy formulation and leadership positions. WfWP works with policymakers and at the grassroots level with local groups, and also acts as an intermediary between national and international stakeholders.

WfWP's networks strive to bridge the gap between internationally agreed-upon principles for sustainable development (the theory) and their application for real-life water management and sanitation projects (the practice). It holds regional conferences and technical working sessions on water management across the world with the participation of women's organizations, local authorities, national parliaments, water and sanitation experts, and international institutions.

WfWP's aim is to give women a voice to influence policy and decision-making at all levels, from the international arena to national and local communities. It advocates for gender-inclusive water management policies and programs.

WfWP continues to focus on core policy areas that promote sustainable development. The main ones include: coordination of inputs of WfWP members in international policy processes; organization of high profile women's preconferences at international water events; lobbying of decision makers; establishment of Water User Associations in partner countries; strengthening of women and water networks at the national level; promotion of women's inclusion in water institutions; and empowerment of women through water interventions and improved gender-responsive provision of services.

WfWP's underlying concept is that the growth in world population, human activities, and lack of consciousness of the water problems are causing major challenges for current and future water needs for drinking, food, energy, health, sanitation, and social and economic activities. The organization believes that water threats are not caused by physical scarcity alone, but also by people's attitudes, disjointed and irresponsible water misuse and pollution activities, and failure to recognize and resolve conflicts among water users ahead of time.

To address these challenges, WfWP emphasizes the need for more effective water governance, women's active participation and insertion in all processes, and greater social equity and cooperation on water management at all levels of society. WfWP works from the premise that women play a central role in the provision, management, and safeguarding of water in many societies, and spend considerable time and effort to meet the water needs of families and communities—and ensure survival. They also believe that better access to water and sanitation is essential to free up women and girls to devote more time to education, income generation, and productive goals.

WfWP's work therefore addresses the issue that in many developing countries men make the most decisions on management and development of water resources at local and national levels, while women still do most of the heavy lifting of water. Hence, the organization's emphasis is on stepping up women's leadership and ownership to bring about water supply and sanitation solutions, and enhance women's productive work (e.g., in agriculture) to raise women, especially female-headed households, out of poverty.

Camille Gaskin-Reyes

Global Activities of the World Bank and Regional Development Banks

While the above-mentioned groups have spearheaded water sector policy initiatives and programs at a smaller or local scale, there are large international players such as the World Bank (WB) and regional development banks (Inter-American, Asian, and African Development Banks) involved in funding large water projects in agriculture, watershed management, energy, and water and sanitation. In addition, there is bilateral aid from developed nations including the United States, Canada, Scandinavia, Australia, and China that provide support to developing countries in the water sector.

By far, the World Bank is the largest external source of financing for water projects in the developing world, amounting to about US$7.5 billion in 2011. Of these funds, 53 percent were allocated to water supply and sanitation, 12 percent to irrigation and drainage, 23 percent to hydropower, and 10 percent to flood protection. World Bank water sector programs also encompassed environmental sanitation programs, partnership programs with other financial institutions, governments and NGOs, and special programs such as the Cooperation in International Waters in Africa Initiative and the Southern Africa Water Initiative.

For the World Bank, water is at the center of economic and social development and a key development platform to foster improvements in health, food supply, energy generation, environmental management, and job creation. Its programs also highlight the linkages between water supply and sustainable development of cities, and aim to strengthen the resilience of rural villages to withstand the impacts of climate change—induced floods or droughts. The WB also pays special attention to the role of women and girls in water, improvement of their educational opportunities, and measures to lessen their water-carrying burdens.

The WB's analyses of water issues guide its policy emphasis and funding for priority areas in the water sector. It believes that water security is emerging as a main global risk for sustained national and international development and supports countries to comply with the water and sanitation-related goals of the United Nations Sustainable Development Goals (SDGs).

The WB has estimated that currently 70 percent of global water withdrawals are used for agriculture, and the world will face a 40 percent shortfall between forecasted demand and available water supply by 2030. It also forecasts that feeding 9 billion people in the next three decades would require a 60 percent increase in agricultural production and a 15 percent expansion of water drawdown—with significant consequences for water resources.

The World Bank also points to the fact that due to rising global demand for energy, the world will need more water for hydroenergy generation (1.3 billion people in the world still lack access to electricity). The WB estimates that more than half of the world's population lives in urban areas, which will intensify the demand for water. It estimates that, currently, poor sanitation, unsafe water, and inadequate hygiene practices cause about 675,000 premature deaths annually around the world, and estimates that by 2025 at least 1.8 billion people will be living in regions or countries affected by water scarcity.

According to WB analyses, climate change will make water shortages more unpredictable; in a worst-case scenario the earth's temperature would rise by 4° Celsius. It points to the high vulnerability of 500 million people already living in low-lying monsoon and delta areas in the world, the 700 million people without access to safe drinking water, and the 2.4 billion people who lack acceptable sanitation standards. Even though the World Bank has a large water sector funding program, the financing provided in loans and grants is not sufficient to match demand and developing countries' capacity to take and repay loans.

Camille Gaskin-Reyes

Work of Regional Development Banks

The regional development banks are smaller financial players than the World Bank in the water sector, but they nonetheless support important regional initiatives in different continents. Under the framework of the Sustainable Development Goals (SDGs) objectives, the areas supported by the Inter-American, Asian, and African Development Banks are similar.

These regional banks assist member countries in Asia, Africa, and Latin America to achieve universal coverage in water, sanitation, and solid waste management. They promote equal access to and improvement of the quality of water services. They are involved in enhancing water sector governance and more efficient water service delivery, private sector participation, and financial sustainability of programs. They also support improved sanitation services to low-income populations and leadership roles of women in the water sector.

Camille Gaskin-Reyes

REFERENCES AND FURTHER READING

African Development Bank Water and Sanitation Sector. http://www.afdb.org/en/about-us/organisational-structure/complexes/sector-operations/water-sanitation-department-owas/.

Asian Development Bank Water and Sanitation Sector. http://www.adb.org/sectors/water/main.

Charity Water. http://www.charitywater.org.
Climate Action Network. http://www.climatenetwork.org.
Gleick, Peter H. "The Toyota Prius and Climate Change: Good Technology versus the Misuse of Science—An ENN Commentary." Environmental News Network. May 31, 2005.
Gleick, Peter H., ed. *Water in Crisis: A Guide to the World's Fresh Water Resources.* New York: Oxford University Press, 1993.
Inter-American Development Bank Water and Sanitation Sector. http://www.iadb.org/en/sector/water-and-sanitation/overview,18357.html.
Pacific Institute. http://pacinst.org/.
United Nations Framework Convention on Climate Change. http://unfccc.int.
Water Aid. http://www.wateraid.org
Women for Water Partnership (WfWP). http://www.womenforwater.org.
World Bank (WB). "Water and Sanitation Program." http://water.worldbank.org/related-topics/water-and-sanitation-program.

OUTLOOK AND A CALL TO ACTION

The previous sections recapped the main themes of the book and presented examples of water management issues and actions. As indicated, there are countless agencies and institutions involved in addressing the planet's water problems, and as many approaches as institutions.

Some agencies fund microprojects; some support megaprojects. Some apply cookie cutter approaches for different countries while others attempt more innovative, or country-customized, initiatives. Many funding agencies have disjointed efforts, while numerous nations struggle for political will, internal commitment, and collaboration among decision-makers, water agencies, and local communities in the same country or across borders.

No Magic Bullet

The experience of many initiatives, programs, and the work of thousands of groups, including United Nations agencies, has taught us there is no magic bullet to bring about solutions. Goals are hard to achieve if countries lack funding, technical know how, governance skills, political commitment, and the will to change.

Concerted, multifaceted action is difficult to come by in such a complex sector as water. The absence of a global water governing or enforcement authority means that all nations have sovereign rights to decide on their course of action—and how little or much they implement on policy statements, commitments, and pledges.

Perspectives: Windows of Opportunity for the Future

Notwithstanding the difficulties in the water sector, this chapter outlines some windows of opportunity to address freshwater problems, the fallout of climate change, ocean pollution, and the destruction of marine resources and coastal ecosystems. Every stakeholder has a role to play: governments, decision-makers, policy analysts, the private sector, communities, multilateral and bilateral aid agencies, development banks, the United Nations, consumers, NGOs, water managers, special interests, and power brokers. The challenge is how to work together in the interest of the common good.

Assuring more equitable water supply, allocation, and distribution mechanisms goes to the core of human values. All groups and sectors in society have roles and responsibilities to protect ecosystems and safeguard them for the welfare of future generations. If humans are to capitalize on the lessons of the past, they need to commit to conserving, recycling, and protecting our most precious resource. The water dilemma we face is not only a crisis of water scarcity or pollution; it is also a crisis of governance and political will.

Ideas for Action

The most urgent actions to respond to global climate change, water crises, and conflicts on land and the high seas rely on international commitments or treaties, e.g., to reduce GHG emissions, promote multilateral or bilateral water-sharing arrangements with neighboring states, and enhance national policies and programs with community or stakeholder involvement at the local level.

Priority Areas

The following priority areas and opportunities for human action cover a range of actions: some are technical, some are institutional, and others are policy-oriented or political, requiring complicated and time-consuming consensus building processes. They include:

- Reduction of the water footprint of industries and other water-using activities, and structural shifts in the strategic planning, production, business models, and supply chains of agricultural, industrial, energy, and service enterprises to reduce GHG emissions and water consumption
- Development of innovative forms of water conservation technologies to address freshwater scarcity, aquifer depletion, and surface water pollution and runoff
- Increase in wastewater recycling and innovative technologies to cleanse wastewater for other uses, including drinking water
- Improvement and/or redesign of irrigation methods to waste less water and achieve better crop outcomes, including the use of drought resistant crops
- Attention to more equitable allocation, distribution, and sharing of water resources, and commitment to water as a human right for millions of poor people without water systems, reliable drinking water, and sanitation services
- Research and development of lower-cost desalinization plants and the use of energy-efficient and low fossil fuel–generating (solar and wind) technologies
- Promotion of educational and community water awareness programs, community involvement, partnership among stakeholders, funding agents and local authorities, and enhancement of women's leadership roles in water management

- Development of more effective policies, regulations, and incentives to curb pollution, and enforcement of the "polluter pays" concept through identification of source polluters and monitoring of water quality and biodiversity changes
- Improvement of water and sanitation infrastructure, and to green building standards, and sustainable responses to urban demands for water
- Avoidance of cross-contamination of water pipes with sewage, reduction of water pollution through human and solid wastes, and decrease of waterborne diseases
- Transfer of water conservation and other technologies and state-of-the-art research and development tools to developing countries to help them to cope with the consequences of water scarcity and climate change
- Promotion of sustainable water conservation practices at national, regional, and local levels, including safeguarding of community water supplies, introduction of water quality monitoring practices, and enforcement of penalties for violations
- Commitment to the SDGs, active participation in the United Nations Framework Convention on Climate Change (UNFCCC) meetings, and implementation of pledges on targets and other actions to reduce emissions of greenhouse gases
- Improvement of water catchment, storage, and harvesting methods to address the impacts of prolonged drought and more equitable and efficient allocation and distribution of water resources, especially at regional, community, or local level
- Reduction of ocean pollution and loss of marine biodiversity, and enforcement of regulations to protect fisheries and marine organisms and curb illegal trafficking

Camille Gaskin-Reyes

REFERENCES AND FURTHER READING

Page, Edward. *Climate Change, Justice and Future Generations*. Cambridge, MA: MIT Press, 2006.

Richter, Brian. *Chasing Water: A Guide for Moving from Scarcity to Sustainability.* Washington, DC: Island Press, 2014.

Sachs, Jeffrey D. *Common Wealth: Economics for a Crowded Planet*. New York: Penguin Press, 2008.

Sachs, Jeffrey D. *The Age of Sustainable Development*. New York: Columbia University Press, 2015.

ANNOTATED DOCUMENT

Conference of the Parties Twenty-First Session Paris, COP 21, November 30–December 11, 2015: Adoption of the Convention Applicable to All Parties, Agenda Item 4(b)

Nations Come Together in 2015 to Pledge GHG Reductions

Restoring climatic and global water imbalances requires tough decisions and tangible action. Endless conferences without commitments or piecemeal or disjointed activities are not likely to resolve the structural issues facing the planet.

However, the most recent conference on climate change in Paris, France, in December 2015 (COP 21) provided a glimmer of hope that the world's nations could come together and make binding pledges to reduce GHG emissions. The conference highlighted the need for shifts in current energy and industrial production models based on fossil fuels to renewable, cleaner sources of energy such as wind, solar, and other green sources.

It ended with calls to nations to honor their commitments to reduce GHG emissions. While all nations were called upon to follow up on their pledges, many discussions emphasized the situation of the most vulnerable states and poorest communities around the world, which stood to bear the highest costs of climate change and had the most to lose. Funds were committed to assisting affected developing countries to adapt to climate changes, but many vulnerable countries maintained they would require more funding.

The deal reached in Paris, if implemented, would only cut global GHG emissions by half of the amount needed to stave off an increase in atmospheric temperatures of 4° Celsius above the preindustrial average of the late 19th century. Some critics consider these targets to be insufficient, but the Paris pledges of individual countries were a step in the right direction that could pave the way for shifts in global economic policies and energy markets, as well as improved management of water resources.

The road ahead appears to be clear, backed by scientific evidence. But can we take it? Can we tap into the same courage, ingenuity, and resourcefulness that humans harnessed in past centuries to tame rivers, develop civilizations, build boats, and plot courses across unknown and uncharted oceans, seas, rivers, and land boundaries? Can we resolve this millennium's challenges and seize the opportunities to improve the situation for the generations to come? Only time will tell.

Camille Gaskin-Reyes

Excerpts from The Paris COP/21 Agreement

The Conference of the Parties,

Recalling decision 1/CP.17 on the establishment of the Ad Hoc Working Group on the Durban Platform for Enhanced Action,

Also recalling Articles 2, 3 and 4 of the Convention,

Further recalling relevant decisions of the Conference of the Parties, including decisions 1/CP.16, 2/CP.18, 1/CP.19 and 1/CP.20,

Welcoming the adoption of United Nations General Assembly resolution A/RES/70/1,

"Transforming our World: the 2030 Agenda for Sustainable Development," in particular its goal 13, and the adoption of the Addis Ababa Action Agenda of the third International Conference on Financing for Development and the adoption of the Sendai Framework for Disaster Risk Reduction,

Recognizing that climate change represents an urgent and potentially irreversible threat to human societies and the planet and thus requires the widest possible cooperation by all countries, and their participation in an effective and appropriate international response, with a view to accelerating the reduction of global greenhouse gas emissions,

Also recognizing that deep reductions in global emissions will be required in order to achieve the ultimate objective of the Convention and emphasizing the need for urgency in addressing climate change,

Acknowledging that climate change is a common concern of humankind, Parties should, when taking action to address climate change, respect, promote and consider their respective obligations on human rights, the right to health, the rights of indigenous peoples, local communities, migrants, children, persons with disabilities and people in vulnerable situations and the right to development, as well as gender equality, empowerment of women and intergenerational equity,

Also acknowledging the specific needs and concerns of developing country Parties arising from the impact of the implementation of response measures and, in this regard, decisions 5/CP.7, 1/CP.10, 1/CP.16 and 8/CP.17,

Emphasizing with serious concern the urgent need to address the significant gap between the aggregate effect of Parties' mitigation pledges in terms of global annual emissions of greenhouse gases by 2020 and aggregate emission pathways consistent with holding the increase in the global average temperature to well below 2° C above pre-industrial levels and pursuing efforts to limit the temperature increase to 1.5° C,

Also emphasizing that enhanced pre-2020 ambition can lay a solid foundation for enhanced post-2020 ambition,

Stressing the urgency of accelerating the implementation of the Convention and its Kyoto Protocol in order to enhance pre-2020 ambition,

Recognizing the urgent need to enhance the provision of finance, technology, and capacity-building support by developed country Parties, in a predictable manner, to enable enhanced pre-2020 action by developing country Parties,

Emphasizing the enduring benefits of ambitious and early action, including major reductions in the cost of future mitigation and adaptation efforts,

Acknowledging the need to promote universal access to sustainable energy in developing countries, in particular in Africa, through the enhanced deployment of renewable energy,

Agreeing to uphold and promote regional and international cooperation in order to mobilize stronger and more ambitious climate action by all Parties and non-Party stakeholders, including civil society, the private sector, financial institutions, cities and other subnational authorities, local communities, and indigenous peoples,

I. ADOPTION

1. Decides to adopt the Paris Agreement under the United Nations Framework Convention on Climate Change (hereinafter referred to as "the Agreement") as contained in the Annex;

2. Requests the Secretary-General of the United Nations to be the Depositary of the Agreement and to have it open for signature in New York, United States of America, from 22 April 2016 to 21 April 2017;

3. Invites the Secretary-General to convene a high-level signature ceremony for the Agreement on 22 April 2016;

4. Also invites all Parties to the Convention to sign the Agreement at the ceremony to be convened by the Secretary-General, or at their earliest opportunity, and to deposit their respective instruments of ratification, acceptance, approval or accession, where appropriate, as soon as possible;

5. Recognizes that Parties to the Convention may provisionally apply all of the provisions of the Agreement pending its entry into force, and requests Parties to provide notification of any such provisional application to the Depositary;

6. Notes that the work of the Ad Hoc Working Group on the Durban Platform for Enhanced Action, in accordance with decision 1/CP.17, paragraph 4, has been completed;

7. Decides to establish the Ad Hoc Working Group on the Paris Agreement under the same arrangement, mutatis mutandis, as those concerning the election of officers to the Bureau of the Ad Hoc Working Group on the Durban Platform for Enhanced Action;

8. Also decides that the Ad Hoc Working Group on the Paris Agreement shall prepare for the entry into force of the Agreement and for the convening of the first session of the Conference of the Parties serving as the meeting of the Parties to the Paris Agreement;

9. Further decides to oversee the implementation of the work programme resulting from the relevant requests contained in this decision;

10. Requests the Ad Hoc Working Group on the Paris Agreement to report regularly to the Conference of the Parties on the progress of its work and to complete its work by the first session of the Conference of the Parties serving as the meeting of the Parties to the Paris Agreement;

11. Decides that the Ad Hoc Working Group on the Paris Agreement shall hold its sessions starting in 2016 in conjunction with the sessions of the Convention subsidiary bodies and shall prepare draft decisions to be recommended through the Conference of the Parties to the Conference of the Parties serving as the meeting of the Parties to the Paris Agreement for consideration and adoption at its first session;

II. INTENDED NATIONALLY DETERMINED CONTRIBUTIONS

12. Welcomes the intended nationally determined contributions that have been communicated by Parties in accordance with decision 1/CP.19, paragraph 2(b);

13. Reiterates its invitation to all Parties that have not yet done so to communicate to the secretariat their intended nationally determined contributions towards achieving the objective of the Convention as set out in its Article 2 as soon as possible and well in advance of the twenty-second session of the Conference of the Parties (November 2016) and in a manner that facilitates the clarity, transparency, and understanding of the intended nationally determined contributions;

14. Requests the secretariat to continue to publish the intended nationally determined contributions communicated by Parties on the UNFCCC website;

15. Reiterates its call to developed country Parties, the operating entities of the Financial Mechanism, and any other organizations in a position to do so to provide support for the preparation and communication of the intended nationally determined contributions of Parties that may need such support;

16. Takes note of the synthesis report on the aggregate effect of intended nationally determined contributions communicated by Parties by 1 October 2015, contained in document FCCC/CP/2015/7;

17. Notes with concern that the estimated aggregate greenhouse gas emission levels in 2025 and 2030 resulting from the intended nationally determined contributions do not fall within least-cost 2°C scenarios but rather lead to a projected level of 55 gigatonnes in 2030, and also notes that much greater emission

reduction efforts will be required than those associated with the intended nationally determined contributions in order to hold the increase in the global average temperature to below 2°C above pre-industrial levels by reducing emissions to 40 gigatonnes or to 1.5°C above pre-industrial levels by reducing to a level to be identified in the special report referred to in paragraph 21 below;

18. Endorsed by decision 2/CP.18, paragraph 2. FCCC/CP/2015/L.9418. Also notes, in this context, the adaptation needs expressed by many developing country Parties in their intended nationally determined contributions;

19. Requests the secretariat to update the synthesis report referred to in paragraph 16 above so as to cover all the information in the intended nationally determined contributions communicated by Parties pursuant to decision 1/CP.20 by 4 April 2016 and to make it available by 2 May 2016;

20. Decides to convene a facilitative dialogue among Parties in 2018 to take stock of the collective efforts of Parties in relation to progress towards the long-term goal referred to in Article 4, paragraph 1, of the Agreement and to inform the preparation of nationally determined contributions pursuant to Article 4, paragraph 8, of the Agreement;

21. Invites the Intergovernmental Panel on Climate Change to provide a special report in 2018 on the impacts of global warming of 1.5° C above pre-industrial levels and related global greenhouse gas emission pathways;

Source: Adoption of the Paris Agreement. Conference of the Parties, 21st Session, Paris. FCCC/CP/2015/L.9. Available online at https://unfccc.int/resource/docs/2015/cop21/eng/l09.pdf.

BIBLIOGRAPHY

Ansohn, Albrecht, and Boris Pleskovic, eds. *Climate Governance and Development*. Washington, DC: World Bank Publications, 2011.

Baker, Judy L., ed. *Climate Change, Disaster Risk, and the Urban Poor: Cities Building Resilience for a Changing World*. Washington, DC: World Bank, 2012.

Baland, Jean-Marie, Pranab Bardhan, and Samuel Bowles, eds., *Inequality, Cooperation, and Environmental Sustainability*. Princeton: Russell Sage Foundation, 2007.

Barlow, Maude. *Blue Future: Protecting Water for People and The Planet Forever*. New York: The New Press, 2013.

Barlow, Maude, and Tony Clarke. *Blue Gold: The Fight to Stop the Corporate Theft of the World's Water*. New York: New Press, 2002.

Barnett, Cynthia. *Rain: A Natural and Cultural History*. New York: Crown Publishers, 2015.

Bergreen, Lawrence. *Magellan's Terrifying Circumnavigation of the World: Over the Edge of the World*. New York: Harper Collins, 2003.

Biswas, Asit K., Olcay Unver, and Cecilia Tortajada, eds. *Water as a Focus for Regional Development*. New Delhi: Oxford University Press, 2004.

Biswas, Asit K., Zuo Dakang, James E. Nickum, and Liu Changming, eds. *Long-Distance Water Transfer*. Tokyo: United Nations University Press, 1983.

Black, Maggie. *The No-Nonsense Guide to Water*. Oxford: New Internationalist Publications, 2004.

Bowermaster, Jon, ed. *Oceans: The Threats to Our Seas and What You Can Do to Turn the Tide*. New York: Participant Media, 2010.

Byers, Alton, and Jorge Recharte. "As Glacial Floods Threaten Mountain Communities, A Global Exchange Is Fostering Adaptation." *News Security Beat* (blog). Environmental Change and Security Program, Woodrow Wilson Center. April 14, 2015. https://www.newsecurity beat.org/2015/04/glacial-floods-threaten-mountain-communities-global-exchange-fostering-adaptation/.

Calow, Roger, Eva Ludi, and Josephine Tucker, eds. *Achieving Water Security: Lessons from Research in Water Supply, Sanitation, and Hygiene in Ethiopia*. Sterling, VA: Stylus Publications, 2015.

Carey, Mark. *In the Shadow of Melting Glaciers: Climate Change and Andean Society*. Oxford, UK: Oxford University Press, 2010.

Cato Institute. *Water for Sale: How Business and the Market Can Resolve the World's Water Crisis*. Washington, DC: Cato Institute, 2005.

Chellaney, Brahma. *Water, Peace, and War: Confronting the Global Water Crisis*. Lanham, MD: Rowman & Littlefield, 2013.

Coudrain, Anne, Bernard Francou, and Zbiginiew W. Kundzewicz. "Glacier Shrinkage in the Andes and Consequences for Water Resources." *Hydrological Sciences Journal* 50, no. 6 (2005): 925–932.

Craig, Robin Kundis. *Comparative Ocean Governance: Place-Based Protections in an Era of Climate*. Cheltenham, UK: Edward Elgar Publishing, 2012.

Davies, Kate. *The Rise of the U.S. Environmental Health Movement*. Lanham, MD: Rowman & Littlefield Publishers, 2015.

Diacu, Florin. *Megadisasters*. Princeton, NJ: Princeton University Press, 2010.

Diaz, Henry F., and Barbara J. Morehouse, eds. *Climate and Water: Transboundary Challenges in the Americas*. Boston: Kluwer Academic Publishers, 2003.

Dow, Kirsten, and Thomas E. Downing. *The Atlas of Climate Change: Mapping the World's Greatest Challenges*. Berkeley: University of California Press, 2006.

Elert, Emily, and Michael D. Lemonick. *Global Weirdness: Severe Storms, Deadly Heat Waves, Relentless Drought, Rising Seas, and the Weather of the Future*. New York: Penguin Random House, 2012.

Ennis-McMillan, Michael C. *A Precious Liquid: Drinking Water and Culture in the Valley of Mexico*. Belmont, CA: Thomson Wadsworth, 2006.

Gifford, Douglas. *Warriors, Gods, and Spirits from Central & South American Mythology*. Glasgow, UK: William Collins, 1983.

Grover, Velma, ed. *Water: Global Common and Global Problems*. Enfield, NH: Science Publishers, 2006.

Handwerk, Brian. "Five Striking Concepts for Harnessing the Sea's Power." *National Geographic*, February 21, 2014. http://news.nationalgeographic.com/news/energy/2014/02/140220-five-striking-wave-and-tidal-energy-concepts/.

Hardoy, Jorge E., Diana Mitlin, and David Satterthwaite. *Environmental Problems in an Urbanizing World*. London: Earthscan, 2004.

Hart, Merriam C. *The Dawn of the World*. Lincoln: University of Nebraska Press, 1993.

Holland, Ann-Christin Sjolander. *The Water Business: Corporations versus People*. Global Issues Series. Chicago: Zed Books, 2005.

Intergovernmental Panel on Climate Change (IPCC). "Synthesis Report of the Fifth Assessment Report." November 2, 2014. http://www.ipcc.ch/report/ar5/syr/.

Jansky, Libor, Martin J. Haigh, and Haushila Prasad, eds. *Sustainable Management of Headwater Resources: Research from Africa and India*. Tokyo: United Nations University Press, 2005.

Jha, Abbas K., Robin Bloch, and Jessica Lamond. *Cities and Flooding: A Guide to Integrated Urban Flood Risk Management for the 21st Century*. Washington, DC: World Bank, 2012.

Khagram, Sanjeev. *Dams and Development*. Ithaca, NY: Cornell University Press, 2004.

Klein, Naomi. *This Changes Everything: Capitalism vs. the Climate*. New York: Simon & Schuster, 2014.

Koch, Wendy. "Denmark Eyes North Pole, but How Much Oil and Gas Await?" *National Geographic,* December 17, 2014. http://news.nationalgeographic.com/news/energy/2014/12/141217/oil-natural-gas-denmark-north-pole-arctic/.

Kolbert, Elizabeth. *The Sixth Extinction: An Unnatural History*. New York: Henry Holt and Company, 2014.

Kristof, Nicholas, and Sheryl Wudunn. *Half the Sky: Turning Oppression into Opportunity for Women Worldwide*. New York: Knopf, 2009.

Kunzig, Robert. *Exploring the World beneath the Waves*. New York: Norton & Company, 1999.

LaHaye, Tim, and John Morris. *The Ark on Ararat*. Nashville: Thomas Nelson, 1976.

Langdon, John. *Mills in the Medieval Economy: England, 1300–1540*. Oxford, UK: Oxford University Press, 2004.

Mann, Charles. *1493: Uncovering the New World Columbus Created*. New York: Random House, 2012.

Maslin, Mark. *Climate Change: A Very Short Introduction*. Oxford, UK: Oxford University Press, 2014.

McCullough, David. *The Path between the Seas: The Creation of the Panama Canal 1870–1914*. New York: Simon & Schuster, 1977.

Murakami, Masahiro. *Arid Zone Water Resources Planning for Peace: Applications of Non-Conventional Alternatives in the Middle East*. Tokyo: United Nations University Press, 1994.

National Geographic. *A Special Issue: Water, Our Thirsty World*. Washington, DC: National Geographic Society, 2010.

Novaresio, Paolo. *The Explorers: From the Ancient World to the Present*. New York: U.S. Media Holdings, 1996.

Overey, Richard, ed. *Hammond Atlas of World History*. Maplewood, NJ: Times Books, 1999.

Payoyo, Peter Bautista, ed. *Ocean Governance: Sustainable Development of the Seas*. Tokyo: United Nations University, 1994.

Pearce, Fred. *With Speed and Violence: Why Scientists Fear Tipping Points in Climate Change*. Boston: Beacon Press, 2007.

Piper, Karen. *The Price of Thirst: Global Water Inequality and the Coming Chaos*. Minneapolis: University of Minnesota Press, 2014.

Ponting, Clive. *A New Green History of the World: The Environment and the Collapse of Civilizations*. New York: Penguin Group, 2007.

Postel, Sandra. *Water: Rethinking Management in an Age of Scarcity*. Washington, DC: Worldwatch Institute, 1984.

Reynolds, Terry S. *Stronger Than a Hundred Men: A History of the Vertical Water Wheel*. Baltimore: Johns Hopkins Press, 1983.

Richter, Brian. *Chasing Water: A Guide for Moving from Scarcity to Sustainability*. Washington, DC: Island Press, 2014.

Rosegrant, Mark W. *Water Resources in the Twenty-First Century: Challenges and Implications for Action*. Food, Agricultural, and the Environment Discussion Paper. Washington, DC: International Food Policy Research Institute, 1997. https://www.ifpri.org/publication/water-resources-twenty-first-century.

Rosemarin, Arno, Jennifer R. McConville, Amparo E. Flores, and Zhu Qiang. *The Challenges of Urban Ecological Sanitation: Lessons from the Erdos Eco-town Project, China*. Sterling,VA: Stylus Publications, 2014.

Rothfeder, Jeffrey. *Every Drop for Sale: Our Desperate Battle over Water in a World about to Run Out*. New York: Putnam, 2001.

Sachs, Jeffrey D. *Common Wealth: Economics for a Crowded Planet*. New York: Penguin Press, 2008.

Sachs, Jeffrey D. *The Age of Sustainable Development*. New York: Columbia University Press, 2015.

Sadoff, Claudia W., and David Grey. "Beyond the River: The Benefits of Cooperation on International Rivers" *Science Direct* 4, no. 5 (2002): 389–403.

Schouten, Ton, Stef Smits, and John Butterworth, eds. *From Infrastructure to Services: Trends in Monitoring Sustainable Water, Sanitation, and Hygiene Services*. Sterling: Stylus Publications, 2015.

Schwartz, Nelson D. "Investors Are Mining for Water, the Next Hot Commodity." *New York Times*, September 24, 2015. http://www.nytimes.com/2015/09/25/business/energy-environment/private-water-projects-lure-investors-preferably-patient-ones.html?_r=0.

Shiva, Vandana. *Water Wars: Privatization, Pollution and Profit*. Cambridge, MA: South End Press, 2002.

Smallwood, Stephanie. *Saltwater Slavery: A Middle Passage from Africa to American Diaspora*. Cambridge, MA: Harvard University Press, 2007.

Solanes, Miguel, and Andrei Jouravlev. *Water Governance for Development and Sustainability*. Santiago, Chile: United Nations Economic Commission for Latin America and the Caribbean, 2006.

Strang, Veronica. "Taking the Waters: Cosmology, Gender and Material Culture in the Appropriation of Water Resources." In *Gender, Water, and Development*, edited by Anne Coles and Tina Wallace, 21–38. Oxford, UK: Berg Publishers, 2005.

Sultana, Farhana, and Alex Loftus. *The Right to Water: Politics, Governance, and Social Struggles*. London: Earthscan Publications, 2011.

Van Koppen, Barbara, Stef Smits, Christina, Rumbaitis del Rio, and John B. Thomas. *Scaling Up Multiple Use Water Services: Accountability in the Water Sector*. Sterling: Stylus Publications, 2014.

White, Sarah C., ed. *Wellbeing and Quality of Life Assessment: A Practical Guide*. Sterling, VA: Stylus Publications, 2014.

Wolf, Aaron T. *Hydropolitics along the Jordan River: Scarce Water and its Impact on the Arab-Israel Conflict*. Tokyo: United Nations University Press, 1995.

World Bank. *Climate Change Impacts in Latin America: Confronting the New Climate Normal*. Washington, DC: World Bank Publications, 2014.

ABOUT THE EDITOR AND CONTRIBUTORS

THE EDITOR

CAMILLE GASKIN-REYES, PhD, is adjunct professor of Latin American studies at Georgetown University. She is a specialist in Latin American and development studies and has more than 28 years of experience in policy development and practice, natural resource management, environmental planning and economics, project management, and monitoring and evaluation in Latin America and the Caribbean. Dr. Gaskin-Reyes is also an international consultant. She received her doctorate at the University of Bonn, Germany.

THE CONTRIBUTORS

ANDREA ARZABA obtained her MA in Latin American Studies with a concentration in development from Georgetown University in Washington, D.C.

MIA BENNETT is a PhD student in the Department of Geography at UCLA. Mia holds an MPhil in polar studies from the University of Cambridge, where she was a Gates Scholar.

PATRICIA BIERMAYR-JENZANO is a sustainable development and agriculture professor at the Center for Latin America Studies and the department of Women and Gender Studies at the Edmund A. Walsh School of Foreign Service, Georgetown University. She is also an international consultant for the Food and Agriculture Organization of the UN.

LAURA BOCALANDRO has worked in international development for more than 20 years at the Inter-American Development Bank. She is the co-founder of P-Lab, a boutique start-up focusing on the creation of public and social value. Ms. Bocalandro holds an LLM from Harvard Law School and a JD from the University of Buenos Aires Law School.

CLAIRE BRENNAN is a lecturer in history at James Cook University in Townsville, Australia. She earned her PhD in history at the University of Melbourne.

MARK TROY BURNETT is an associate professor of geography and geography program coordinator at Mount Royal University in Calgary, Alberta.

TAMAR BURRIS is an independent researcher and writer. She has worked with the Discovery Channel, PBS, and ESPN, as both a writer and educational curriculum expert.

ROBERT D. CRAIG is an emeritus professor of history at Alaska Pacific University.

RENEE DUBIE is a writer and editor at ABC-CLIO. She holds an MA in political science from San Diego State University.

JANET EDMOND is the senior director for peace and development partnerships in the Policy Center for Environment and Peace at Conservation International. She obtained her Master of Public Health from Tulane School of Public Health and Tropical Medicine.

NATHAN EIDEM is an adjunct lecturer in geography at the University of Nebraska, Kearney. He has worked as a consultant for the National Geographic Society, the United Nations, USAID, and the Mekong River Commission.

ANOUR ESA studies peace and conflict resolution and sustainability at GWU's Elliott School of International Affairs in Washington, D.C.

ANGUS M. GUNN was professor emeritus of social and educational studies at the University of British Columbia.

HARALD HAARMANN is vice president and director of the Institute of Archaeomythology in Luumäki, Finland. He holds a PhD in linguistics from Bonn University.

ALBERT C. HINE is a professor in the College of Marine Science at the University of South Florida.

BERNADETTE HOBSON obtained her MA in Latin American studies with a concentration in political economic development from the School of Foreign Service at Georgetown University.

RUTH A. JOHNSTON is a writer and educator. She is author of *All Things Medieval* (Greenwood 2011) and *A Companion to Beowulf.*

1ST LT. JAMES LINK is a scout platoon leader in HHC, 1-41IN 2/4 IN. He obtained his BS in human geography with an emphasis on the Middle East and Africa from the United States Military Academy at West Point.

LIZA LUGO, JD, is a legal scholar and president of her own consulting firm, Monarch Consultants. She is author of *How Do Hurricane Katrina's Winds Blow? Racism in 21st-Century New Orleans* (Praeger 2014).

MAJ. DYLAN MALCOMB is an assistant professor of geography at the United States Military Academy. He is currently serving as the operations officer of an aviation task force in Afghanistan.

JOHN F. MONGILLO is a middle-school science teacher at Mercymount Country Day School in Cumberland, Rhode Island. He is author of *A Student Guide to Energy* (Greenwood 2011).

DENNIS MORAN was senior editor and writer for geography and world cultures at ABC-CLIO. He spent 20 years as a journalist, working for newspapers in Illinois, California, and the Czech Republic.

CORALIE NOËL has been the deputy director of the International Office for Water since 2007. She was responsible for European water policy at the French Ministry of Environment from 2000 to 2003, served as head of the European Affairs Department from 2003 to 2005, and was an adviser to the minister of work and employment from 2005 to 2007.

AMY KRAKOWKA RICHMOND is associate professor of geography at the United States Military Academy. She obtained her PhD in geography from Boston University.

TOM SHATWELL is a scientist specializing in lake ecology at the Leibniz-Institute of Freshwater Ecology and Inland Fisheries in Berlin. He obtained his PhD in biology from Humboldt University in Berlin.

CHARLES SKEETE is the former executive director and senior advisor of Plans and Programs at the Inter-American Development Bank.

ZACHARY SMITH is Regents' Professor of Political Science at Northern Arizona University. A consultant both nationally and internationally on natural resource and environmental matters, he is the author or editor of 20 books and many articles on environmental and natural resource policy topics.

ORI Z. SOLTES teaches theology, philosophy, and art history at Georgetown University. He has authored nearly 250 books, articles, and essays. Dr. Soltes has been interviewed on CNN, the History Channel, and the Discovery Channel.

NANCY L. STOCKDALE is associate professor of history at the University of North Texas.

LEEANN SULLIVAN is a PhD candidate in the human dimensions of natural resources program at Colorado State University.

GRENETTA THOMASSEY is policy director at Tip of the Mitt Watershed Council in Petoskey, Michigan, and holds a PhD in public policy.

JENNIFER C. VEILLEUX is a water geographer and lead researcher for Dr. Aaron Wolf's Transboundary Freshwater Dispute Database at Oregon State University. Jennifer also works as water advisor for The Nile Project.

TIM J. WATTS is a content development librarian at Kansas State University.

JANE WHITMIRE is a researcher who specializes in public policy analysis. She obtained her PhD in political science and her MA in sustainable communities from Northern Arizona University.

INDEX

31192021151749